TECHNOLOGICAL SYSTEMS AND INDUSTRIAL DYNAMICS

Economics of Science, Technology and Innovation

VOLUME 10

The titles published in this series are listed at the end of this volume.

TECHNOLOGICAL SYSTEMS AND INDUSTRIAL DYNAMICS

Edited by

BO CARLSSON

Weatherhead School of Management
Case Western Reserve University

SPRINGER SCIENCE+BUSINESS MEDIA, LLC

Library of Congress Cataloging-in-Publication Data

A C.I.P. Catalogue record for this book is available from
the Library of Congress.

ISBN 978-0-7923-9940-7 ISBN 978-1-4615-6133-0 (eBook)
DOI 10.1007/978-1-4615-6133-0

Printed on acid-free paper

TABLE OF CONTENTS

PREFACE

This volume consitutes a summary of several years' multi-disciplinary research by a group of Swedish researchers. The project 'Sweden's Technological Systems and Future Development Potential' was initiated by the Swedish National Board for Industrial and Technical Development (NUTEK) and has been carried out at the Department of Industrial Management and Economics at Chalmers University of Technology in Gothenburg, the Research Policy Institute at the University of Lund, the Industrial Institute for Economic and Social Research (IUI) in Stockholm, and the Department of Industrial Economics and Management at the Royal Insitute of Technology, Stockholm, under the direction of Bo Carlsson, Case Western Reserve University, Cleveland, Ohio.

The project group decided early on to focus first on the technological system for factory automation – a relatively mature system of great importance to Swedish industry and in which Sweden has reached a leading position internationally – and then to shift the attention to other systems in various stages of development and with varying Swedish strength. The work on factory automation resulted in numerous papers and publications, summarized in a volume published in 1995 (*Technological Systems and Economic Performance: The Case of Factory Automation*, ed. Bo Carlsson. Dordrecht. Boston and London: Kluwer Academic Publishers). The present book constitutes the second volume, summarizing our work on all four technological systems studied thus far (electronics and computers, pharmaceuticals and biotechnology, and powder technology, in addition to factory automation). The book also contains a brief analysis of Sweden's technological profile, a survey of the literature on technological spillovers, a presentation of the results of simulations of the macroeconomic effects of spillovers, and a discussion of key lessons for public policy. A third volume, focusing on the methodological aspects of our work, is currently in preparation.

On behalf of the entire project group I would like to thank NUTEK for generous financial support throughout this project, and Lennart Elg in particular for his interest and involvement as the NUTEK representative on the project team. Financial support from the Swedish Council for Planning and Coordination of Research (FRN) is also gratefully acknowledged. I would also like personally to thank each of my colleagues in the research group for all their efforts to make this project a success. The potential hurdles of geographic distance and different disciplinary backround were cheerfully overcome in a spirit of cooperation which made the research experience enjoyable as well as productive.

Cleveland in March, 1997

Bo Carlsson
Project director

LIST OF CONTRIBUTORS

Bo Carlsson, Associate Dean for Research and Graduate Programs and William E. Umstattd Professor of Industrial Economics, Weatherhead School of Management, Case Western Reserve University, Cleveland, Ohio (Ph.D. in Economics, Stanford University, 1972).

Gunnar Eliasson, Professor of Industrial Economics, Royal Institute of Technology, Stockholm. Formerly President, Industrial Institute for Economic and Social Research, Stockholm (Fil. Dr. and Docent in Economics, University of Uppsala, 1968).

Åsa Eliasson received her Fil.Dr. in Microbiology at the Biomedical Centre (BMC) at the University of Uppsala in 1996. She is currently doing post-doctoral research at IBMP-CNRS in Strasbourg, France.

Anders Granberg, Director and Senior Research Associate, Research Policy Institue, University of Lund (Fil.Dr. in Sociology, University of Lund, 1988).

Staffan Jacobsson, Professor, Department of Industrial Management and Economics, Chalmers University of Technology (D Phil in Economics, University of Sussex, 1985, and docent in Economics, Chalmers University of Technology, 1990).

Joakim Philipson received his M.Sc. from Chalmers University of Technology in 1992 and is currently working for Ericsson Radio Systems in Stockholm.

Rikard Stankiewicz, Professor of Technology Dynamics, Department of Industrial Economics and Strategy, Copenhagen Business School, Copenhagen, Denmark. Formerly Associate Professor, Research Policy Institute, University of Lund (Fil.Dr. in Sociology, 1980, and Docent in R&D Management, 1987, University of Lund).

Erol Taymaz, Professor of Economics, Middle East Technical University, Ankara, Turkey, and former Research Associate, Industrial Institue for Economic and Social Research, Stocholm (Ph.D. in Economics, Case Western Reserve University, Cleveland, Ohio, 1989).

LIST OF TABLES

LIST OF FIGURES

1

INTRODUCTION

Bo Carlsson

BACKGROUND

This is a study of technological innovation as the primary source of economic growth in an advanced economy. The purpose is to show how technological innovation which may occur at many levels - in individual firms or in various groupings of firms - gives rise to economic growth at the aggregate (macro) level. This process involves interaction, both planned and unplanned, via markets as well as via other mechanisms, among a variety of economic agents. In a series of case studies, our focus is on who the various participants in this process are, how their interaction takes place, what the conditions for desirable outcomes are, and how these conditions might be improved in each case.

This book is a summary of the findings in a multi-year, multi-disciplinary effort involving several leading research institutes in Sweden: the Department of Industrial Management and Economics at Chalmers University of Technology, the Industrial Institute for Economic and Social Research, the Research Policy Institute at the University of Lund, and the Department of Industrial Economics and Management at the Royal Institute of Technology. The project has been funded by the Swedish National Board for Technical Development (Nutek, formerly STU) and by the Swedish Council for Planning and Coordination of Research (FRN) and directed by Bo Carlsson, Case Western Reserve University.

There is no shortage of studies of economic growth. What is unique and novel about the work presented here is the approach, viewing the economy as a set of dynamic and complex systems, partially overlapping and interacting with each other. Innovations — new combinations of products, processes, organizations, buyers, and suppliers — are at the core of the analysis. They propel the economy forward. Sometimes they improve and expand existing systems, and sometimes they generate new systems. The technological possibilities are virtually limitless, but the ability of individuals, firms, and other entities to identify and exploit the resulting business opportunities is strictly limited.

The central units of analysis in our study are *technological systems*. We have previously defined technological systems as knowledge and competence networks supporting the development, diffusion and utilization of technology in established or emerging fields of economic activity (Carlsson & Stankiewicz 1991). They consist of networks of firms, research and development (R&D) infrastructures, educational institutions, and policy-making bodies. Technological systems can be seen as mediating structures between the knowledge base of a sector and the firms active in it. Their quality influences the degree to which the firms can exploit the emerging technological opportunities. Hence, the importance of technological systems increases with the complexity and heterogeneity of the knowledge base of innovation.

Why Systems?

According to the *American Heritage Dictionary*, a system may be defined as a group of interacting, interrelated, or interdependent elements forming a complex whole. The idea that the economy may be thought of as a system is certainly not new; it is as old as the economics discipline itself. But why should one study systems rather than their component entities — say, sectors, industries, and firms? What is gained by bringing in the notion of a system, or set of systems?

By viewing economic activity in a systems framework, two things are highlighted. The first is the interdependence of the elements which make up the system. There is important interaction among agents, only some of which takes place in markets and is governed by prices. This market-mediated interaction is typically the focus of conventional industry studies. By taking a systems approach, we explicitly recognize also the importance of nonmarket-mediated interaction, a form of external economies, especially in the form of knowledge spillovers. Such externalities (spillovers) are pervasive; they are not rare exceptions. They constitute an important part of the dynamics of technological systems.[1]

Another implication of taking a systems approach is the explicit recognition of the role of institutions — or what might be called the context or framework of economic activity. Thus, the support systems for R&D, technology and innovations, the educational system, physical infrastructure, the organization of labor, capital, commodity, and service markets, etc., need to be integrated into the analysis.

Systems are not all alike; there is insight to be gained by studying more than one system and looking for differences as well as similarities: patterns and regularities. We started by examining the technological system for factory automation in Sweden. This first phase of our study was summarized in Carlsson (1995). The present volume broadens the analysis to include electronics and computers, pharmaceuticals, and powder technology as well.

---

[1] Nadiri (1978) appears to have been the first to use the term 'spillovers,' but there is now a vast and rapidly growing literature on this (see the survey by Eliasson in ch. 8).

Other Systems Approaches

Various systems approaches have been suggested in the economics literature. The earliest effort incorporating a systems view was made by Erik Dahmén in 1950 in his epic study of '**development blocs**' in Sweden (Dahmén 1950). He defined development blocs as "sequences of complementarities which by way of a series of structural tensions, i.e., disequilibria, may result in a balanced situation" (Dahmén 1989, p. 111). Probably the first example of dynamic, Schumpeterian analysis, Dahmén's study focused on the role of entrepreneurs and entrepreneurship and analyzed the tension that arises between available resources and capabilities and those required to take advantage of new business opportunities identified by entrepreneurs.

After Dahmén's initial study there was a long delay before other systems approaches appeared in the literature. In the late 1980s, Freeman (1988), Lundvall (1988 and 1993), and Nelson (1988 and 1993) launched a series of studies on **national innovation systems**. Here the focus is on the institutions and actors, primarily in science and technology, as well as the role of technology policy, at the national level: R&D, universities, research institutes, government agencies, and government policies. This approach lends itself particularly well to comparisons among countries during a certain time period.

Another approach is Michael **Porter's 'diamond'** described in his 1990 book (*The Competitive Advantage of Nations*), focusing essentially on clusters of industries. The four sides of the diamond are made up of factor conditions (skills, technologies, capital, etc.), demand conditions, links to related and supporting industries, and firm strategies, structure, and rivalry. Each economic activity is viewed as part of a cluster of activities and agents rather than as taking place in isolation.

A similar approach is represented by **'sectoral innovation systems'** (Breschi and Malerba, 1995; see also Malerba and Orsenigo, 1990; 1993; 1995). This concept is based on the idea that different sectors or industries operate under different technological regimes which are characterized by particular combinations of opportunity and appropriability conditions, degrees of cumulativeness of technological knowledge, and characteristics of the relevant knowledge base. Similarly to Porter's analysis, this approach focuses on the competitive relationships among firms by explicitly considering the role of the selection environment.

Yet another approach is built around the concept of **local industrial systems** as represented in AnnaLee Saxenian's study (1994) of the electronics industry in Silicon Valley in California and along Route 128 in Massachusetts. Here the focus is on differences in culture and competition which have led to differences among the two regions in the degree of hierarchy and concentration, experimentation, collaboration, and collective learning which, in turn, have entailed differences in the capacity to adjust to changing market circumstances.

The Technological Systems Approach

Our own approach is akin in many ways to the previous ones, but there are also distinct differences. Technological systems are most closely related to Erik Dahmén's dynamic concept of 'development blocs'. They differ from sectoral innovation systems in that they focus on generic *technologies* rather than industries. In contrast to national innovation systems, there are many (or at least several) technological systems in each country, and national borders do not necessarily form the boundaries of the systems. Although our systems concept can be applied at the local level or regional level, we have not yet done so systematically, thus differing from Saxenian. Compared to Porter, our approach gives more emphasis to the nonmarket-mediated interaction and collaboration among entities, putting somewhat less emphasis on the competitive market processes. In contrast to other approaches (except Dahmén's), our approach emphasizes the fact that systems evolve over time, i.e., the number and types of actors, institutions, relationships among them, etc., vary over time; there is thus a need for longitudinal studies. In addition, our analysis focuses on *technologies* rather than industries.

There are four basic assumptions underlying our studies. The first two have already been indicated. (1) The system as a whole is the primary unit of analysis, not its components; the focus is on interaction and interdependence among actors and institutions; the main function of the system is to capture, diffuse and enhance spillovers; there may be a need to create appropriate institutions.

(2) Dynamics: the systems are not static, given once for all. A newly emerging system has characteristics different from those of a mature system, and the configuration of actors and institutions changes over time within each system.

There are two additional assumptions. (3) The global technological opportunities are practically unlimited. This means that it is impossible to identify *all* possibilities. (4) There is bounded rationality (limited knowledge and information processing capability; the competence of various entities is differentiated but fairly stable and path dependent; and the search for new knowledge is local). Together, these last two assumptions imply that it is necessary to identify and successfully exploit a limited number of business opportunities, that the agent's own economic competence (the ability to identify, exploit, and expand business opportunities — see Carlsson and Eliasson, 1994; Eliasson 1990) is an important constraint, and that the choice of business opportunities can never be 'optimal' but must always be based on limited knowledge and intuition. This choice is not made up of optimization but of experimentation (cf. Eliasson 1987). Contrary to the conventional wisdom, according to which the opportunity set is limited and the decision-makers are completely informed and rational, here the opportunity set is unlimited and the decision-makers boundedly rational. This also means that information is more effectively and efficiently gathered, filtered, processed, interpreted, and implemented via networks than completely independently. A well-functioning network consists of a number of agents, each with a different knowledge base but with a similar problem-orientation. The collective knowledge base is therefore likely to be vastly larger than that of each individual agent.

Technological systems involve market and non-market interaction in three types of network: buyer-supplier (input/output) relationships, problem-solving networks, and informal networks. While there may be considerable overlap between these networks, it is the problem-solving network which really defines both the nature and the boundaries of the system. Buyer-supplier linkages are important, the more so the more technical information is transmitted along with the transactions and less so, the more commodity-like the transactions are. Sometimes the most important technical information comes from sources (e.g. universities and research institutes) separate from buyers and sellers. Sometimes the informal, mostly personal, networks established through professional conferences, meetings, publications, etc., are important channels of information gathering and sharing.

The main carriers of competence, the prime movers, are the key players in each system. They identify the economic potential, define the key technical requirements, and undertake the necessary entrepreneurship. Through their leadership role they create opportunities for others as well, both directly via market interaction and indirectly via spillovers. One of the main functions of technological systems is to capture and enhance technological spillovers, as well as to create favorable conditions for market exchange. The most important features of the systems, therefore, are the characteristics of knowledge and spillover mechanisms which determine the potential spillovers, the receiver competence of each actor (particularly of the prime movers), the connectivity (the degree to which various components in the system are tied together), and the mechanisms which create variety within the system.

THE FOUR CASES

Our study covers four different technological systems. We started with factory automation because it represents a well-established set of products and technologies, being rooted in engineering and metalworking industries which have long constituted nearly half of manufacturing output, employment, and exports in Sweden. This suggested a higher probability relative to other systems that we would be able to identify and describe a full set of economic agents, institutions, and networks. It is also an area in which Sweden is known to be highly advanced across the board (including both users and suppliers). Therefore, this is the system analyzed in the first phase of our project and reported in Carlsson (1995).

In the second phase we extended our analysis to electronics and computers, pharmaceuticals and biotechnology, and powder technology. The first of these fields is in a less mature development phase than factory automation, with rapid technological change still taking place, with strong international linkages, and with Swedish actors in only parts of the system rather than across the board. Digital electronics clearly is a generic technology which penetrates into almost all areas of economic activity. A better understanding of the technological system for electronics and computers in Sweden is likely to contribute to a more solid foundation for assessing future economic growth than would otherwise be the case.

In our original discussions we envisaged a study of the technological system for biotechnology as a system with extraordinary growth potential, but it soon became clear that most commercial applications of biotechnology in Sweden were in pharmaceuticals. Thus, the pharmaceutical industry became the focus rather than biotechnology per se. In the course of the study, the technological base of the pharmaceutical industry has shifted rather dramatically from the previous chemical base to a biological science base. And it so happens that the Swedish pharmaceutical companies were in a particularly advantageous position to learn about and exploit the new opportunities. Thus, the system examined here is not only in an extremely rapid development phase but is also one in which Swedish companies and institutions are playing a much more prominent role than we expected at the outset.

Powder technology is defined as the production of powder and powder-based materials and components for structural (engineering) applications. For all practical purposes, this means that powder technology comprises two types of materials, metals and ceramics, as well as combinations, in the form of composites and compound materials. The technological system is relatively mature with respect to metals, being closely tied to traditional metallurgical activities in Sweden, while the ceramics-based side is still in an early development phase, not only in Sweden but also globally. The main rationale for picking this particular system is to allow the study of a system in an earlier development phase and with more limited scope than the other cases, especially with a view toward how the composition with respect to types of actors and institutions might differ from more mature systems. This system is also focused in the government's technology policy.

In addition to extending our analysis to more cases, we also took a more dynamic approach than in the previous study by examining the evolution of each system over time rather than just "taking a snapshot" at a moment in time. This adds a further dimension to our study, particularly with respect to the policy conclusions.

The four technological systems studied here represent a significant part of the Swedish economy. As shown in table 1-1, total employment in each of the four areas may be estimated to vary between about 100,000 and nearly four million persons. The latter figure corresponds to the total labor force in Sweden and reflects the fact that practically every person alive today is affected by electronics and computers. Of course, it is not possible to simply add together the total number of persons affected by each system; there are considerable overlaps and interdependencies among the systems. This simply reflects the complexity of the economy. Each person or unit is, at least potentially, a part of several technological systems and can be positively or negatively affected by developments in several different areas.

Table 1-1 is based on estimates which are necessarily rough. But even if each individual item may be questioned, it is difficult to avoid the conclusion that the systems studied are quite comprehensive. Many people are influenced by each system.

There are many technological systems besides those studied here. For example, there are systems focused on processing of certain materials (e.g., metals, polymers, forest and agricultural raw materials) and others focused on certain processes or functions (such as energy supply, transportation, communication, construction, and

Table 1-1 Rough Estimate of the Number of Employees in Various Technological Systems

Number of employees in:	Factory automation	Electronics & computers	Pharma-ceuticals	Powder technology
Users	386,000[a]	3,600,000[e]	306,000[h]	100,000[m]
Suppliers	10,000 - 15,000[b]	93,000[f]	11,000[i]	5,400[n]
Consultants	28,000[c]	28,000[c]	...[j]	...[o]
Universities/Board of Education	180[d]	600[g]	10,000 - 12,000[k]	40[p]
Industry associations and public agencies	400	400	500[l]	20[q]
Total	425,000	3,600,000	330,000	105,000

[a] Refers to the engineering industries (SNI 38).
[b] Refers primarily to SNI 3823 (metal- and woodworking machinery).
[c] Refers to SNI 38230 (data processing, systems design, programming, and computer consulting services).
[d] Refers to employees at 11 academic institutions, including doctoral students and technical/administrative personnel.
[e] Refers to the entire labor force in Sweden.
[f] Refers to SNI 3825 (computers and office machines), 383 (electrical machinery) and 385 (instruments, photographic and optical equipment, and watches and clocks).
[g] Estimated as follows: 1 instructor, 4 doctoral students, and 1 secretary are assumed to be associated with each professorship in the relevant subjects (about 85; see chapter 4).
[h] Refers to the entire health care sector, including 23,000 physicians.
[I] About half of these are employed abroad.
[j] Not possible to estimate. The most important group, physicians engaged in clinical testing of drugs, is included in the figure for "users."
[k] Refers to persons working in R&D and education at medical, pharmacological, and veterinary faculties. (The full-time equivalent number is about 7,000 persons.) If specialists, e.g., chemists and certain biologists (such as zoophysiologists) at science faculties, as well as certain engineers (in chemistry and biotechnology) whose expertise is needed in R&D in pharmaceuticals are also included, the number rises to 13,000 - 15,000 persons.
[l] Difficult to estimate. 500 persons is probably an upper limit.
[m] If one defines "users" as "units which in at least some of their products use powder technology-based components" and computes the total number of employees in these units, the total becomes very large, probably exceeding 100,000. If one defines as users "units which not only use but also produce or design PT-based components," the number would shrink considerably. If one were to count only those persons whose primary occupation is design and production of PT components, the number would shrink further, perhaps to 10-20 persons.
[n] Refers to producers of materials, ca. 1988. Hard metals dominate heavily.
[o] There may be a few other individuals who could be referred to as "consultants."
[p] Refers to persons in whose work PT research constitutes an important part (ca. 1990).
[q] These are found primarily in two cooperative research institutes (IM and SCI), ca. 1990.

combustion). Within each system, several subsystems may be identified (e.g. different modes of transportation within the transportation system). Also, there may be partial overlaps between systems, e.g. transportation and energy supply systems.

OVERVIEW OF THE BOOK

As a prologue to the chapters that follow, we begin by providing an overview of how the technological systems studied here relate to Sweden's overall technological profile. Thus, in chapter 2, Staffan Jacobsson and Joakim Philipson analyze the technological specialization in Sweden in comparison with that in other OECD countries, including recent trends in Sweden's share of OECD R&D efforts and patenting activity. The results are interesting from both a substantive and a methodological standpoint. Sweden's technological specialization is measured using both so-called revealed comparative advantage and Sweden's share of the OECD's technological activities. Both of these indicators show a heavy orientation towards metals and mechanical engineering and relative weakness in electronics, computers, and chemicals (other than pharmaceuticals). This picture does not seem to have changed much over the last two or three decades.

The authors also contribute to the methodological debate on technology indicators by comparing patents and R&D as technology indicators and discussing the degree to which they provide a consistent view of the Swedish technological landscape and changes therein. The authors find that the two indicators are consistent as regards Swedish strength in mechanical engineering and weakness in electronics and computer science. The indicators diverge, however, with respect to Sweden's position in both pharmaceuticals and, most importantly, the total technological activities of the OECD countries. In both cases, the performance in terms of patenting activity turns out to be weak while R&D data indicate strength. These divergences illustrate the danger of relying on only one indicator when assessing the technological position of firms and countries. The conflicting results can only be resolved through further research.

The next several chapters present the results of our investigations of each of the technological systems. Chapter 3 (by Bo Carlsson and Staffan Jacobsson) summarizes the findings on factory automation previously presented in more detail in Carlsson (1995). As mentioned already, this technological system was picked precisely because of Sweden's strength in engineering industries in general and factory automation in particular. The main purpose of the analysis is therefore to examine the reasons for this strength by analyzing the technological system supporting it, and to assess the contribution of the system to the Swedish macro economy.

The most important features of the factory automation system may be grouped under three broad headings: (1) The competence of various agents within the system, particularly users and suppliers, is a prominent feature. In factory automation, the buyer competence of "first echelon" or leading users such as Volvo and ABB (which have often been first movers) is of paramount importance. The role of the lead users is to define the technical requirements of new products, thus providing useful information and stimulus to suppliers. The existence of competent domestic suppliers of most factory automation products is also important and can be largely related to intensive user-producer interaction. (2) Another important feature is the institutional infrastructure which consists of the organization of industrial research and development, the academic infrastructure, other institutions, and government policy. Among these, the most important contribution comes from academic institutions in the form of training of engineers and also a strong user (applications) orientation of academic research. (3) A third component is knowledge/competence networks, the most important features of which are "bridging" institutions which form linkages between academic research, industrial firms (both users and suppliers), and government agencies, strong user-supplier linkages which stimulate new product development and diminish the risks, and sufficient density (critical mass) of actors for sustained interaction to take place.

Another question addressed in chapter 3 is what difference it makes in macroeconomic terms that Sweden has a strong technological system in factory automation. It is shown that the engineering sector in Sweden has maintained its share of total industrial activity over the last two decades, similarly to the development in major competitor countries. But Sweden's share of OECD exports of machinery and transport equipment has declined since the mid-1970s (although less so if the shipbuilding sector is excluded). It is argued, however, that the decline would have been even worse in the absence of a high degree of automation, particularly in view of the high labor costs in Sweden. Also, the increased role of trade internally within the European Community and the growing overseas activities (particularly in Europe) of Sweden-based multinational firms have contributed to the declining export share. It appears as though the technological position of Swedish firms is considerably stronger than that which is indicated by their export performance. We conclude that the poor macroeconomic performance in Sweden in recent decades is attributable to factors external to the technological system for factory automation (but internal to the domestic economy).[2]

In chapter 4, Staffan Jacobsson analyzes the Swedish technological system for electronics and computer technology. Using data on production and international trade as well as science and technology indicators of the performance of the Swedish electronics industry since 1975, he finds that the Swedish electronics industry is small in international comparison, that the net export ratio has deteriorated from slightly positive to slightly negative for the sector as a whole, and that the largest net imports are in consumer electronics, office machinery, electronic components, and computers, while there are net exports in telecommunications, radio communications, medical electronics, and recently also in industrial electronics. Thus, the performance is decidedly mixed within the system. In addition, Swedish R&D in electronics is not growing very fast, and the country's share of U.S. patenting as well as its revealed comparative advantage are declining. Sweden's scientific activities have been focused on other areas.

Several factors explain this relatively poor performance. First, there is strong path dependence. Large Swedish firms have continued to specialize in their traditional

---

[2] For an in-depth analysis of the reasons for the poor Swedish macroeconomic performance in recent decades, see Andersson *et al.* (1993).

(mostly mechanical) areas of strength, and even when they have moved into electronics, they have done so primarily to improve traditional product areas rather than going into new product lines. The exceptions are firms such as Ericsson, ABB, and SAAB which already in the 1960s were technologically close to electronics.[3] In each of these cases there has also been close collaboration with a competent buyer (the telecommunications agency, the State Power Board, and the Defense Procurement Agency, respectively — all government agencies, as it happens). There has not been much formation of new technology-based firms either, eliminating another potential source of industrial renewal.

Secondly, the academic sector has played a reactive rather than proactive role. It was slow to increase the capacity to train engineers in electronics and computer science, and academic research in relevant areas was long neglected.

Thirdly, during the 1970s the technology policy of the government contributed to preserving the traditional specialization pattern of Swedish industry but has since been reoriented towards increasing awareness of electronics technology in industry and raising the research and training capabilities of the universities in electronics.

Chapters 5 and 6 provide two perspectives on the technological system for pharmaceuticals in Sweden and its development in recent decades. In chapter 5, Rikard Stankiewicz explains how the Swedish pharmaceutical industry was transformed within a mere 15 years from a marginal player in Sweden's economy into one of its two main high-tech industries (the other being telecommunications). He does this by examining in depth a particular case of innovation during the 1960s and 1970s, namely the development of a group of cardiovascular medicines known as beta blockers, and by placing this innovation within the technological system, company (Astra), and subsidiary (Hässle) in which it took place.

Stankiewicz describes the evolution of the Swedish pharmaceutical industry and then traces the development of beta blockers in an international perspective. He examines both the failure of several large would-be innovators to exploit their early discoveries and the somewhat serendipitous success at both ICI in the U.K. and Hässle. The question then becomes how it was possible for a small company like Hässle with very limited resources to do so well in competition with large firms. How could an R&D team with practically no experience of developing genuinely new drugs select and effectively execute a highly innovative drug development strategy?

Hässle's success can be explained in terms of factors both external to the company and specific to Hässle. As far as the company itself and its relationships to its environment are concerned, there are three main components. (1) The first is the particular style of innovation management and research culture at Hässle — the fact that Hässle was unencumbered by an established tradition made it relatively easy to create a genuinely new research organization and develop innovative approaches. (2) Secondly, the company's limited knowledge base made it necessary to seek contacts

---

3 Among these companies, SAAB has played an especially important role in diffusing technology and generating new business areas in electronics in Sweden, including parts of Ericsson's business which derived from the aircraft industry. See Eliasson (1995).

with the universities. The resulting close relations to academia provided Hässle with important product ideas, problem-solving capability, highly qualified manpower, and a contact surface with the best international science. Thus, Hässle was able to tap into the global technological system. But several aspects of the Swedish system were also important to Hässle: the high standard and large volume of clinical research in Sweden was important in testing and documenting drugs, and the research orientation of a large fraction of the Swedish medical profession made the Swedish doctors less conservative and better equipped to evaluate and use new drugs than were many of their foreign colleagues. (3) Thirdly, the relation between the R&D program at Hässle and the parent company's (Astra's) corporate leadership played an important role. Hässle managed to develop a competent team which was well balanced in terms of chemical, clinical and biological/pharmacological competence; it combined an academic-analytical attitude with an industrial-entrepreneurial spirit. The decentralized structure of Astra allowed its subsidiaries a high degree of autonomy and prevented the bureaucratization and over-management which has been characteristic of many large pharmaceutical firms.

The general properties of the Swedish technological system for pharmaceuticals which were of greatest importance in the development of beta blockers were the existence of a well-developed public biomedical R&D system, a highly competent and innovation-oriented health care system, and strong linkages between the health care system and the universities.

Stankiewicz also discusses the changing dynamics of pharmaceutical innovation in general, both in Sweden and globally. He notes the change in the knowledge base from chemical products and processes to molecular biology, biotechnology, and information technologies and observes that this tends to increase rather than decrease the costs of pharmaceutical innovation. It also calls for a high degree of flexibility and responsiveness in the R&D infrastructures as well as in the organization and orientation of corporate R&D. High R&D costs and large risks seem to indicate a need for large companies which can absorb technical and financial risks. Yet large companies may not be the best organizations to carry out the kind of experimental research that is required. Also, the need for large firm size has severely reduced the number of companies (from about a dozen to only two in the Swedish case) via mergers and acquisitions, thus increasing the system's vulnerability through dependence on a very limited number of actors. The changes in the regulation and organization of health care in several countries in the last decade have weakened the technological system of the global pharmaceutical industry. The expansion of the knowledge base, the changing corporate dynamics, and the new economic and regulatory regimes are transforming the structure of the industry. This is illustrated by the increasing consolidation and globalization of the industry and the growing complexity and heterogeneity of both the industry and its technological system.

Several of the same themes emerge also in chapter 6 in which Gunnar and Åsa Eliasson provide a systematic competence-oriented perspective on the technological system for pharmaceuticals and biotechnology. Whereas chapter 5 examines the development at Astra-Hässle of beta blockers which appeared on the market in the mid-1970s, chapter 6 focuses on the development, also at Astra-Hässle, of the peptic ulcer drug Losec, introduced on the market in 1988 and currently (1997) the world's best-selling drug. Thus, the time periods covered in these chapters overlap partially, but the main focus in chapter 6 is on the late 1970s and 1980s. During this period there was a continued significant shift in the technology base toward biotechnology. Therefore, the principal discovery processes involved in biotechnological innovative activity, the formation of the biotechnological competence bloc, and the diffusion of related technology are discussed. The Losec case illustrates both the cross-disciplinary nature of commercially oriented biotech research and the need to focus research efforts across disciplinary boundaries and methods. It also emphasizes the importance of experimentation and the role of serendipity and good luck in successful research.

Innovations, by definition, cannot be planned. Looking for particular innovations involves screening by known standards which are likely to screen out radically new solutions. Innovative activity is experimental, and new products and ideas are *discovered.* Great discoveries often start as a mistake or something thought of as an error that someone with a creative mind decides to take a closer look at. The art is to recognize new and superior solutions. Thus, the organization of innovative activity and the criteria for innovative choice are decisive for success.

The Losec story brings out four things that relate to innovative discovery processes in general. *First,* for new ideas to be recognized, significant and diverse (broad-based) competence has to be present among the resource providers (the venture capitalists). *Second,* the discovery process may be extremely complex and the gestation period very long (sometimes decades). *Third,* whereas the conventional policy role of government is to step in when the market has failed, when the market is populated by large firms that hold the resources, the proper public policy focus may be to correct private non-market failures *within* the large business organizations. Finally (*fourth*), the complexity of biotech and pharmaceutical discovery, stretching across several diverse academic and industrial fields and reaching beyond the capacity of individual teams and hierarchies to comprehend, makes it necessary for them to be organized over the market to achieve a sufficient foundation of diverse competencies and alternative solutions.

The authors therefore address a general problem when it comes to organizing corporate R&D. They recognize that there are two types of errors associated with innovative activity: (1) losing ventures are allowed to run for too long, and (2) winners are rejected. Managers in big companies with large financial resources are often excellent at terminating bad projects early, but they also lack the creative understanding of explorative research necessary to come up with radically new ideas. They often have to work through expert reference teams which are much less creative than the business people. Top management teams seldom include people who have worked with and managed the kind of people that populate R&D laboratories. By contrast, small firms cannot afford to support losers for very long, but they also lack the resources to take full advantage of potential winners. The authors conclude that the optimal organization to exploit the competence base of an industry is to have many small players and many sources of capital (esp. venture capitalists) trying it out in the market.

In chapter 7, Anders Granberg examines the technological system for powder technology. As indicated already, one of the reasons for selecting this particular system is that it offers an opportunity to study a system which is at least partially in an early development phase. Thus, one aim of this chapter is to put forward a conceptual approach to the exploration of certain cognitive and institutional aspects of the development of modern, science-based technologies. The basic elements of the approach are the same as in the other systems studied. They include the general view of technology as knowledge and the notions that technology advances by problem solving; that technologies can be decomposed into clusters of problems and related competencies; that problem solving takes place in an institutional entity - the technological system - comprised of partially interconnected organizations or actors; and that national support programs targeting a given technology can be interpreted as interventions in the corresponding national system.

Powder technology (PT) is defined as the field of technical knowledge and competence underlying the development and production of powders and powder-based materials for engineering applications requiring advanced structural materials properties. Powder technology comprises two major subfields - powder metallurgy (P/M) and engineering ceramics - with broadly similar production, and with overlaps in terms of possible materials combinations. But the two subfields are at different stages of development. Powder metallurgy is a fairly mature, commercially well-established industrial technology which has evolved as an outgrowth from the heavily dominant industrial tradition in Sweden of metallic materials and metals-based design. Engineering ceramics, by contrast, is still, by and large, a field of research and experimental development, of prototype trials and minor niche products.

Given this composition of the technological system, it is not surprising that powder metallurgy and ceramics subsystems are indeed found to be institutionally connected, with several units active in both fields. The institutional links are strongest in the scientific and technological infrastructure. The distribution of actors is 'imbalanced' in the sense that there is a heavy concentration of industrial actors on the metallurgy side and a concentration of infrastructure actors on the ceramics side. Thus, a relatively high level of industrial activity is paired with a low level of infrastructure activity in metallurgy while in ceramics the opposite is true. This reflects the fact that Sweden has old-established metals and engineering-industry traditions and that several materials firms were early in recognizing the potential of the new technology. Thus, drawing on their existing competencies in iron and steel metallurgy, Swedish firms have been able to build strong niche positions in powder metallurgy, with a significant presence in powders, processes, and finished products. These efforts have been essentially industry-initiated and industry-driven, with relatively little infrastructure support.

On the ceramics side, by contrast, the initiatives have come largely from the infrastructure, particularly in the form of government policies designed to broaden and reinforce the scientific and technological infrastructure. A series of national R&D programs have been implemented. In parallel with the public-sector developments, a few Swedish firms have engaged in monitoring, materials testing, and exploratory development, but active industrial involvement has been limited to a few potential leading users and materials- or process-oriented firms.

The major strengths of the system are the following. The first is simply that the Swedish materials industry has developed a significant presence in the field - that a number of firms have acquired advanced competencies and built strong commercial positions in certain areas or product niches; that the R&D and production activities engaged in are broadly extended over the PT chain; and that the actors are partially interlinked and supported by user-supplier as well as problem-solving networks, with ties also to the national infrastructure and the international environment, in particular to foreign users. For domestic users, actual and potential, the existence of a large and diversified engineering industry sector must be considered a major asset. Also to be mentioned are the R&D and educational capabilities of the infrastructure - notably on the ceramics side, which has been a favored area in terms of technology policy support - and the valuable bridging functions performed by the institutes in relation to industry and academia. A final positive feature that partly reflects the small-country status of Sweden is that an extensive informal network has evolved which facilitates the development of collaborative relations of a more formal nature, in R&D projects, national technology programs, and commercial undertakings.

The weaknesses and limitations of the system are also, in part, attributable to the constraints typically faced by small industrialized countries. Thus, it is hardly surprising that the Swedish PT suppliers are able to satisfy only a minor portion of the product and technology needs of the national engineering industry, nor is it surprising that the home market merely accounts for a minor portion of the sales of most suppliers. Accordingly, there is limited scope for the growth of industrial networks or development blocks within the system. Such growth is further constrained by a particular feature of the Swedish PT system, viz. its narrow 'components waist' in the key area of sintered products. A significant strengthening of domestic user-supplier links would thus require a broader and more R&D-intensive materials industry base in that area. Efforts to expand the national supplier base beyond its present areas of (largely P/M) specialization and to develop the PT competencies of the user industry would, however, be hampered by another structural characteristic of the system. This is the 'diagonal pattern', noted earlier, where a concentration of industrial activity on the P/M side is combined with a concentration of infrastructure support on the ceramics side. Attempts to branch out from existing technological strongholds and product niches into other areas of the P/M field would clearly benefit from a more evenly balanced infrastructure. Finally, a general weakness of the infrastructure - not only in powder technology, but in other fields as well - lies in the small size and limited internal research funds of most academic units. Lacking adequate resources and 'critical mass', it is difficult for these units to maintain a standby capacity in areas which are not targeted in current programs, to effectively engage in broad-spectrum monitoring, and to respond in a flexible and timely manner to new possibilities and demands.

Chapter 8 (by Gunnar Eliasson) surveys the recent literature on technological spillovers, particularly from so-called high-tech industries to other industries. The results of econometric studies show that such spillover effects are quite significant, but the interpretations of the results vary. This paper takes the interpretation down to the micro level, within firms, to study the channels through which technology is diffused.

R&D spending in manufacturing consists largely of the cost of acquiring (*imitating* and *introducing*) new technology developed in leading-edge (high-tech) firms. Such firms therefore often serve other firms as 'higher institutes of technical education and research' and suppliers of complex applications-oriented production and product development know-how which regular technical universities are often incapable of providing. Therefore, the focus here is on the diffusion of advanced production and product development know-how in large-scale production.

Diffusion of new technology occurs through the application of best-practice techniques by advanced firms in the production of other firms in four ways: (1) through movement of competent people among firms, (2) through the growth of applications-oriented competence blocs around particularly advanced firms, (3) through the introduction of new technologies in other firms, and their effects on existing production and (4) learning in subcontracting firms through competent purchasing or vice versa, from advanced specialist subcontracting firms or engineering consultants. To a high degree, advanced technology consists of an international pool of opportunities available only to those able to access it. To be successful, firms must develop the *receiver competence* needed to transform externally available technology to internal industrial production competence.

Most econometric studies on spillovers are based on industry data, not firm data. The results reviewed in chapter 8 are reported as they appear in the original source, but they are also reinterpreted from the micro-to-macro perspective. Information technology and the aircraft industry are used for illustration of integrated production. The Swedish aircraft industry represents frontier technology in mature engineering industries which constitute the industrial backbone of the advanced economies. The aircraft industry already today employs the tools, materials and product technologies that engineering firms generally will have to know in the future. The analysis shows the importance of other forms of knowledge capital than R&D investment, and of dynamic market mechanisms to support the diffusion of new technology. The most important such mechanism is the market for competence embodied in people, i.e., the labor market.

One implication of the analysis is that the educational performance and the capacity of labor markets to allocate competence are decisive for the growth and economic welfare of nations. The educational attainment is intimately tied to the technological performance, the most advanced firms providing valuable educational services to all other firms. Spillovers are technical and commercial as well as organizational. Only the internationally most competitive firms diffuse internationally competitive technology to domestic firms, and efficient learning by firms can only occur in competitive markets. Firms improve their performance in response to challenges from the best competitors, and they do it predominantly through organizational change. Leading-edge firms are created only in direct confrontation with the best producers in international markets. An important conclusion is that to become and stay excellent in an industry, such as mechanical engineering, the country must have some of the internationally most advanced producers of that industry represented domestically, as educational and research institutions.

In chapter 9, Bo Carlsson, Gunnar Eliasson and Erol Taymaz make an attempt to bring together many of the ideas in the previous chapters by modeling the types of technological spillovers which occur in technological systems and estimating the quantitative effects. This is done through simulations on the Swedish micro-to-macro simulation model. The basic idea is to specify how innovation takes place within firms, how technological spillovers to other firms take place via networks, and how various characteristics of networks influence performance at the macroeconomic level.

Innovative activity within firms is modeled in the following way. At any time, the technology used in each sector of the economy can be characterized as a set of technological trajectories. These trajectories may be thought of as historically developed alternative technologies (similar to the notion of vintages). Each trajectory consists of a set of techniques. The best-practice technology of a trajectory (the "global technology") defines the best combination of techniques. The technological level of each firm is determined by the degree of correspondence between the global technology and the technology employed by the firm. In each time period, the firm can try to improve its technological position in a variety of ways. It can invest in physical capital, in general knowledge (thereby improving its ability to learn from others and absorb new ideas), in specific skills (through training), and it can innovate. In innovating, it can choose either incremental innovation (by selecting a better combination of techniques within the technological trajectory it currently employs), or radical innovation (by moving to a new trajectory). Which strategy it will choose and how successful it will be depends on a number of factors, including the firm's general stock of knowledge, its current technology level, and its ability to network with other firms. Firms can learn about the global technology by recombining their own techniques (experimentation), by recombining their sets of techniques with those of other firms (imitation), or by mutations (radical innovation).

Networks involve both connectivity and receptivity. Connectivity determines the amount of knowledge and R&D capital that a firm emits to other firms and depends, of course, on whether or not the firm belongs to a network. If it does, it gives off more information to others than it would if it did not belong, and it gives off more to firms within the network than to those outside. The level of receptivity determines how much a firm can use the information emitted by other (network as well as non-network) firms. Receptivity depends on the stock of general human capital within the firm. If a firm does not have general human capital, it cannot benefit from being a member of a network.

Thus, there are two benefits from belonging to a network. One is increased receptivity: the effective stock of knowledge (receiver competence or absorptive capacity) of the firm increases because it can draw on the knowledge and R&D in other firms (especially in the same network). Therefore, the probability of engaging in external search increases, the learning efficiency is improved, and the probability of successful innovation (both radical and incremental) increases. The second benefit is that through increased connectivity, firms become more aware of what other firms in the network are doing; in effect, the opportunity set increases. In its innovative efforts, the firm can therefore draw from a larger pool. This enhances both incremental and radical innovation.

Two different sets of simulations were run: one in which the main thrust was to explore the implications of different make-up of network membership, and one in which the intensity of receptivity and/or connectivity is reduced. In the first group of experiments, firms form networks either randomly or on specific criteria, within their respective sectors or across industries. The results show that the best economic performance (in terms of economic growth rates and rates of return) is obtained when the technological spillovers are the most intensive (i.e., in the first group of experiments). It seems to matter less how the network is made up; the important thing is that networks expand the range of options available to each participant and increases the experience base upon which decisions can be made. The performance of the economy declines dramatically when both connectivity and receptivity are reduced.

Chapter 10 (by Bo Carlsson) tries to summarize what we have learned in our case studies about technological systems in general. Thus, rather than just summarizing our findings, an attempt is made to systematize and reformulate our observations according to a more formal and fully elaborated scheme than employed initially. In particular, the dynamic aspects of each system are more systematically integrated than in the original case studies: the evolution over several decades of each technological system as distinct from a static view at a moment in time. Asking time-related questions forces us to examine the time frame and initial conditions of each system, to analyze who the various actors have been, what role they have played, how these roles have shifted over time, and how the institutional framework has evolved. Other aspects which are also given more emphasis and systematic coverage than in the preceding chapters are the role of capital (especially the sources of venture capital), international linkages, and geographic proximity.

Our analysis show that there are four features of technological systems which determine their performance: (1) the nature of knowledge and spillover mechanisms; (2) the receiver competence (or absorptive capacity) on the part of various actors, (3) the connectivity between various parts of the system, and (4) the vigor of variety creation and selection mechanisms.

The nature of knowledge in each field determines both the potential and the mechanisms for spillovers. If knowledge in a particular system is tacit by nature (as distinct from codified or articulated), spillovers are likely to take place primarily via personnel transfers. Thus, physical proximity and high receiver competence are likely to be important. If the crucial knowledge is tied to individual components of the system rather than more architectural (over-arching) in character, the carriers of knowledge are likely to be more numerous and dispersed. If knowledge is embodied in artifacts rather than disembodied, the receiver competence required for successful transfer may be much lower.

Receiver competence refers to the ability to tap into the global technology set. In each system, a few leading edge firms — prime movers — who are also major (potential) users have formed the link to the global opportunity set. They have achieved a high level of receiver competence via investment in R&D, most of which is oriented towards locating the technology frontier and identifying who is on it. Leading-edge firms also generate significant spillovers, (a) by setting high standards in the form of specific technical requirements for their suppliers, and (b) by training their own labor force, thus supplementing the public education system.

The nature and importance of spillovers are determined by the degree of connectivity. A technological system which exhibits a dense network of links among its participants is more robust and benefits more from spillovers than one which has weak and few links. The higher the density in a given geographical area, the greater the potential for spillovers.

The vigor (or survival, for that matter) of the system over the long haul is determined also by the extent to which it is continually rejuvenated by being exposed to new challenges. The market is often an exceedingly tough selection environment; failure and exit are inherent in the competitive process. The more path dependent and resistant to change the actors are, the greater are the risks of a gradual reduction in the number of players and hence the number of links into the global opportunity set. Mechanisms to create more variety are needed in order to avoid the gradual erosion and eventual collapse of the system.

Even if one limits the the analysis to these four features, the number of dimensions is so large that it is impossible to generalize on the basis of only the systems studied here. But a few general observations may be offered.

The main function of technological systems is to capture, diffuse, and magnify spillovers, some of which are the result of intentional acts of individual actors, while some occur as a result of largely unintended interaction.

A common feature in all the systems studied thus far is their evolutionary character. The systems are continually changing, some slowly and more or less predictably, others dramatically and unpredictably. They emerge over several decades, but not out of thin air: they are strongly path dependent. Initial conditions in the form of the pre-existing technology base, actors, and networks play an important role. The nature and locus of knowledge in each field place important requirements on the entire system and its components, on the relationships among various actors, and on the institutions which facilitate their interaction. Therefore, the traditional role and specialization of the main actors and the culture of which they are a part constitute important features which shape, and are shaped by, the system.

Our study has brought out the importance of certain types of interaction within the networks, particularly the identification of specific technical requirements. This takes place in the interface between competent customers and other actors within the system. Entrepreneurial activity is crucial but seems to be successful only when the technical requirements are well understood. User-supplier interaction is often a good vehicle, but the important part is the technical problem-solving rather than the buying and selling of inputs.

Institutions play an important role in each system, but the make-up of the institutional framework varies from one system to another, and from one time period to another. Given the small size of the Swedish economy, it is perhaps inevitable that international linkages figure prominently in each system. The ability to tap into the global system may be the single most important institutional feature in each system studied thus far. Going abroad for talent, information, and capital is clearly an effective and quick way to bypass domestic constraints in the educational and financial systems. This is undoubtedly one of the contributing factors to the strong reliance on a few well-established and well-connected domestic firms in each area. These firms function as channels through which the local system is connected to the global system and through which domestic competence is acquired or enhanced.

In the final chapter (chapter 11), Bo Carlsson and Staffan Jacobsson draw out the key lessons and issues for policy makers, both public and private. The theoretical foundation for public policy intervention rests on the notion of market failure. Market failures may be caused by the presence of increasing returns to scale and scope, externalities, missing markets, coordination problems, uncertainty, etc. — all of which are present in abundance in each technological system. But the market represents only a part of the context for technological innovation and diffusion. Institutions and networks are also part of the system; the nature and functioning of institutions and networks influence the processes of selection and discovery. Therefore, in addition to market failures, institutional and network failures also need to be considered as part of the domain of public policy.

The policy discussion addresses two main issues: strengthening or preserving existing systems on the one hand and creating new systems on the other. The latter is by far the most difficult and is therefore the primary focus in the discussion.

Improving the receiver competence in the system is a concern of both private and public actors. The receiver competence of private actors determines the ability of firms to exploit new technological opportunities and is raised through research and development, hiring of competent personnel, training of personnel (via both formal education and on-the-job training), and accumulation of experience (learning by doing). Public policy can supplement and enhance private efforts in two ways: via the educational system and university R&D, and via policies to avoid the risks of 'lock-in' in traditional areas of activity (path dependence) and to create positive externalities through various forms of spillovers. In large part, this relates to policy towards prime movers, the key actors who raise awareness of a new technology, undertake investment in it, give it legitimacy, and diffuse it locally. Policies may be needed to build or strengthen such prime movers and to make sure that they have strong linkages to the rest of the economy. In part, this refers to the academic sector which needs to be both proactive with respect to new technology and have the ability to 'change gears' in the supply of engineers and scientists in a new field.

Increase connectivity. Increasing the degree to which various components in each technological system are tied together is a 'natural' area for public policy. Our study has shown the important role played by bridging institutions, many of which are public or semi-public agencies. The main function is to increase the number and intensity of linkages among businesses, between business and academia, among these and policy agencies, and especially to stimulate international linkages of all agents. The role of public policy should be to promote linkages which do not come about without intervention, not least those between academic institutions and business, and to promote system integration. Considerations of physical proximity and high density (critical mass) may stimulate transfer of knowledge, through various mechanisms, between firms.

Given the strong path dependence and the powerful market mechanisms which weed out most new initiatives, perhaps the most important area of all for public policy is to find mechanisms which *sustain and increase variety*. The main idea here is to promote the creation and continued existence of multiple players in each system. Often this is accomplished without public policy intervention, but sometimes it is not. Public policies may be especially useful in two areas, namely to broaden and diversify the capital market and to promote the creation of new technology-based firms (NTBFs). The absence of a well-functioning venture capital market is one of the most important reasons for Sweden's (and other European countries') poor performance in the 'high tech' industries, particularly the electronics industry.

REFERENCES

Andersson, Thomas, Pontus Braunerhjelm, Bo Carlsson, Gunnar Eliasson, Stefan Fölster, Lars Jagrén, Eugenia Kazamaki Ottersten, and Kent Rune Sjöholm, 1993. *Den långa vägen — den ekonomiska politikens begränsningar och möjligheter att föra Sverige ur 1990-talets kris* (The Long Road — The limitations and Possibilities of Economic Policy to Bring Sweden Out of the Crisis of the 1990s). Stockholm: Industriens Utredningsinstitut (IUI).

Breschi, Stefano and Franco Malerba, 1995. "Sectoral Innovation Systems: Technological Regimes, Schumpeterian Dynamics and Spatial Boundaries," paper prepared for the Systems of Innovation Research Network Conference, Söderköping, Sweden, 7-10 September.

Carlsson, Bo (ed.), 1989. *Industrial Dynamics: Technological, Organizational, and Structural Changes in Industries and Firms.* Boston and Dordrecht: Kluwer Academic Publishers.

Carlsson, Bo (ed.), 1995. *Technological Systems and Economic Performance: The Case of Factory Automation.* Boston and Dordrecht: Kluwer Academic Publishers.

Carlsson, Bo and Gunnar Eliasson, 1994. "The Nature and Importance of Economic Competence," *Industrial and Corporate Change*, 3 (1): 687-711. Also published as ch. 3 in Carlsson (1995).

Carlsson, Bo, Gunnar Eliasson, and M. Ishaq Nadiri, (eds.), 1978. *The Importance of Technology and the Permanence of Structure in Industrial Growth. IUI Conference Reports 1978:2.* Stockholm: Industriens Utredningsinstitut (IUI).

Carlsson, Bo and Rikard Stankiewicz, 1991. "On the Nature, Function, and Composition of Technological Systems," *Journal of Evolutionary Economics*, 1 (2), 93-118. Also published as chapter 2 in Carlsson (1995).

Dahmén, Erik, 1950. *Svensk industriell företagarverksamhet* (Swedish Industrial Entrepreneurial Activity). Stockholm: Industriens Utredningsinstitut. An English translation is available: Erik Dahmén, 1970. *Entrepreneurial Activity and the Development of Swedish Industry, 1919-1939.* American Economic Association Translation Series. Homewood: Richard D. Irwin.

Dahmén, Erik, 1989. "'Development Blocks' in Industrial Economics," in Carlsson (1989), 109-121.

Dosi, Giovanni, Freeman, C., Nelson, R.R., Silverberg, G., and Soete, L. (eds.), 1988. *Technical Change and Economic Theory.* London: Francis Pinter.

Eliasson, Gunnar, 1987. *Technological Competition and Trade in the Experimentally Organized Economy.* IUI Research Report No. 32. Stockholm, IUI.

Eliasson, Gunnar, 1990. "The Firm as a Competent Team," *Journal of Economic Behavior and Organization*, 13 (3), 273-298.

Eliasson, Gunnar, 1995. *Teknologigenerator eller nationellt prestigeprojekt? Exemplet svensk flygindustri* (Technology Generator or a National Prestige Project? The Example of the Swedish Aircraft Industry). Stockholm: City University Press.

Freeman, Christopher, 1988. "Japan: A New National System of Innovation?" in Dosi *et al.* (eds.), 330-348.

Lundvall, Bengt-Åke , 1988. "Innovation as an Interactive Process: From User-Supplier Interaction to the National System of Innovation", in Dosi *et al.* (eds.), 349-369.

Lundvall, Bengt-Åke (ed.), 1992. *National Systems of Innovation: Towards a Theory of Innovation and Interactive Learning.* London: Pinter Publishers.

Malerba, Franco and Luigi Orsenigo, 1990. "Technological Regimes and Patterns of Innovation: A Theoretical and Empirical Investigation of the Italian Case," in Arnold Heertje and Mark Perlman (eds.), *Evolving Technology and Market Structure.* Ann Arbor: Michigan University Press, 283-305.

Malerba, Franco and Luigi Orsenigo, 1993. "Technological Regimes and Firm Behavior," *Industrial and Corporate Change,* 2 (1), 45-71.

Malerba, Franco and Luigi Orsenigo, 1995. "Schumpeterian Patterns of Innovation," *Cambridge Journal of Economics.*

Nadiri, M. Ishaq, 1978. "A Dynamic Model of Research and Development Expenditure," in B. Carlsson, G. Eliasson, and M.I. Nadiri (eds.), 51-71.

Nelson, Richard R., 1988. "National Systems of Innovation: Preface" and "Institutions Supporting Technical Change in the United States" in Dosi et al.,, 309-329.

Nelson, Richard R. (ed.), 1993. *National Systems of Innovation. A Comparative Analysis.* Oxford: Oxford University Press.

Porter, Michael E., 1990. *The Competitive Advantage of Nations.* New York: The Free Press.

Saxenian, AnnaLee. 1994. *Regional Advantage. Culture and Competition in Silicon Valley and Route 128.* Cambridge, MA: Harvard University Press.

2

SWEDEN'S TECHNOLOGICAL PROFILE*

Staffan Jacobsson and Joakim Philipson

INTRODUCTION

Competitive advantage is to a growing extent based on the ability of firms and nations to generate, diffuse and utilize knowledge, primarily scientific and technological. Grupp (1990), for example, points out that there is a quite consistent correlation between the ranking of nations according to the state of their civilian-market technologies and their trading positions.

An understanding of the magnitude and distribution across technologies, firms and industries of scientific and technological efforts ought therefore to be central to any analysis of competitive advantage and economic growth. Indeed, Rappa *et al.* (1992, p. 133) stress that "the ability of managers and policy makers to comprehend the pace and direction of technological advancement will largely determine their firm's or nation's competitive performance in world markets in the next century."[1]

The main purpose of this chapter is to attempt to analyze the Swedish 'technological landscape,' focusing on two questions: first, what are the characteristics of technological specialization in Sweden and, second, what trends can be discerned regarding Sweden's share of the technological efforts in the OECD countries? In other words, we will discuss the orientation and magnitude of Sweden's technological activities in an international perspective.

A subsidiary objective is to contribute to the methodological debate on technology indicators. Frequently, the issues raised above are analyzed using one indicator only, usually patents (e.g. IVA 1993, 1993a). Whilst the availability of data makes patents

---

* A previous version of this paper has been published in *Technovation*, 16 (5), pp. 245-253. We are grateful to Bo Carlsson, Anders Granberg and Lennart Stenberg for useful comments on earlier drafts.

1 Rosenberg (1991, p. 335) also argues for an improved understanding of scientific and technological change; "...in a world in which the economic role of science may reasonably be expected to grow over time, and in which policy-making will need to be based on a more sophisticated understanding of the ways in which science and technology interact and influence each other, a better roadmap of the science/technology landscape is vitally necessary."

easy to use, their inherent weaknesses (Pavitt 1988) makes sole reliance on patents data a risky venture. We will therefore supplement patents with R&D as technology indicator and discuss the degree to which these two measures provide a consistent picture of the Swedish 'technological landscape.'

The chapter is structured as follows. We begin, in section 2, with a brief recapitulation of the advantages and disadvantages of the two indicators used; patents and R&D data. Section 3 analyses Sweden's technological specialization, as well as changes therein, whilst section 4 focuses on Sweden's share of the technological efforts in the OECD, and changes therein. The final section summarizes the main conclusions.

ADVANTAGES AND DISADVANTAGES OF PATENTS AND R&D AS TECHNOLOGY INDICATORS

A number of technology indicators has been used extensively, e.g. patent statistics (Schmookler 1966, Griliches 1957, Grupp and Hohmeyer 1986, Grupp 1990, Archibugi and Pianta 1992, Pavitt 1985, Pavitt 1988, Patel and Pavitt 1993); bibliometric techniques (e.g. Walsh 1982, Granberg 1986) and R&D statistics (e.g. Kodama 1986).[2] Each of these indicators exhibits its own particular characteristics in terms of strengths and weaknesses and what type of activity is measured. Below, we will briefly review the literature on R&D and patent statistics.

R&D expenditures reflect formal expenditures on research and development as these are reported to the Central Bureau of Statistics. The main advantage of this indicator is that data are available at the firm, industry and national level.

The main disadvantages are:

• as data are normally collected according to the firms' principal product area, R&D data neglect both the range and specific mix of technologies in firms and industries (Patel and Pavitt 1994);

• the data do not reflect R&D expenditures which are not reported to the Government and may therefore underestimate the technological activities of smaller firms (since these may not have a formal R&D department and, depending on the sampling procedure, may not be identified by the Bureau of Statistics).

A *patent* application is evaluated using three criteria. First, a patent has to be novel to the world. This suggests that patents reflect scientific and technological activities (mainly the latter, according to Archibugi 1992) which are 'leading-edge.' Second, the invention described in the patent application needs to be technically reproducible and industrially exploitable. This suggests that a patent is filed in order to create a business impact.[3] It may also suggest that patents are a particularly appropriate indicator to

---

2 Diffusion data on products (e.g. robots) incorporating new technologies (e.g. electronics) (see Edquist and Jacobsson 1988) and data on industrial output dividing industries into low, medium and high tech industries (OECD 1986) can also be regarded as technology indicators.

3 Defensive patenting often occurs, however.

capture proprietary technical change (Archibugi 1992).[4] Third, a patent must contain solutions that are not obvious to the average practitioner.

The main advantages of patents are:

- that they cover virtually every field of technology with the major exception of software not linked directly to technical processes and products;
- that detailed information is available;[5] and
- that information is today easily obtained and rapidly disseminated through information technology.

The disadvantages include:

- variance among firms and countries in the propensity to patent;
- variance among technologies and industries in the costs of developing a patentable solution; and
- inertia in the patent system. It takes a while before a new technology is given its own class in the patent system. Until then, it is hidden within an already existing class. Thus, for a new technology to become visible, a prior alteration in the classification system is required. Such alterations are not instantaneous with respect to the emergence of new technologies.[6]

REVEALED TECHNOLOGICAL COMPARATIVE ADVANTAGE

This section analyzes the technological specialization of Sweden, as well as changes therein in the past decade. Technological specialization is measured by calculating the Revealed Technological Comparative Advantage (RTCA). This is the share of Swedish R&D (or patents) in a particular technology (or set of technologies) divided by the Swedish share of all R&D (or patents) in the OECD countries.[7]

As far as R&D[8] is concerned, two features can be discerned from the data given in table 2-1. First, Swedish industry exhibited a high degree of specialization in metalworking & mechanical engineering[9] and pharmaceuticals, in both 1981 and 1989.

---

4 However, patents are not the only way to exploit firm-specific technology and hinder imitation. Numerous other ways can be found, including relying on secrecy, imitation lags and firm specific skills and know-how (Pavitt 1988, Archibugi 1992). Despite this, empirical work demonstrates that a large share (66-87 percent) of all patentable inventions are patented (Archibugi 1992).

5 They provide information not only on the amount, but also the composition and direction of inventive and innovative activity at a very detailed level (Archibugi 1992, Mogee 1991). Patents also include a lot of other useful information such as year of invention, assignee and citations which can be used for numerous analyses at the technology, firm, industry and national levels.

6 International comparisons are made difficult by different patent systems in various countries. This has led to the frequent use of patenting in a 'third country', often the U.S.

7 The details are given in Table 2-1.

8 R&D is classified according to principal product area, as was mentioned in section 2.

9 See appendix 2-1 for a description of the databases used and how the aggregation took place.

The revealed comparative advantage in electrical engineering, electronics, and computer science[10] as well as chemicals (other than pharmaceuticals), was relatively weak. Second, while the position improved in chemicals (both with and without pharmaceuticals) during the period studied, that of electronics, etc., remained weak.

Table 2-1 Revealed Technological Comparative Advantage (RTCA*) of Sweden Using R&D data, 1981 and 1989

	RTCA 1981	RTCA 1989
Metalworking and mechanical engineering	1.98	1.95
Electrical engineering, electronics and computers	0.91	0.87
Chemicals	0.93	1.11
pharmaceuticals	1.70	1.83
other chemicals	0.72	0.79

* the share of Swedish R&D in each of the three areas divided by the Swedish share of all business-level R&D in a subset of OECD countries. The countries included in the analysis were: Belgium, Canada, Denmark, Finland, France, Germany, Italy, Japan, Norway, Spain, Sweden, U.K and U.S. Note that the sum of R&D in the three areas does not add up to the total business-level R&D in the countries. Some sectors were not classified into the three areas; agriculture, mining, aerospace, textile and clothing, stone, clay and glass; wood, cork and furniture; other manufacturing, utilities and the service sector. In 1981 and in 1989, R&D in the three areas accounted for 80 percent of the total *manufacturing* R&D in the mentioned countries. See Appendix 2-1 for the method of classifying the other sectors into our three areas.

Sources: Elaboration on OECD (1991, 1993).

The growing strength of R&D in the chemical field, including pharmaceuticals, is also reflected in Sweden's *scientific* specialization (based on bibliometric data). It is well-known that Sweden's scientific activity has a relative strength in the biomedical sciences whilst the engineering field is relatively weak (Persson 1991). What is, perhaps, less well-known is that *within* the relatively weak engineering field, the 'winners' in the period 1970-1990 were to be found in chemistry, chemical processes, biotechnology, and environmental and waste technology, areas which would appear to be relatively close to the fields of natural science and medicine (see appendix 2-2, table A2-1). On the other hand, scientific activity in the electronics and computer science field (as well as optics) is still relatively weak in Sweden. In addition, several of the metals and mechanical engineering areas have suffered a major deterioration in their revealed scientific advantage. A shift is therefore occurring from metals and mechanical engineering over to more natural science-linked fields (Persson 1991).

10 Electrical engineering is merged with electronics and computers for two reasons. First, and most importantly, it is difficult to separate these two areas, especially with the growing importance of power electronics within electrical engineering. Second, as is evident in appendix 2-2, figure A2-1, the two areas exhibit the same trend with respect to RTCA.

Turning to patents, Table 2-2 underlines the prior observation concerning Sweden's strength in metals and mechanical engineering and relatively weak (and deteriorating) position in electronics, etc. However, the two indicators show different results in the field of pharmaceuticals. While the R&D data suggest a strong and growing specialization in that field, the patent data indicate a weak position in the 1980s.

Table 2-2 Revealed Technological Comparative Advantage[a] in Swedish Industry, Using Patent Statistics, 1963-1990

	1963-1968	1969-1974	1975-1980	1981-1986	1987-1990
Metalworking and mechanical eng.	1.36	1.32	1.47	1.53	1.55
Electrical eng./ electronics/ computers	0.89	0.78	0.74	0.71	0.63
Chemicals	0.47	0.50	0.52	0.49	0.61
pharmaceuticals	1.12	1.12	1.02	0.66	0.88
other chemicals	0.42	0.45	0.42	0.45	0.53

[a] the share of Swedish patents in the U.S. patent system in each of the three areas divided by the Swedish share of all patents taken in the U.S.A. in each time period. For the method of aggregation, see appendix 2-1.

Source: Elaboration on data supplied to the Science Policy Research Unit, by the U.S. Department of Trade and Commerce, Patent and Trademark Office.

Thus, the two technology indicators confirm our picture of Swedish firms' strength in metals and mechanical engineering and a continued relative weakness in electronics (see chapter 4 in this volume). More thought-provoking is the wide discrepancy in the message that they give us regarding the pharmaceutical field.

SWEDEN'S SHARE OF THE TECHNOLOGICAL EFFORTS IN THE OECD

Revealed technological comparative advantage measures the relative strength of technological efforts in different fields. But it says nothing about the share of the absolute level of technological activities, which, after all, is what matters in linking technological efforts to economic performance.[11] We turn, therefore, to a study of how the development of Sweden's technological efforts compares with that of other OECD countries in the three technological fields.

[11] There may, of course, be significant differences in efficiency in the use of the R&D resources among firms and countries.

Table 2-3 Sweden's Share of R&D* in the OECD Countries 1981 and 1989** (in percent)

	1981	1989
All manufacturing	1.15	1.26
Metalworking/mechanical	2.28	2.45
Electrical/Electronics/Computers	1.05	1.09
Chemicals	1.07	1.40
Pharmaceuticals	1.95	2.31
Other chemicals	0.83	1.00

* See appendix 2-1 for sources and for the method of aggregation.

** See table 2-1 for the list of countries included.

In table 2-3 we can see that Sweden increased its share of R&D in the OECD[12] from 1.15 percent in 1981 to 1.26 percent in 1989. Indeed, Sweden's share of R&D in the OECD grew in all areas, but only very marginally for electronics, etc.[13]

Sweden's strength varies greatly, however, from one area to another. While Sweden has quite a large share of the R&D in metalworking/mechanical engineering and pharmaceuticals, it is quite small in electronics, etc., and in the chemical industry, other than pharmaceuticals.

In terms of the market share of *patents* granted in the U.S. (see table 4), we note that, as in the case of R&D statistics, Sweden has a substantial share of the technological efforts in metalworking and mechanical engineering and a weak position in electronics, etc., and in chemicals other than pharmaceuticals. Again, however, the two indicators show a big discrepancy as regards pharmaceuticals. Whilst the R&D data suggest a strong Swedish position, the patent data indicate the opposite for the 1980s.

Equally importantly, there is a disturbing discrepancy in the *trends* of Sweden's share of the OECD countries' technological efforts between R&D and patent data. In the case of patents, a fairly positive trend may be seen from 1963-1968 to 1975-1980. During this period, Sweden's share (for all patents) increased from 0.86 percent to 1.30 percent. After 1980, our share declined in all areas, and quite dramatically in electrical engineering/electronics/computer science. Sweden's share of R&D develop-ed quite differently, though; it *increased* for all categories (although only very slightly for

12 See table 2-1 for the list of countries included.

13 A closer look at the electrical/electronics data reveals that the increase in Swedish R&D is less than that of most countries; it is the relatively slow growth of the United States which improves our share; see appendix 2-2, table A2-2. Thus, in spite of an increase in the number of engineers and scientists (see appendix 2-2, table A2-3) as well as in R&D, the electrical/electronics/computer science field is not managing so well in an international perspective.

electrical engineering/electronics/computer science) during the same period.[14],[15]

These figures may have a whole set of explanations.[16] One important explanation is probably connected to the changing behavior of Japanese firms. While Japan increased its share of R&D performed by a set of OECD countries[17] from 32.6 to 39.5 percent in 1981-1989, its share of patents in the U.S.[18] increased from 33.4 to as much as 51.1 percent. Hence, the Japanese pattern is the opposite to the Swedish.

Table 2-4 Sweden's Share of Patents Granted in the United States, 1963-1990 (in percent)

	1963-1968	1969-1974	1975-1980	1981-1986	1987-1990
All patents[a)]	0.83	1.11	1.30	1.16	0.96
Metalworking and mechanical eng.	1.13	1.47	1.91	1.77	1.49
Electrical eng./ electronics/ computers	0.74	0.87	0.96	0.82	0.60
Chemical industry	0.39	0.56	0.67	0.57	0.59
Pharmaceuticals	0.93	1.24	1.32	0.76	0.84
Other chem. ind.	0.35	0.50	0.55	0.52	0.51

a the sum of the 34 classes included in the analysis. The data have been aggregated as in appendix 2-1.

The Japanese pattern could, for instance, reflect an increase in the propensity to

14 This discrepancy appears to have been created in the 1980s. Sweden's share of United States patenting as well as OECD R&D amounted to about 1.15 percent in 1981. (However, the R&D data are for a subset of all countries patenting in the U.S. but included as much as 86 percent of all non-U.S. patents in 1979 - see table 3 in Freeman and Hagedoorn 1992.) Subsequently, Sweden's share of U.S. patents decreased to 0.88 percent in 1989 whilst the share of OECD R&D increased to 1.26 percent (elaboration on data supplied to the Science Policy Research Unit, by the U.S. Department of Trade and Commerce, Patent and Trademark Office for the patent data and by the OECD (1991, 1993) for the R&D data).

15 One source of error may be that United States statistics on patents overrepresent the share of the U.S. (Freeman and Hagedoorn 1992). Another source of error may be found in the R&D statistics which cover fewer countries than do the patent data. This, of course, automatically increases Sweden's market share. If we exclude the R&D by U.S. firms as well as patenting in the U.S. by U.S. firms, and compare exactly the same set of countries (the countries included were Belgium, Canada, Germany, Denmark, Finland, France, Spain, Britain, Italy, Japan, Norway and Sweden), we find that Sweden's share of U.S. patenting decreased from 3.5 percent in 1981 to 2.1 percent in 1989 while the R&D share remained at 2.5 percent. The same trend therefore emerges; indeed, it is even more pronounced.

16 Differences among countries in the ability of the statistical agencies to statistically capture the R&D undertaken in the nation may, of course, also affect the ratio between markets shares of R&D and patents.

17 See footnote 15 for a list of the countries included in the analysis.

18 Japan's share of patents granted to the sub-set of countries mentioned in footnote 15.

patent abroad and/or a growing tendency for Japanese firms to apply for 'easier' patents (although this is unlikely, see IVA 1993); both reflecting two disadvantages of patents as technology indicator (see section 2).

Another set of contributory factors might be found in particular features of Swedish industry. Much of the increase in R&D in the 1980s took place in a very limited number of firms, e.g. Ericsson, chiefly in telecommunications; Astra as well as Kabi-Pharmacia in pharmaceuticals; and Volvo in automobiles. This concentration of the growth in formal R&D expenditures implies that the inherent 'patentability' of the R&D efforts of these firms[19], as well as their patent strategies, will have a significant effect at the national level. Again, this illustrates the weakness of patents as a technology indicator.

Until we know more about the reasons for the discrepancy between the two indicators, we cannot exclude the possibility that patent statistics may underestimate Sweden's share of the OECD countries' technological activities.[20] Reliance on patent statistics alone for analyzing the technological performance of Sweden should therefore be avoided.

CONCLUSIONS AND IMPLICATIONS FOR FURTHER RESEARCH

In this final section, we will summarize the main results regarding the magnitude and orientation of technological efforts in Sweden and point to two mysteries which deserve more work.

First, we measured the specialization of Sweden's technological activities using both Sweden's revealed technological comparative advantage and its share of the OECD's technological activities. What comes out of both analyses is that Sweden's technological specialization is heavily oriented towards metals and mechanical engineering and is relatively weak in electronics, etc., and in chemicals other than pharmaceuticals. In this, there seems to be little room for doubt.

However, the R&D and patent data differ widely in regard to pharmaceuticals. The R&D data suggest a strong specialization in pharmaceuticals whilst the patent data do not. Here is the first mystery that needs to be solved by future research.

Second, at the scientific level, the dominance of the biomedical area is very strong. Within the engineering field, it would seem as if those areas most closely connected to biology and chemistry are gaining at the expense of metals and mechanical engineering. Although the patent analysis suggests a poor performance in chemistry, there is some room for arguing that there is a growing Swedish scientific and technological strength

---

19 Sweden may specialize in software as opposed to hardware in electronics and computer science.

20 To the extent that there are technology-specific differences reflected in this discrepancy, patents may also wrongly portray the relative distribution of such activities (indeed, the greatest discrepancy between R&D and patents is in pharmaceuticals). Such possible biases should then supplement those revealed in Jacobsson *et al.* (1994) where it is shown that in relation to educational statistics, patents tend to overstate the specialization towards metal and mechanical engineering at the expense of the chemical industry.

in natural science-related areas. Moreover, as in the case of our technological specialization, electronics and computer science is not improving its position.

Third, whilst patent and R&D data are consistent in what they suggest to be Sweden's technological specialization, with the great exception of pharmaceuticals, they are not so consistent when it comes to analyzing the trend in Sweden's share of the OECD's technological efforts. While a patent analysis would indicate that Sweden's technological performance deteriorated rapidly in the 1980s, an analysis of R&D data would indicate otherwise. Indeed, Sweden's share of the OECD's technological efforts increased in all our fields in the period studied. Here is the second mystery that needs to be solved by future research.

Clearly, the two mysteries are quite disturbing, from both methodological and policy perspectives, and illustrate the dangers of relying on only one indicator when assessing the technological behavior of firms and nations.[21]

Take, for instance, an example where we cross-tabulate, using patent data, the RTCA of Sweden in various technological fields and the rate of growth of these fields (see Appendix 2-2, figure A2-1). The latter variable is supposed to capture future economic significance of the patents. As Archibugi (1992, p. 366) explains: "... the fields of rapid expansion of patents today are at the technological frontier, and will represent the common technologies of future economic systems". That is, a quickly growing patent class is assumed to correspond not only to original developments in scientific and technological knowledge but also to a growing number of applications of that knowledge.

We would then conclude that our RTCA is declining fastest in the most rapidly growing fields, namely pharmaceuticals as well as electronics and computer science. This is, of course, an extremely important observation which, if valid, should give rise to much concern (IVA 1993a). However, the considerable discrepancy between R&D and patent data suggests that it would be wise not to draw definite conclusions from that observation.

More work clearly remains to be done to solve these two mysteries. Until that is done, great caution needs to be exercised in drawing conclusions not only about Sweden's technological profile but, most importantly, about how well Sweden performs in an international perspective.

REFERENCES

Archibugi, Daniele, 1992. "Patenting as an Indicator of Technological Innovation: A Review," *Science and Public Policy*, 19 (6), 357-368.

Archibugi, Daniele, and Mario Pianta, 1992. *The Technological Specialization of Advanced Countries. A Report to the EEC on International Science and Technology Activities.* Dordrecht, Boston and London: Kluwer Academic Publishers and The Commission of the European Communities.

---

[21] On this issue, see also Jacobsson *et.al.* (1994).

Edquist, Charles and Staffan Jacobsson, 1988. *Flexible Automation. The Global Diffusion of New Technology in the Engineering Industry*. Oxford: Basil Blackwell.
Freeman, Christopher and John Hagedoorn, 1992. "Convergence and Divergence in the Internationalization of Technology," paper prepared for the MERIT Conference on Convergence and Divergence in Economic Growth and Technical Change, 10-12 December, Maastricht.
Granberg, Anders, 1986. "A Bibliometric Survey of Fibre-Optics Research in Sweden, West Germany and Japan," in Hariolf Grupp (ed.), *Problems of Measuring Technological Change*. Köln: Verlag TÜV Rheinland, 207-241.
Griliches, Zvi, 1957. "Hybrid Corn: An Exploration in the Economics of Technological Change," *Econometrica*, 25, 501-522.
Grupp, Hariolf, and Olav Hohmeyer, 1986. "The ISI-Approach to Technometrics: Outline of the Conceptual Framework and Assessment of Technological Standards," in Hariolf Grupp (ed.) *Problems of Measuring Technological Change*. Köln: Verlag TÜV Rheinland, 146-170.
Grupp, Hariolf, 1990. "Technometrics as a Missing Link in Science and Technology Indicators," in Jon Sigurdsson (ed.), *Measuring the Dynamics of Technological Change*. London: Frances Pinter Publishing, 57-76.
IVA, 1993. *Profit from Innovation. A comparison of Swedish and Japanese Intel-lectual Property Management*. Stockholm: Royal Swedish Academy of Engineering Sciences.
IVA, 1993a. *Halkar Sverige efter? Sveriges position i det globala patentlandskapet*. Stockholm: Royal Swedish Academy of Engineering Sciences.
Jacobsson, Staffan, Christer Oskarssson and Joakim Philipson, 1994. "Indicators of Technological Activities - A Comparison of Educational, Patent and R&D Statistics in the Case of Sweden," mimeo., Department of Industrial Management and Economics, Chalmers University of Technology, Gothenburg.
Kodama, Fumio, 1986. "Technological Diversification of Japanese Industry," *Science*, 233, July 18, 291-296.
Mogee, Mary Ellen, 1991. "Using Patent Data for Technology Analysis and Planning," *Research - Technology Management*, 34, (4), 43-49.
OECD, 1986. *OECD Science and Technology Indicators*. Paris: OECD.
OECD, 1991. *Basic Science and Technology Statistics*. Paris: OECD.
OECD, 1993. *Basic Science and Technology Statistics*. Paris: OECD (on diskettes).
Patel, Parimal and Keith Pavitt, 1994. "The Continuing, Widespread (and Neglected) Importance of Improvements in Mechanical Technologies," *Research Policy*, 23 (5), 533-545.
Pavitt, Keith, 1985. "Patent Statistics as Indicators of Innovative Activities: Possibilities and Problems". *Scientometrics*, 7 (1-2) 77-99.
Pavitt, Keith, 1988. "Uses and Abuses of Patent Statistics," in Van Raan (ed.), *Handbook of Quantitative Studies of Science and Technology*. Amsterdam: Elsevier North-Holland, 509-536.
Persson, Olle, 1991. *Forskning i bibliometrisk belysning*. INUM, Umeå.

Rappa, Michael, Koenraad Debackere and Raghu Garud, 1992. "Technological Progress and the Duration of Contribution Spans". *Technological Forecasting and Social Change,* 42, 133-145.
Rosenberg, Nathan, 1991. "Critical issues in Science Policy Research," *Science and Public Policy*, 18 (6), 335-346.
Schmookler, Jacob, 1966. *Invention and Economic Growth.* Cambridge, MA: Harvard University Press.
Walsh, Vivien, 1982. "The use of Patents and Other Indicators in the Study of Invention and Innovation in the Chemical Industry," Conference Paper at STIU/OECD, Paris.

APPENDIX 2-1

We have used two databases in our work. First, the patent data stem from the Science Policy Research Unit's data on patenting in the United States. This database was kindly made available to us by Professor Keith Pavitt and Mr. Pari Patel. The R&D data come from the OECD (1991, 1993). The patent and R&D data are aggregated to five main technological fields: metalworking and mechanical engineering, electrical engineering/electronics/computers and chemistry. Chemistry is then divided into pharmaceuticals and chemistry other than pharmaceuticals.

The **R&D data** from the OECD (1991,1993) were aggregated as follows:

Metalworking and Mechanical Engineering
Motor vehicles, ships, other transport equipment, ferrous metals, non-ferrous metals, fabricated metal products, and machinery, n.e.c.

Electrical Engineering, Electronics, and Computers
Electrical machinery, electronic equipment and components, instruments, and office machines and computers. For 1981, we have assumed 30 million U.S. dollars' worth of R&D in office machinery & computers in Sweden (in 1985 prices and purchasing power parity).

Chemical Engineering
Chemicals, drugs, petroleum refineries, food, drink & tobacco, rubber & plastics, paper & printing. Paper was included in this group simply because paper and pulp manufacturing is to a large degree a chemical process, as indicated by the fact that nearly three-quarters of the engineers working in the three largest firms in Sweden are chemical engineers. Drugs are assumed to by synonymous with pharmaceuticals.

As far as **patents** are concerned, we aggregated the initial 34 classes into three categories.

Metalworking and mechanical Engineering
Metallurgical and other mineral processes, Apparatus for chemicals, food, glass, etc., General non-electrical machinery, Non-electrical special, industrial equipment, Metallurgical and metalworking equipment, Assembling and material handling apparatus, Nuclear reactors and systems, Power plants, Road vehicles and engines, Other transport equipment, Mining and wells machinery and processes and Miscellaneous metal products.

Electrical Engineering, Electronics, and Computers
General electrical industrial apparatus, Electrical devices and systems, Instruments and controls, Photography and photocopy, image and sound equipment, Calculators, Computers and office equipment, Semiconductors and Telecommunications.

Chemical Engineering
Inorganic chemicals, Organic chemicals, Agricultural chemicals, Chemical processes, Hydrocarbons, Drugs and bio-engineering, Mineral oils and fuels, etc., and Bleaching, dyeing and disinfecting, Plastic and Rubber products. Drugs and bio-engineering are assumed to be synonymous with pharmaceuticals.

APPENDIX 2-2

Table A2-1 Revealed Scientific Advantage within the Engineering Field in Sweden, 1970-1975 and 1985-1990

	1970-1975	1985-1990
Electrical machinery	0.72	0.82
Electronics and telecommunications	0.90	0.73
Computer science	0.71	0.75
Optics	0.98	0.91
Material handling	0.68	0.90
Railroad equipment	0.70	0.74
Shipbuilding	0.90	0.81
Hydraulics	0.52	0.75
Machinery	1.35	0.69
Metallurgy, metals	1.81	1.01
Metallurgy, process	1.67	1.06
Solid mechanics	1.25	0.93
Aerospace	0.33	0.73
Transport	0.95	1.23
Chemical industry, general	1.05	1.30
Chemical, process-industry	0.92	1.20
Biotechnology	1.79	2.18
Environmental & waste treatment	1.39	1.81
Agriculture and food	0.90	1.28

Source: Persson (1991, p. 91) based on online study of Compendex.

Table A2-2 Annual Cumulative Growth Rates of R&D in the Electronics Industry in a Set of OECD Countries, 1981-1989, in Constant 1985 USD and Purchasing Power Parity

Country	Growth rate (percent)
Spain	19.0
Japan	12.1
Finland	9.9
Italy	9.7
Norway	9.6
Germany	8.7
Belgium	8.5
Sweden	6.9
ALL COUNTRIES	6.4
France	6.0
USA	4.3
UK	1.9

Source: Elaboration on OECD (1991,1993).

Table A2-3 Distribution of Engineers and Scientists* in Swedish Industry and Industry-Related Services over Four Areas, 1985 and 1990, in Percentage of the Total Stock**

	1985	1990	Annual Growth rate (percent)
Metalworking and mechanical eng.	41.1	35.4	5.6%
Electrical eng., electronics & computer science	36.7	39.1	10.2%
Chemical industry***	14.8	16.6	11.2%
Medicine	7.4	8.9	12.8%

* We have excluded physicists, industrial economists and civil engineers from the engineers and natural scientists other than chemists.

** These include railroad traffic, harbor services, air transport, telecommunications, computer consultancy (ISIC 8323), technical consultancy (ISIC 8324), other consultancy (ISIC 8325, 83292,83299), Defence, Research and Development (ISIC 932).

*** Natural scientists with a specialization in chemistry are included in this group.

Source: Elaboration on data supplied by the Swedish Central Bureau of Statistics.

Figure A2-1 Sweden's RTCA in Relation to the Rate of Growth of Patenting in Five Technology Areas, 1963/68 - 1987/90

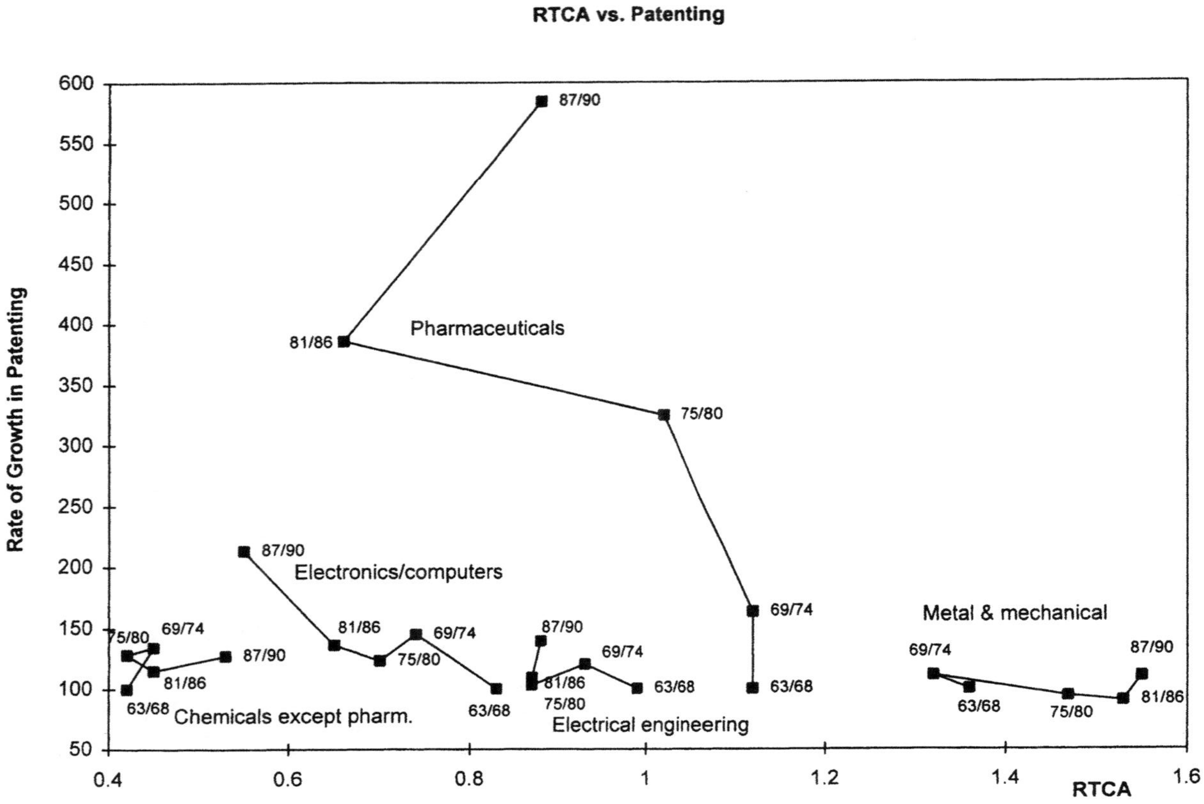

Source: Elaboration of data supplied by the Science Policy Research Unit and the U.S. Department of Commerce

3

THE TECHNOLOGICAL SYSTEM FOR FACTORY AUTOMATION

Bo Carlsson and Staffan Jacobsson

INTRODUCTION

In the *Fortune* 1992 list of the world's 500 largest corporations, Sweden had 14 entries (15 if the Swiss-Swedish-owned Asea Brown Boveri, ABB, is included), the highest number in relation to GDP of any country (Fortune 1992, p. 226; Jagrén 1993, p. 78). Half of these Swedish companies are in the engineering industries: Asea Brown Boveri (ABB) in electrical equipment, Volvo and Saab in motor vehicles, Electrolux in household appliances, Ericsson in telecommunication equipment, SKF in ball bearings, and Sandvik in metal products.

One of the features shared by all of these companies (besides being large engineering firms with vast multinational operations) is that they have a common technological base: they are all major users of factory automation technologies such as numerically controlled machine tools, industrial robots, and flexible manufacturing systems (FMS). In fact, even though these companies have located substantial manufacturing operations outside Sweden in recent years, it is nonetheless true that Swedish industry is among the most highly automated in the world. Thus, while these large companies are clearly not dependent solely on domestic manufacturing capability, neither is it the case that the Swedish engineering industry is totally dependent on them. But there are certainly strong and mutually beneficial links not only among these firms but also between them and other firms in Sweden; they are all part of a technological system which is defined by the basic set of technologies used in the engineering industries.

Why is it that a small country in a relatively remote corner of Europe has been able to establish such a strong base in such a technically highly complex field? The purpose of the present chapter is to examine some of the reasons for the Swedish strength in engineering industries by analyzing the technological system which supports them, particularly in the areas of production and design.

Factory automation is the term used in this study to refer to the cluster of technologies in a complex and diversified technological domain focused on solving certain technical problems involving the enhancement, assistance, and (sometimes) substitution of human labor through the use of certain techniques or machines in manufacturing operations.[1] The technology consists of both knowledge (constituent disciplines and specialties) and artifacts (products such as numerically controlled machine tools, industrial robots, etc.). Thus, the technological system for factory automation consists of the network of actors (individuals, firms, and other organizations) interacting to generate, diffuse, and utilize knowledge and artifacts pertaining to the solutions of these technical problems.

In a previous volume (Carlsson 1995), the characteristics of the Swedish technological system for factory automation have been explored in detail. The object of the present chapter is merely to summarize the salient features of that system.[2] Several dimensions have been identified which are particularly important in factory automation. These may be grouped under three broad headings, namely (1) the competence of various agents within the system, particularly users and suppliers, (2) the institutional infrastructure, particularly in the areas of education and the role of government, and (3) the clustering of resources in the form of networks and/or development blocs which hold the whole system together.

The chapter is organized as follows. In the next section, the salient features of the technological system for factory automation in Sweden are summarized. This is followed by an assessment of the contribution of the factory automation system to Swedish economic growth and competitiveness. The conclusions are stated in the final section.

SALIENT FEATURES OF THE SWEDISH TECHNOLOGICAL SYSTEM FOR FACTORY AUTOMATION

Overview

The main types of actors in the technological system for factory automation in Sweden are shown in Figure 3-1. In addition to users/buyers and suppliers there is infrastructure in the form of research institutions (academic departments and independent research institutes), policy and supporting institutions, and "bridging" institutions which form links among various actors in the system. The users are domestic industrial companies ranging from international giants such as ABB and Volvo to small independent companies. The suppliers are both domestic and foreign (but increasingly foreign)

---

1 Thus, office automation is not included in the study.

2 Questions relating to the internal organization of firms are not dealt with here.

companies which sell factory automation products. The main products are computer-aided design (CAD), flexible manufacturing systems (FMS), numerically controlled and computer numerically controlled (NC/CNC) machine tools, industrial robots, and automatic materials handling (AMH) equipment. Supporting this cluster of technologies is a wide range of academic disciplines (see Figure 3-2) and a set of policy and other supporting organizations.

Figure 3-1 Actors in the Technological System for Factory Automation in Sweden

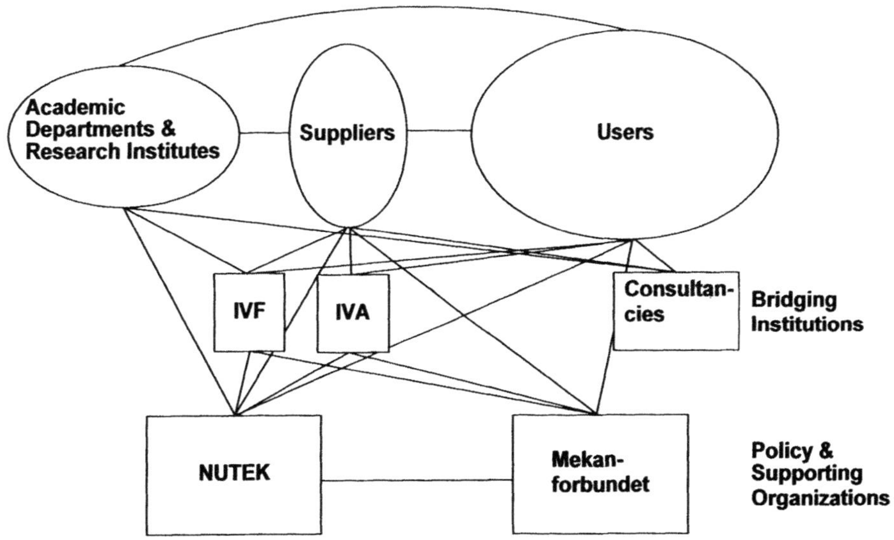

Competence

There are two aspects of economic competence (defined as the ability to identify, expand, and exploit business opportunities — see Carlsson and Eliasson 1994) which seem particularly important. The first and foremost of these is *buyer competence*, i.e. the existence of a sufficiently large set of highly competent buyers and/or users of factory automation technology. These users are referred to as "first echelon" or leading users. They are strongly internationally oriented firms such as ABB, Electrolux, SAAB, and Volvo. There are about 10-15 such firms in the field of factory automation in Sweden. These users have the capability to identify and articulate technical problems in such a way that the solutions are generalizable. Often they have the capability to solve the problems entirely on their own but choose not to do so in order not to venture too far from their core competencies. Such firms are not only in a good position technically vis-à-vis their competitors; they are also attractive as partners for technology- or equipment-supplying firms in developing new technology which may then be made available generally to other customers. The more complex and customized the technology, the greater the required interaction between supplier and customer for successful transfer and usage. Customized solutions inherently involve adaptation and/or improvement of an existing product or process and normally require far more competence on the part of the buyer both in procurement and in use than do standard products. Collaborative relationships of this sort are not limited to domestic suppliers, but without a sufficient competence base among domestic users it is much more difficult for domestic suppliers to thrive. They would then have to overcome the difficulty of working with more remote partners instead (see further Carlsson and Jacobsson 1991).

Figure 3-2 Academic Disciplines Supporting the Factory Automation Technology Cluster

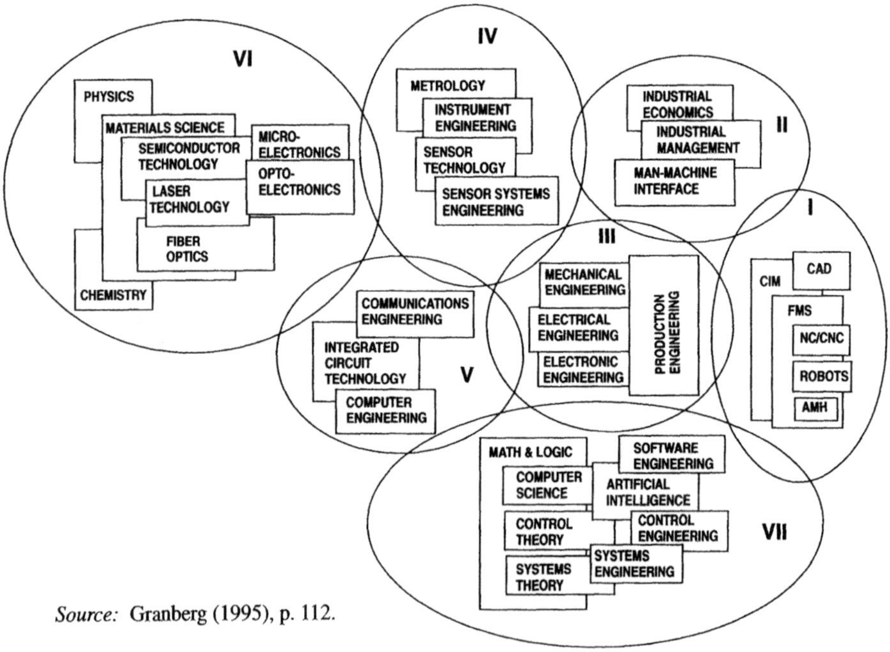

Source: Granberg (1995), p. 112.

In addition to these most advanced users there are second and third echelon users. Second echelon users ("technological followers") operate slightly behind the technological frontier. They do not have the same competence or attractiveness for specialist automation suppliers — especially distant suppliers — as do the most advanced users. Lacking the ability to articulate technical problems of sufficient generality, and hence to come up with their own solutions or to persuade suppliers to generate new solutions for them, they are to a greater extent forced to rely on already existing solutions (or rather variants of these) to similar (though not necessarily exactly the same) problems.

Third echelon firms, finally, may be subcontractors to large firms or they may be small, independent firms. Characteristic of them is that they lack the ability to identify or articulate the technical requirements; it is difficult for them to attract suppliers interested in helping them find customized solutions; and they are therefore forced to rely on standardized (off-the-shelf) technology or equipment.

Table 3-1 Density* of Flexible Automation Techniques in Various Countries

Country	NC Machine Tools* (1984)	Industrial Robots (1989)	FMS** (1988)	CAD (1985)
France	n.a.	3.98	n.a.	2.89
W. Germany	11.38	5.84	19.2	2.62
Italy	n.a.	8.57	n.a.	0.31
Japan	22.40	43.50	31.7	0.72
Sweden	**22.18**	**9.35**	**108.1**	**3.76**
U.K.	10.51	2.87	43.7	3.17
USA	11.73	4.64	17.6	6.33

* Number of units per 1,000 employees in the engineering industry (for NCMTs, industrial robots, and FMS) and in the manufacturing sector (for CAD)

** per million employees in the engineering industry

Sources: *Numerically controlled (NC) machine tools*: Edquist and Jacobsson (1988), p. 104. *Industrial robots:* Karlsson (1991) and OECD (1989). *FMS:* Ranta (n.d.) and OECD (1989). *CAD:* Åstebro (1991).

Buyer competence is, of course, difficult to measure. The high degree of diffusion of factory automation in Sweden is one indicator. See Table 3-1, which presents an overview of the usage of flexible automation techniques per employee (referred to as "density") in various countries. In density of numerically controlled (NC) machine tools, Sweden and Japan are world leaders. The same is true with regard to industrial robots and flexible manufacturing systems (FMS), with Japan clearly ahead in industrial robots and Sweden in FMS. In the area of computer-aided design (CAD), the United States has a significant lead, followed by Sweden, the U.K., France, and West Germany, while Japan and Italy lag far behind. Thus, Sweden has attained a strong position in the utilization of all of these technologies. The high diffusion rates suggest that the diffusion is not limited to the most advanced users but that second and third echelon firms are also involved.[3]

Another indicator of buyer competence is the relatively extensive distribution and service networks of foreign suppliers in Sweden (in spite of the relatively small market size), the international success and reputation of the leading Swedish firms in mechanical products, and the observations made in connection with the interviews and other empirical work for this project. All of this evidence points to the high degree of competence of many user firms, particularly the first-echelon firms. In addition, there is a substantial layer of second-echelon firms who have considerable buyer competence, though not as high as that of the leading users.

Another aspect which appears strong but not quite as strong as buyer competence is *supplier competence*. This refers to the existence of a sufficiently large number of suppliers of the whole range of equipment or technologies which make up factory automation. At least until recently there have been Swedish suppliers of most, if not all, of the major factory automation products. The most pervasive of these are numerically controlled machine tools (NCMTs). Swedish machine tool firms became involved relatively early in manufacturing them and had a high percentage of their output accounted for by NCMTs in the late 1960s and early 1970s (Carlsson 1983). But the subsequent development has not been as favorable. When Japanese machine tool firms began to apply microprocessors in their numerical controllers, producing small, standardized machines for a much broader category of users than before, they revolutionized the industry. They were thus able to take advantage of economies of scale in purchasing components, in machining and assembly, and in marketing and distribution. This gave them a substantial cost advantage over their Western competitors who were strongly oriented towards large, complex machines for sophisticated but few users. Price competition and economies of scale entered the competitive picture to an extent not known in this industry before. This made Swedish producers with a small domestic market particularly vulnerable (Carlsson 1989). Their strategy of product differentiation in the end proved non-viable in competition with the cost leadership strategies of the Japanese firms (for further analysis of company strategies in the machine tool industry, see Ehrnberg and Jacobsson 1995). The largest Swedish suppliers of both NCMTs and their successors, flexible manufacturing systems (FMSs), have been forced out of business in recent years. Thus, a portion of supplier competence which used to be strong in Sweden is now becoming an area of weakness.

In other areas, particularly industrial robots, the domestic supplier competence remains strong. ABB Robotics, working closely both with customers within the Asea group and with other customers, has been able to develop a successful strategy. It is

---

[3] A high degree of automation is not automatically reflected in high productivity, however. As illustrated by Carlsson, Taymaz and Tryggestad (1994), investments in automation may be undertaken for environmental or health and safety and other reasons, or may simply fail to achieve desired goals.

now one of the two largest suppliers in the world (the other being GM Fanuc).[4] The Swedish factory automation industry has also been successful in automatically guided vehicles (BT Carrier Systems), measurement systems (C.E. Johansson), and CNC grinding machines including systems (Lidköpings Mekaniska Verkstad).

Institutional Infrastructure

There are four components of the institutional infrastructure: (1) the organization of industrial research and development (R&D); (2) academic infrastructure, particularly the technical universities; (3) other institutions; and (4) government policy. We will limit ourselves here to discussing the academic infrastructure. Government policy and other institutional infrastructure seem to have few unique features in Sweden, if each aspect is treated individually. However, they do seem to be distinctive if treated as a system. Therefore, these components are dealt with under knowledge/competence networks below, as is the organization of R&D.

Academic Infrastructure.[5] The educational system, in particular the technical universities, plays a critical role in the process of generating and diffusing knowledge. It is a dual role in that the output of the educational system is both training of people and research results. The latter expand the set of technological opportunities, while training increases the ability to take advantage of the opportunities. Indeed, the quality and quantity of the engineers graduated are among the main determinants of a nation's innovative and absorptive capacity.[6]

Until quite recently, the Swedish educational system in the field of factory automation was largely confined to the training of engineers rather than research. The advanced Swedish research capabilities have rested primarily in firms and in the non-commercial research sector. Before the 1980s, little research was done at the universities, and the curricula lacked a factory automation orientation.[7] It was only during the second half of the 1980s that the universities began to play an important role in research and then chiefly on the basis of external funding from the Swedish National Board for Technical Development (STU, later NUTEK). This is quite remarkable, since the diffusion of microelectronics-based factory automation technology began to be highly visible already in the mid-1970s. Although we do not have data on the adjustment pattern in other OECD countries, it is pertinent to raise the question if a

---

4 For an analysis of industrial dynamics and corporate strategies in the robot industry, see Dahlin (1995).

5 Unless otherwise indicated, this section is based on Granberg (1995a).

6 For a discussion of the notion of absorptive capacity in firms and the dual role of industrial research and development as representing both innovation and learning, see Cohen and Levinthal (1989, 1990).

7 A relevant question, to which we do not yet know the answer, is that if little research was done in this rapidly expanding area, what was the quality of the education provided? How important is first-class education for application (as distinct from generation) of technology?

whole decade needs to pass between the point when leading-edge firms[8] identify the new technological opportunities and when the academic system adjusts (or begins to adjust). The persistent shortage of application engineers in the robotics field in the 1980s suggests that a faster adjustment would have been warranted. Indeed, the lack of application engineers is still today claimed to be a major obstacle to a faster diffusion of robots.

Moreover, the academic part of the Swedish technology system is still small. The expenditures on factory automation training and research in the Swedish technical universities represent no more than approximately one-half of the R&D budget of ABB Robotics (a subsidiary of Asea Brown Boveri)! Additionally, in spite of the expenditure increase in the 1980s, the academic infrastructure in this area is still underfinanced. The smallness of the system means that it has increasing difficulties to cover the whole range of competencies (including both basic technology and applied knowledge) required for the eventual design and implementation of factory automation systems which are becoming ever more complex while, at the same time, not eroding its competence base in other areas. This is, of course, an inherent problem in a small nation, but the underfinanced status of academia makes matters worse.

The academic part of the technology system for factory automation uses its skills primarily to educate engineers, and it also has a strong user orientation in its research.[9] This research is, significantly, not simply oriented to machine design but concerns itself with the broader, practical issues of the implementation of factory automation. This orientation of the research presumably means that the awareness of the newly graduated engineers of these issues is heightened. The applied character of the research strengthens the diffusion orientation of the Swedish technological system for factory automation in spite of the above-mentioned weaknesses of the academic part.[10]

---

[8] The President of the highly successful Japanese firm Fanuc (global leader in both CNC units and robots) explained once that 'We applied the technical innovations in the semiconductor field to machine tools earlier than the computer industry' (*Metalworking Production*, 1975, cited in Jacobsson 1986). In particular, Fanuc introduced microprocessor-based CNC units in 1975, a fact of great significance in explaining Fanuc's global leadership today.

[9] One of the reasons for the emphasis on training rather than research may be the decentralization of Swedish universities via growth of new universities which took place during the 1970s and 1980s. It is possible that this policy was more beneficial to technological diffusion than a more centralized, research-oriented strategy might have been.

[10] There may be another reason for the relatively applied character of research in factory automation in Sweden. Henderson and Clark (1990) have suggested a distinction between "component knowledge" and "architectural knowledge." In the area of factory automation it may well be that it is primarily the "component knowledge" (i.e. how a machine or process functions in detail — the realm of the engineer rather than the researcher) which is important. In other fields, such as biotechnology, "architectural knowledge" may be more important, and hence also the role of academic research.

Knowledge/Competence Networks (Clustering)

There are three aspects of networks which seem noteworthy in regard to Sweden's technological system for factory automation, namely "bridging" institutions, user-supplier linkages, and critical mass.

"Bridging" Institutions. As indicated earlier, the academic sector has played a supportive rather than a leading role. Its emphasis has been on training rather than research, and its research emphasis has been on application/implementation rather than basic research.

By contrast, the business sector has distinguished itself by what seems to be the highest level of R&D expenditure in relation to sales or value added anywhere in the world. According to OECD statistics, Sweden's R&D intensity has been the highest in the world in recent years. The vast bulk of these expenditures are in the engineering industry (although they are probably more product- than process- [factory automation] oriented). But it is inherent in such spending that it is highly skewed: a few firms are responsible for most of the R&D. This is true everywhere, not just in Sweden. In our view, what may be of considerable importance in explaining the relatively fast diffusion in Sweden — although more international comparisons are needed for conclusive evidence — is the existence of well-functioning networks consisting not only of these few large R&D spenders but also of smaller firms, academic institutions, government agencies, and perhaps even more importantly, institutions which provide bridges among these various types of units.[11]

These "bridging institutions" are NUTEK and its various sub-units, as well as Mekanförbundet and IVF. NUTEK[12] is the main government agency for implementation of technology policy. Mekanförbundet is the branch organization of the Swedish engineering industry. In addition to the duties normally performed by a branch organization, Mekanförbundet plays a significant role in the diffusion of technology as well as in the formulation of technology policy in Sweden. IVF (the Swedish Institute of Production Engineering Research), a private organization, is a collective research

---

[11] This is not to say that other factors are not important for explaining the high degree of diffusion of factory automation in Sweden, only that we have elected to focus here on what we deem to be most important, namely institutional aspects. More conventional explanatory factors have been explored elsewhere. For example, Carlsson *et al.* (1979, pp. 21-23) suggest that the conditions for factory automation may have been particularly favorable in Sweden in the late 1970s because of a combination of factors (including a wage explosion exacerbated by wage subsidies to declining industries) which reinforced the persistent labor shortage (reflecting both demographics and a strong preference among young people for non-industrial jobs) in Sweden in the postwar period. In addition, the relatively high level of education of the labor force and the favorable attitude of labor unions towards automation constituted positive factors. On the other hand, one can show that neither the distribution of value added by industry in manufacturing nor the size distribution of firms can provide much explanatory power.

[12] On July 1, 1991, the Swedish National Board for Technical Development (STU) was merged with the Board for Industrial Development (SIND) and the Energy Agency (Energiverket) to form NUTEK, the National Board for Industrial and Technical Development.

institute financed jointly by Mekanförbundet and NUTEK (Stankiewicz 1995). It conducts technology scanning, monitoring, adaptation, and diffusion, as well as contract research and testing for individual firms.

There is cross-representation of all these organizations on numerous committees and task forces, and small as well as large firms are represented. Academic institutions are also represented, primarily in NUTEK and IVF. Thus, IVF and NUTEK provide links between academia and business, while Mekanförbundet bridges the gap between government and industry. Through their organization and interaction they scan the world for new technology, evaluate it and rapidly disseminate information on it, thus fostering a high level of technical awareness at all levels of Swedish industry, an awareness which is central to explaining a rapid diffusion of technology. In addition, the bridging institutions help to accumulate and integrate the results of innovative activities which otherwise tend to be highly firm-specific and make them useful and available to other firms as well, thus internalizing them within the technological system.

The role of these organizations is not only to disseminate know-how but also to provide a compensating mechanism for weaknesses within other parts of the technological system, particularly the academic system. IVF may serve as an example here: It has an important monitoring role; it builds up specialized skills very early in the technology's life; it collaborates with the leading edge firms in the development and application of new technology, and it acts strongly as a diffuser of the new technology to small and medium-sized firms. By playing this role, IVF has, for instance, compensated for the lack of a strong Swedish supplier of CAD/CAM (see Stankiewicz 1995).

Typically, the prime role of these 'bridging institutions' is to speed up the diffusion process while strong user-supplier relationships (see next section) are more important with respect to the innovation process. However, the innovation and diffusion processes are strongly interdependent. Innovations in the form of new generations and varieties of a product may be a prerequisite for a fast diffusion. To a degree, these bridging institutions, in particular IVF, therefore participate in the innovation process as a means of speeding up the diffusion process. The following example illustrates how the system works.

> In 1983/84, NUTEK (then STU) started a five-year support program in CAD/CAM designed to stimulate the construction of reference systems by supporting technical development with regard to application within user firms. CAD/CAM was not a new technology when the program started, but it was not widely diffused. During the operation of the program, there was rapid technical development of the equipment, making it difficult for firms interested in the technology to "link up". At the same time, there were early users who wanted to proceed from an experimental phase to more thought-out, long-run investments.
>
> The design of the program reflected the strong need for both increased user competence and technical development of the systems on the supplier side. As a result, the practical application has been speeded up by several years in most

participating companies. In many cases, the program was the decisive factor influencing investment decisions even in large firms like SKF. Participating firms report having had a valuable exchange of ideas and experiences helpful in their own development. The participating group of firms initiated technology diffusion to other firms (via 26 conferences on various topics where participating firms presented their experiences), arranged seminars for teachers of technical subjects, and influenced standardization (via IVF) at the international level. Through the participation of academic researchers in company projects, a basis was established for exchange of know-how and experience across the academic research-industry line.

The program was directed by a person from IVF, enabling IVF to further develop, expand, and disseminate its know-how within the CAD/CAM area. For instance, in order to reach smaller tool-making firms whose managers rarely take the opportunity to visit organizations such as IVF, a diffusion program was designed in which IVF instead went to the firms. They filled a bus with CAD (and later CAM) equipment and simply went to the firms and demonstrated the technology. Within a short time, the bulk of the firms visited had acquired the technology. (*Teknik i tiden*, No. 1, 1991, p. 20; interview with a representative of IVF.)

User-Supplier Linkages. It is clear that in spite of the large and growing international trade in factory automation products, access to new technology is not the same for all potential customers. In the early stages of a product's life cycle, particularly in the case of highly complex products, only the most advanced users (such as Volvo) can expect to attract distant suppliers. These distant suppliers normally use their initially limited supply capacity to sell to users with which they have long-term relationships and refrain from selling to distant markets due to the great economic uncertainties of supplying complex and untried technology to such customers. All firms other than the technologically most advanced are therefore dependent (and increasingly so with declining technical competence on the part of the users) on domestic suppliers — until the technology is mature and supply restrictions and uncertainties are eliminated. A local, and advanced, automation industry therefore has a critical role to play in the diffusion of new technology in the local market. This strategic role is further underlined by the role domestic suppliers can play by diffusing information and using their locally dense service networks for new products to reduce the perceived uncertainty about the new products by local firms (Carlsson and Jacobsson 1991).[13]

The strategic role of the automation industry implies that it is important that it be broad in the sense of covering many product areas. At the same time, firms in the industry need to pursue strategies which emphasize product development as the prime source of competitive strength, in contrast to emphasizing low costs.

---

13 For a general analysis of user-supplier relationships in innovation, see Lundvall (1988).

As indicated earlier, the Swedish factory automation industry has been very successful thus far. In part, its success rests on the interaction (market and non-market mediated) that it has had with a set of very advanced local user firms which possess significant problem-identifying and problem-solving capabilities in the field of new production technology as well as in systems integration. In some cases, as in robotics and automatically guided vehicles, and to some extent also in CNC grinding machines, the new technology was developed internally within an advanced user firm (ASEA, Volvo, and SKF, respectively) and then spun off to a new division, a new independent firm, or an existing subsidiary. While such a close relationship between buyer and supplier is sometimes helpful, the crucial aspect is not the financial or ownership link but rather close collaboration with a technically advanced buyer. As we have shown in a previous paper (Carlsson and Jacobsson 1991), it appears that the existence of technologically advanced domestic users is a prerequisite for an advanced and healthy supplier industry. Jointly, these technologically leading firms and the local factory automation industry form a development bloc (Dahmén 1950, 1989) which has the strength to influence positively the diffusion of new production technology to the very large number of small and medium-sized firms.

Public policy has influenced the workings of this development bloc in several ways. First, as reported by Stankiewicz (1995), a large number of demonstration plants have been partly financed by NUTEK and SIND. The diffusion of FMS technology to small and medium-sized firms was directly stimulated by this policy, which also had a demonstration effect. Moreover, links between the Swedish factory automation industry and Swedish users were stimulated, thus improving the connectivity of the development bloc. Finally, the Swedish suppliers had an opportunity to gain valuable experience at a fairly early stage of the product's life cycle.

Second, two of the leading machine tool firms were acquired by the state in periods of financial crisis. However, as indicated by Ehrnberg and Jacobsson (1995), the economic competence of the state has proven to be quite questionable, and the future is now precarious in some central areas of Swedish machine tool industry, particularly lathes and machining centers as well as systems built around these, as mentioned above. Holes in the hitherto well-functioning development bloc have thus begun to open up.

Critical Mass. In order for a set of relationships such as a network or a technological system to function as a development bloc capable of generating economic activity, two things are needed: an entrepreneur who can formulate a vision of potential economic activity and who can bring it into being, and a certain density or intensity of the relationships among the various agents. We refer to the latter aspect as critical mass. (See further Carlsson and Stankiewicz 1991.) The basic idea is that there is a certain minimum number and intensity of interactions required for sustained economic activity to take place. Thus far, Sweden has been in the fortunate position of having a sufficient number of highly advanced users (e.g. Volvo, ABB (formerly ASEA), SAAB, and Electrolux) to sustain a healthy domestic supplier industry. The close networking among various agents in Sweden has made for high technical awareness and rapid diffusion of new technologies. There are, however, signs that the critical mass problem is increasing in importance as a result of rising complexity of technology (i.e. requiring expertise in more fields than before) and weakening domestic manufacturing capability of advanced users as they become more and more internationally spread out and/or globally oriented. Intriguing questions for future research are what is "sufficient" and "close" — in other words, what is required to achieve critical mass? International comparisons could shed light on this.

HOW WELL DOES THE SWEDISH SYSTEM PERFORM?[14]

As shown in Table 3-1, Swedish industry is among the most highly automated in the world. But what difference does it make that Sweden has a high density of flexible automation, reflecting a strong technological system in factory automation? This is the question to which we now turn.

Because of the size and complexity of technological systems and the variety of ways in which they interact with other systems both domestically and abroad, this is a difficult question. Only a partial answer can be expected, even if several different performance indicators are used. Put differently, the technological system is only one, albeit an important one, of many determinants of economic performance. This is one of the main points illustrated in Carlsson and Taymaz (1995) and Carlsson, Taymaz and Tryggestad (1994) in which the contribution of factory automation to economic growth in Sweden is analyzed, using a micro-based simulation model. As expected, the findings indicate that the contribution is positive. More importantly, however, the microeconomic mechanisms through which the contribution is made are spelled out: changes in microeconomic behavior and organization. These changes involve such things as lower operating expenses due to reductions in indirect labor input and inventories, increased flexibility as a result of speeding-up the production process coupled with a reduction in work-in-process inventories, improved managerial and working processes which enhance the output obtainable from given capital input, investment in new capital, and improved quality of output. The macroeconomic impact of such changes includes increased rates of output and productivity. The most consistent result in our simulations occurs at the microeconomic level in the form of an increase in the rate of return (particularly in the engineering industries), suggesting increased competitiveness.

Another way to gauge the impact of factory automation is by way of an international comparison. The question is whether in some sense Sweden is better off because of a strong (national) technological system in factory automation than it would be otherwise. Two different indicators are used: the relative size of the industries supported by factory automation, primarily the engineering industries; and the development of world market shares.

---

14 This section is based on Carlsson (1995a).

Figure 3-3 Shares of OECD Exports of Machinery and Transport Equipment (SITC 7) for Sweden, Germany, the European Community, Japan, and the United States, 1970-1992. Index, 1970 = 100

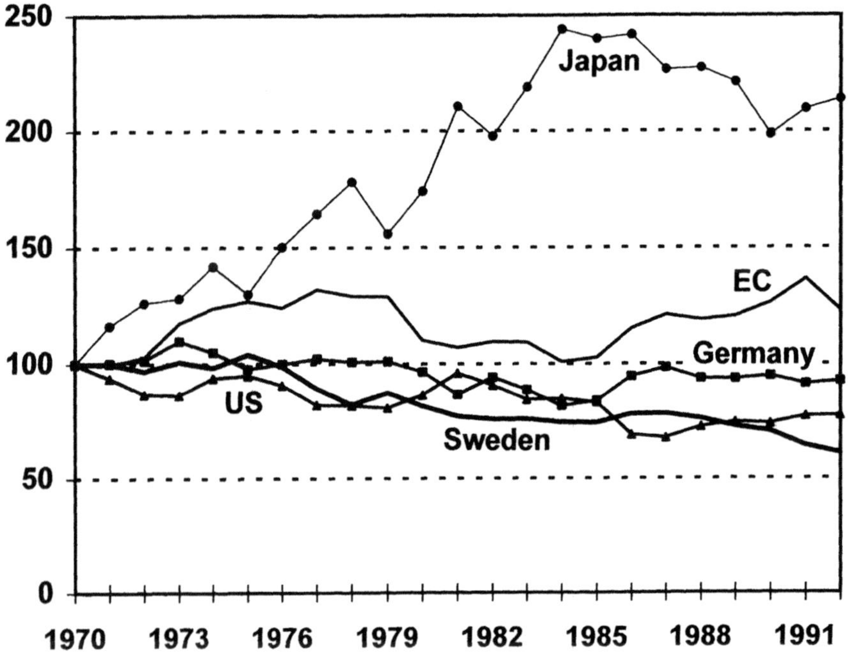

Sources: OECD, *Foreign Trade by Commodities*, various years (Paris: OECD); United Nations, *Yearbook of International Trade Statistics, Vol. 1*, various years (New York: United Nations).

With respect to the former, we found that the engineering industries represented 39.0 percent of the total value added in manufacturing in Sweden in 1970, compared to 38.6 percent in Germany, 41.8 percent in Japan, and 40.7 percent in the United States (Carlsson 1995a, p. 458). Thus, the engineering industry was of roughly comparable size in all four countries at that time. It should be pointed out, however, that all of these countries have a larger engineering industry than do most other industrial countries. It turns out also that the engineering industry grew faster than other manufacturing industries in all four countries. In 1991, its share of manufacturing value added had increased to 45.8 percent in Sweden, while the comparable figures for Germany, Japan, and the U.S. were 50.4, 48,7, and 42.1, respectively (Carlsson 1995a, p. 458). It is evident that the growth of the Swedish engineering industry relative to other manufacturing industries (its share rising from 39.0 to 45.8 percent) is not unique, nor is its size particularly large compared to that in the other countries. Also, industrial output grew considerably more slowly in Sweden than in the other countries throughout the whole period. It grew by only 25 percent between 1970 and 1991, compared to 56 percent in Germany, 74 percent in the United States, and 126 percent in Japan (*ibid.*, p.

458). This means that the expansion of the Swedish engineering industry took place in a slow-growth environment, i.e., that this was one of the few industries that grew at all. A high degree of factory automation was presumably one reason why this industry grew faster than most others.

The data on international trade performance give a similarly inconclusive result. Sweden's overall share of industrial country exports declined by 30 percent between 1970 and 1992, from 3.1 to 2.1 percent (IMF 1993). Surprisingly, its share of OECD exports of engineering goods (represented by machinery and transport equipment, SITC 7) declined even more: by nearly 40 percent (from 3.5 to 2.1 percent) over the same period. See Figure 3-3. The figure shows that Sweden's market share has declined almost continuously since the mid-1970s. The United States has suffered a similar, although not quite as precipitous, decline, while Germany has maintained its share. The fact that the market share of European Community (EC) exports has increased is probably attributable largely to growing internal trade within the Community. The decline in market share of countries outside the EC (e.g. Sweden and the U.S.) would not be as great if the data were adjusted for this internal trade. The rapidly increasing Japanese market share until the early 1980s, of course, had a similar depressing effect on the market shares of other countries. But it remains clear that Sweden's share of OECD exports of engineering products declined more than that of some of its main competitors over the last two decades.

The picture is somewhat less bleak if we examine only the ten largest product categories ("Top 10") at the 3-digit SITC level.[15] According to Figure 3-4, Sweden's share of industrial country exports of machinery and transport equipment declined by 27 percent over the period 1976-1991 if only the 10 largest product groups in Swedish exports are included. The corresponding figure for all SITC 7 goods was 35 percent. (This means that the market share of other SITC 7 goods declined even more, by 45 percent, as shown in Figure 3-4.) Thus, Sweden was more successful in maintaining its export shares in its traditionally strong product lines than in others.

However, much of the decline in export share outside the Top 10 groups is due to the virtual demise of the Swedish shipbuilding industry in the late 1970s. Ships and Boats (SITC 793) accounted for 16.4 percent of Swedish SITC 7 exports in 1976 but only 1.0 percent in 1991; some 15 percent of Swedish exports simply disappeared. This

---

[15] The largest product group in Swedish exports of SITC 7 goods at the 3-digit level in 1976 was Ships and Boats, etc. (SITC 793). Because of the special problems afflicting the shipbuilding industry, this product group was treated separately in the analysis here. Excluding this group, the "Top 10" product groups, ranked in descending order in terms of total exports in 1976, were Passenger Motor Vehicles except Buses (SITC 781); Telecommunications Equipment, Parts, and Accessories NES (SITC 764); Lorries, Special Materials Vehicles NES (SITC 782); Motor Vehicle Parts and Accessories NES (SITC 784); Mechanical Handling Equipment (SITC 744); Other Machinery for Special Industries (SITC 728); Internal Combustion Piston Engines (SITC 713); Nonelectrical Machinery Parts and Accessories NES (SITC 749); Nonelectrical Machinery and Tools NES (SITC 745); and Heating & Cooling Equipment (SITC 741). It is noteworthy that the ranking among the Top 10 product groups in 1991 was identical to that in 1976, except that Nonelectrical Machinery and Tools (SITC 745) jumped from 9th to 6th position.

Figure 3-4 Sweden's Share of Industrial Country Exports of Machinery and Transport Equipment, Various Product Groups, 1976-1991. Percent

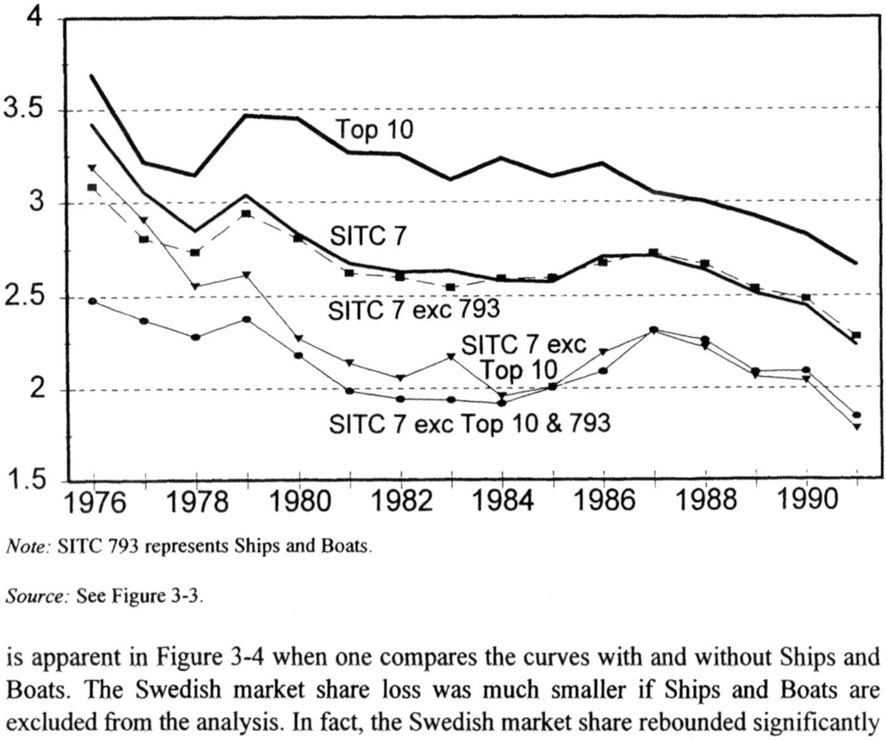

Note: SITC 793 represents Ships and Boats.

Source: See Figure 3-3.

is apparent in Figure 3-4 when one compares the curves with and without Ships and Boats. The Swedish market share loss was much smaller if Ships and Boats are excluded from the analysis. In fact, the Swedish market share rebounded significantly during 1985-1987 outside the Top 10 product groups. Those gains have since been lost, however. The disturbing conclusion is that no matter which definition is used, Sweden suffered substantial market share losses in engineering products over the last 15-20 years.

In interpreting this development, it is important to keep two things in mind. One is the increased role of trade internally within the European Community which has already been mentioned. In 1992, the EC share of OECD SITC 7 exports was 49.6 percent, up from 40.3 percent in 1970.[16] If, say, two-thirds of that increase is due to increased internal trade, the external trade share gain of the EC is much less impressive.

---

16 Sources: OECD, *Foreign Trade by Commodities* (Paris: OECD), various years; United Nations, *Yearbook of International Trade Statistics, Vol. I* (New York: United Nations), various years.

Secondly, the role of multinational firms cannot be ignored, particularly in the case of Sweden which is more dependent on large multinational firms than any other industrial country in the world. As Bergholm and Jagrén (1985) have shown for Sweden and Blomström and Lipsey (1989) for the United States, while both countries lost about 20 percent of their shares of world exports of manufactures between the mid-1960s and mid-1980s, the export shares of their multinational firms stayed fairly stable or even increased. The shares of U.S. and Swedish parent firms' exports as a percentage of total home-country exports also increased. In the Swedish case this occurred while the Swedish firms with production facilities abroad expanded their foreign production even faster than they increased their exports from Sweden (Swedenborg *et al.* 1988, p. 35). As shown by Andersson (1993, p. 58), the world market share of Swedish firms actually increased during the 1980s if not only their domestic but also their overseas production is taken into account.[17]

The implication here is that the world export share of a country may or may not reflect the country's international competitiveness, depending on how one views the role of its multinational firms.[18] To the extent that Swedish firms exploit strong technological systems at home by expanding their activities abroad, a declining Swedish world market share does not necessarily indicate weak domestic technological systems. Even though Sweden's loss of world export share has actually been larger in engineering products than in manufactured goods as a whole, it is part of a more general decline of Swedish competitiveness during the last 20 years whose roots go far beyond the engineering industry. It is not merely a problem of cost: despite several devaluations of the Swedish krona (accumulating to approximately 50 percent between 1977 and 1982), the Swedish market share continued to fall until another devaluation (of about 25 percent) in 1992. It is difficult to say whether the rapid export growth since then can be sustained over the longer haul.

How should these findings be interpreted? It is difficult to argue that the relatively slow growth of the Swedish engineering industry and the loss of world export share in engineering products (ignoring the growing overseas activities of Swedish multinational firms) provide direct, convincing evidence that the technological system for factory

---

17 Further illustrating the point, Encarnation (1992) has shown that the foreign sales by majority-owned subsidiaries of U.S. multinationals were approximately three times larger than U.S. exports in 1988 (six times larger in the European Community than U.S. exports to the EC) and in the case of Japan almost twice as large. Calculations for Sweden, based on material from Swedenborg *et al.*, suggests that the corresponding Swedish figure is 0.7. The lower Swedish figure reflects the fact that the ratio of exports to total domestic manufacturing output is considerably higher than those in Japan and the United States, i.e., that *all* firms (not just multinationals) export a substantial share of their output.

18 Surely a country is better off if it has domestically based multinational firms with strong positions abroad than it would be in the absence of such firms — provided that there are domestic spin-offs in the form of technology transfer, both direct and indirect domestic employment, and build-up of domestic infrastructure. There are strong indications that the impact on the Swedish economy of direct foreign investment by Sweden-based multinational firms was largely positive through the 1970s but that the linkages between the foreign and domestic activities of MNFs have since weakened (Andersson 1993).

automation has contributed positively to the international competitiveness of Swedish engineering industries. On the other hand, neither do these developments indicate the opposite conclusion. Apparently, forces outside the technological system for factory automation were stronger determinants. On the other hand, if the technological system had not been as strong as it is, it is likely that the Swedish performance would have been even worse. It has been shown in previous studies using United States data[19] that the use of flexible manufacturing technology made a statistically significant contribution to explaining the composition of net exports of U.S. engineering industries in 1983. Industries characterized by large plants with flexible production, a high degree of automation, and a large number of products were found to have the most successful export performance (Carlsson 1991). Another study (Carlsson and Taymaz 1993) showed that U.S. engineering industries were negatively affected by the relative decline in the technological capabilities of the U.S. machine tool industry (particularly involving computer numerical control) during the 1980s. Thus, these studies indicate that both the production technology in engineering industries and the support system for that technology are important determinants of export performance.

One of the determinants of competitiveness outside the technological system which is worth commenting on here is the narrowing of the Swedish industrial base which has been noted in previous studies (e.g. Carlsson *et al.* 1979). Among the main characteristics of factory automation is that it has greater impact on process innovation than on product innovation. This is one of the main reasons why our simulation results (reported partly in Carlsson 1995, chapters 12 and 13, and partly in chapter 9 below) indicate that factory automation reduces input requirements more than it increases output and hence increases profitability. In some further simulations (see chapter 9) we compared the impact of factory automation with that of the introduction of new firms and new economic activities. We found that the macroeconomic consequences of new entry were considerably larger than those attributable to increased factory automation. It is also consistent with the finding in a recent OECD study (OECD 1994, p. 218) that Sweden gained market share between 1980 and 1990 in only four commodity groups in "dynamic markets" (pharmaceuticals, aerospace, instruments, and chemicals), while in other groups it either lost market share in dynamic markets (e. g. computers, electronics, motor vehicles) or gained market share in less dynamic ones (e. g. ferrous metals, fabricated metals, and nonelectrical machinery), the overall result being a loss of market share. To further illustrate the point, it may be noted that even in telecommunications equipment and parts, where Ericsson has made indisputable market share gains within its product lines, Sweden has lost market share (from 6.25 percent in 1976 to 3.75 percent in 1991). Thus, exports grew more rapidly in other product areas than those represented by Swedish manufacturers. Similar phenomena are probably occurring in other industries as well.

---

[19] Unfortunately, similar data on the use of flexible manufacturing equipment are not available for Sweden.

Another consideration is that the technology base supporting the country's economic growth must be larger, the greater the role of domestically based multinational firms which draw upon that base not only for their domestic activities but also for their overseas production. The reliance upon the domestic base for industrial R&D (although not necessarily for other corporate functions) seems to persist and dominate in multinational firms, notwithstanding the continuing globalization of their operations overall (Patel 1993). In view of this, it is not surprising that Sweden's share of industrial R&D is increasing.

Table 3-2 shows the development of Sweden's shares of OECD R&D, patents, and exports in manufacturing industry as a whole and in engineering industries during the 1980s. An interesting picture emerges. Sweden's share of R&D in the OECD countries increased in both manufacturing as a whole and in engineering industries, while its share of patents fell.[20] As noted already, its share of OECD exports also fell, more so in engineering goods than in all manufactured products. As a result, the revealed comparative advantage (RCA)[21] in engineering was almost unchanged if measured by R&D but declined according to both the patent and export measure. The Swedish specialization as measured by RCA is much stronger (over 1.9) in R&D than in patenting (1.55 in 1987-90) and seems to have disappeared completely in exports.

Table 3-2 Measures of Sweden's Specialization in Engineering Products, 1970-1992

	Sweden's share (%) of total OECD						
	R&D expenditures		Patents		Exports		
	1981	1989	1975-80	1987-90	1970	1980	1992
All manufacturing	1.15	1.26	1.30	0.96	3.1	2.4	2.1
Engineering products	2.28	2.45	1.91	1.49	3.5	2.8	2.1
Revealed comparative ad vantage*	1.98	1.94	1.67	1.55	1.13	1.17	1.00

* Revealed comparative advantage is computed as the ratio between the Swedish shares of the OECD total for Engineering products (line 2) and the corresponding share for All manufacturing (line 1).

Sources: Jacobsson and Philipson (Ch. 2 in this volume); OECD (various years); IMF (1994).

---

20 Due to a sharp increase in Japanese patenting activity during the 1980s, many countries have found similar declines in their shares of patents.

21 Revealed comparative advantage (RCA) is measured as the country's share of activity in a particular field compared with its share in all fields combined.

The differences among the three measures are noteworthy. Sweden's share of patents declined in spite of the increase in the R&D share, but its export share declined even more. A possible explanation of this apparent anomaly is that the Swedish multinational firms continue to do most of their R&D in Sweden but that the results are shared with their foreign subsidiaries who may reap more than their share of patents and sales.

The question that arises is whether, in spite of its growth, the domestic R&D base in engineering industries is large enough: According to table 3-2, the Swedish R&D share in engineering products is only slightly larger than the Swedish export share which, in itself, has declined by about 40 % over the last two decades. If the total sales of Swedish firms are roughly twice the size of their domestic production, it could be argued that the R&D in relation to domestic sales should be substantially larger than it is. It is difficult to say to what extent the decisions by Swedish multinational firms to locate their expansion abroad are attributable to insufficient capacity of the domestic technological system to support both their domestic and their overseas operations and to what extent they are attributable to other factors. Our guess would be the latter, i.e., that those decisions have been made more *in spite of* the existence of a strong technological system in factory automation at home than *due to* any major deficiency in that system.

CONCLUSION

What, then, can we conclude from this overview? Let us first summarize what we have learned.

Sweden appears to have a fairly strong and robust technological system in factory automation and performs well in terms of both R&D and results of R&D (as measured by patents) in the engineering industries. In spite of this, Sweden has had a dismal growth record during the last two decades and has lost international market share. This is presumably due to problems in the general domestic economic environment having to do more with institutions than technology, and (until 1995) lack of EC membership. The former (e.g. a taxation system which is not conducive to growth, insufficiently developed capital markets, insufficient supply of engineers, too much of the economy being guided by non-market principles, and corporativistic structures have caused a lack of entrepreneurship and growth of small and medium-sized enterprises (and hence too much dependence on a handful of MNFs). The lack of EC membership has strongly influenced foreign direct investment of Swedish firms. As a result, the market share of Sweden-based firms and Swedish exports combined has held up much better than Swedish exports alone. For similar reasons, Sweden has not been an attractive environment for inbound direct foreign investment. We have also raised the question whether the technological system supporting the engineering industry is large and robust enough to support the global (not just domestic) activities of Sweden-based multinational firms and to attract foreign-based multinationals to Sweden.

REFERENCES

Andersson, Thomas, 1993. "Den svenska industrins expansion i utlandet" (The Foreign Expansion of Swedish Industry), chapter 2 in Andersson, T., P. Braunerhjelm, B. Carlsson, G. Eliasson, S. Fölster, L. Jagrén, E. Kazamaki Ottersten and K.R. Sjöholm, *Den långa vägen — den ekonomiska politikens begränsningar och möjligheter att föra Sverige ur 1990-talets kris*. Stockholm: IUI.

Andersson, T., P. Braunerhjelm, B. Carlsson, G. Eliasson, S. Fölster, L. Jagrén, E. Kazamaki Ottersten and K.R. Sjöholm, 1993. *Den långa vägen — den ekonomiska politikens begränsningar och möjligheter att föra Sverige ur 1990-talets kris*. Stockholm: IUI.

Åstebro, Thomas, 1991. "Computer Aided Design," in R. Ayres, W. Haywood, E. Merchant, J. Ranta, H-J. Warnecke (eds.): *Computer Integrated Manufacturing; Vol 2. The Past, Present and the Future.* London: Chapman & Hall.

Bergholm, Fredrik and Lars Jagrén, 1985. "Det utlandsinvesterande företaget — en empirisk studie" (The Firm Investing Abroad — An Empirical Study) in G. Eliasson, F. Bergholm, E.C. Horwitz and L. Jagrén, *De svenska storföretagen — en studie av internationaliseringens konsekvenser för den svenska ekonomin.* Stockholm: IUI.

Blomström, Magnus and Robert R. Lipsey, 1989. "The Export Performance of U.S. and Swedish Multinationals," *Review of Income and Wealth*, 35 (3), 245-264.

Carlsson, Bo, 1983. "The Machine Tool Industry - Problems and Prospects in an International Perspective." Paper presented at the European Association for Research in Industrial Economics (EARIE) Conference, Bergen, Norway, August 24.

Carlsson, Bo, 1989. "Small-Scale Industry at a Crossroads: U.S. Machine Tools in Global Perspective," *Small Business Economics*, 1 (4), 245-261.

Carlsson, Bo, 1990. "Technology and Competitiveness: The Micro-Macro Links - A Comparison between the United States and Sweden," paper presented to the OECD Conference on Technology and Competitiveness, Paris, June 24-27.

Carlsson, Bo, 1991. "Flexible Manufacturing and U.S. Trade Performance," *Weltwirtschaftliches Archiv*, 127 (2), 300-332.

Carlsson, Bo (ed.), 1995. *Technological Systems and Economic Performance: The Case of Factory Automation.* Dordrecht, Boston and London: Kluwer Academic Publishers.

Carlsson, Bo, 1995a. "The Technological System for Factory Automation: An International Comparison," ch. 15 in Carlsson (*op. cit.*), 441-475.

Carlsson, Bo and Gunnar Eliasson, 1994. "The Nature and Importance of Economic Competence," *Industrial and Corporate Change*, 3 (1), pp. 687-711. Also published as ch. 3 in Carlsson (1995), *op. cit.*, 57-87.

Carlsson, Bo and Staffan Jacobsson, 1991. "What Makes the Automation Industry Strategic?" *Economics of Innovation and New Technology*, 1 (4), 257-269. Also published as ch. 8 in Carlsson (1995), *op. cit.*, 241-261.

Carlsson, Bo and Rikard Stankiewicz, 1991. "The Nature, Function and Composition of Technological Systems," *Journal of Evolutionary Economics*, 1 (2), 93-118. Also published in revised form as ch. 2 in Carlsson (1995), *op. cit.*, 21-56.

Carlsson, Bo and Erol Taymaz, 1993. "Technological Capabilities and International Competitiveness in the Engineering Industries," *Review of Industrial Organization*, 8 (3), 293-313.

Carlsson, Bo and Erol Taymaz, 1995. "The Importance of Economic Competence in Economic Growth: A Micro-to-Macro Analysis," ch. 12 in Carlsson (1995), *op. cit.*, 359-389.

Carlsson, Bo, Erol Taymaz, and Kjell Tryggestad, 1994. "Factory Automation and Economic Performance: A Micro-to-Macro Analysis," in Ove Granstrand (ed.), *Economics of Technology* (Amsterdam: Elsevier), 37-63. Also published in revised form as ch. 13 in Carlsson, *op. cit.*, 391-416.

Carlsson, Bo, Erik Dahmén, Anders Grufman, Märtha Josefsson and Johan Örtengren, 1979. *Teknik och industristruktur - 70-talets ekonomiska kris i historisk belysning* (Technology and Industrial Structure - The Economic Crisis of the 70's in Historical Perspective). Stockholm: Industriens Utredningsinstitut and The Royal Swedish Academy of Engineering Sciences.

Cohen, Wesley M. and Daniel A. Levinthal, 1989. "Innovation and Learning: The Two Faces of R&D," *The Economic Journal*, 99 (397), 569-596.

Cohen, Wesley M. and Daniel A. Levinthal, 1990. "Absorptive Capacity: A New Perspective on Learning and Innovation," *Administrative Science Quarterly*, 35 (1), 128-152.

Dahlin, Kristina, 1995. "Diffusion and Industrial Dynamics in the Robot Industry," ch. 11 in Carlsson, *op. cit.*, 323-357.

Dahmén, Erik, 1950. *Svensk industriell företagarverksamhet. Kausalanalys av den industriella utvecklingen 1919-39* (Swedish Industrial Entrepreneurial Activity. Causal Analysis of the Industrial Development 1919-39). Stockholm: IUI.

Dahmén, Erik, 1989. "'Development Blocks' in Industrial Economics" in Bo Carlsson (ed.), *Industrial Dynamics: Technological, Organizational, and Structural Changes in Industries and Firms.* Boston: Kluwer Academic Publishers.

Dosi, Giovanni, Christopher Freeman, Richard Nelson, Gerald Silverberg and Luc Soete (eds.), 1988. *Technical Change and Economic Theory.* London: Pinter Publishers.

Edquist, Charles and Staffan Jacobsson, 1988. *Flexible Automation: The Global Diffusion of New Technology in the Engineering Industry.* Oxford: Basil Blackwell.

Ehrnberg, Ellinor and Staffan Jacobsson, 1995. "Technological Discontinuities and Company Strategies — Machine Tools and Flexible Manufacturing Systems," ch. 10 in Carlsson (1995), *op. cit.*, 273-322.

Encarnation, Dennis J., 1992. *Rivals beyond Trade: America versus Japan in Global Competition.* Ithaca: Cornell University Press.

Fortune, 1992. "The Global 500," July 27, 176-232.

Granberg, Anders, 1995. "Mapping an Evolving Technology Cluster: The Composition and Structure of Factory Automation," ch. 4 in Carlsson, *op. cit.*, 89-117.
Granberg, Anders, 1995a. "The Academic Infrastructure of Factory Automation," ch. 5 in Carlsson, *op. cit.*, 119-164.
Henderson, Rebecca M. and Kim M. Clark, 1990. "Architectural Innovation," *Administrative Science Quarterly*, 35 (1), 9-30.
IMF (International Monetary Fund), 1994. *International Financial Statistics.* Washington, D.C.: IMF.
Jacobsson, Staffan, 1986. *Electronics and Industrial Policy. The Case of Computer Controlled Lathes.* London: Allen and Unwin.
Jagrén, Lars, 1993. "De dominerande storföretagen" (The Dominant Large Companies), ch. 3 in Andersson *et al., op. cit.*, 77-90.
Karlsson, Jan, 1991. *A Decade of Robotics.* Tyresö, Sweden: Mekanförbundets förlag.
Lundvall, Bengt-Åke, 1988. "Innovation as an Interactive Process: From User-Producer Interaction to the National System of Innovation," in Giovanni Dosi *et al.* (eds.), *op. cit.*
OECD, 1989. *Industrial Structure Statistics.* Paris: OECD.
OECD, 1994. *Science and Technology Policy. Review and Outlook, 1994.* Paris: OECD.
OECD (various issues). *Foreign Trade by Commodities.* Paris: OECD.
Patel, Parimal, 1993. "Localised Production of Technology for Global markets," mimeo., Science Policy Research Unit, University of Sussex, February.
Ranta, Jukka, n.d. "Economics and Benefits of Flexible Manufacturing Systems: Conclusions for Practice." Mimeo, IIASA, Laxenburg, Austria.
Stankiewicz, Rikard, 1995. "The Role of the Science and Technology Infrastructure in the Development and Diffusion of Industrial Automation in Sweden," ch. 6 in Carlsson (1995), *op. cit.*, 165-210.
Swedenborg, Birgitta, Göran Johansson-Grahn and Mats Kinnwall, 1988. *Den svenska industrins utlandsinvesteringar, 1960-1986* (Foreign Investment by Swedish Industry, 1960-1986). Stockholm: Industriens Utredningsinstitut.
United Nations (U.N.), various years. *Industrial Statistics Yearbook.* New York: United Nations.
United Nations (U.N.), various years. *Yearbook of International Trade Statistics, Vol. 1.* New York: United Nations.

4

SWEDEN'S TECHNOLOGICAL SYSTEM FOR ELECTRONICS AND COMPUTER TECHNOLOGY*

Staffan Jacobsson

INTRODUCTION

Electronics and computer technology (henceforth electronics technology) is today becoming as pervasive as mechanical technology; fewer and fewer economic areas remain untouched. This development can be traced back to the transistor in 1948 and the integrated circuit in 1960. Since that time we have seen a set of key innovations which have greatly increased the technological and commercial potential of electronics. Some of these were the microprocessor (1969); the semiconductor laser (end of the 1960s); the dynamic memory (beginning of the 1970s) and optical fibers (with their first commercial applications in telecommunications in the middle of the 1970s). Just as important was the addition of a continuous stream of computer languages and programs to these hardware developments.

Furthermore, new ways have been developed to apply these technologies in both old and new products. The first electrically driven and electronically controlled robot was put on the market in 1973; the first completely digital electronic telephone switchboard came in the second half of the 1970s and the personal computer in the late 1970s, just to mention a few of the rapidly growing number of products based on electronics technology.

In this chapter we analyze the Swedish technological system for electronics technology. A system of this kind can be delineated in three dimensions: the technological (where we distinguish between, for example, electronics, or some of its sub-technologies, and genetic engineering technology), the industrial or sectoral (where we

---

* I am grateful to the following individuals for helpful comments on previous versions of this chapter: Erik Bohlin, Bo Carlsson, Charles Edquist, Ellinor Ehrnberg, Lennart Elg, Anders Granberg, Ove Granstrand, William Ingberg, Anders Lundgren, Keith Pavitt, Sven-Ingmar Ragnarsson, Sören Sjölander, Rikard Stankiewicz and Jan Torin. I am also grateful to Keith Pavitt and Pari Patel for their generosity in sharing the SPRU Data Base on U.S. patenting with us.

distinguish between, say, the automobile and the electronics industries); and the spatial dimension (regional, national, global levels). We will restrict ourselves to analyzing how well the **Swedish** technological system for **electronics** has worked with respect to creating an internationally competitive and dynamic **electronics industry**.[1]

The chapter is structured as follows. The performance of the system is discussed in section 2. In section 3, we discuss three central components of a technological system: the performance of the academic system, the character of economic competence, and government policy. Finally, in section 4, we summarize the main results and argue that it is mainly these factors which, separately, as well as through feedback mechanisms, have shaped a technological system for electronics which is not as strong as it ought to be.

THE PERFORMANCE OF THE SWEDISH TECHNOLOGICAL SYSTEM FOR ELECTRONICS TECHNOLOGY

The performance of the Swedish technological system for electronics technology will be assessed using two different types of indicators. The first consists of production and trade indicators from the Swedish electronics industry. The second type of indicators are science and technology related.

Production and Trade Performance of the Swedish Electronics Industry

The Swedish electronics industry[2] is quite small in relation to value added in manufacturing (ISIC 3). In Sweden, the combined value added of the computer, heavy electrical machinery, telecommunications and precision instrument industries constituted only 8.3 percent of manufacturing value added in 1988. This is on a par with Finland (7.3), Norway (5.9) and Denmark (7.6) but significantly lower than West Germany (15.9), the Netherlands (14.4), USA (15.1) and Japan (16.1) (OECD 1991).

Of course, Sweden (and Finland) has a large resource based industry, but the smallness of the Swedish electronics industry is revealed also if we look at its share of the engineering industry. While the figure for Sweden was 18.9 percent (Finland 24.1,

---

1 Electronics is, of course, a set of technologies used by a whole range of actors, including those which normally are seen to form the 'electronics industry.' Electronics technology can also be of vital importance to firms outside the electronics industry.

2 Three definitions of the electronics industry are used in our work. First, the Swedish Central Bureau of Statistics (SCB) uses a definition which is based on detailed assessment of the actual output of firms. This definition is employed when we discuss production, exports and imports of electronics products in Sweden only. Second, where international comparisons of production are made, we use an approximation of the industry by aggregating the following ISIC numbers: 3825, 383 and 385. The overlap with the SCB definition is great but not complete. Third, where international trade comparisons are made, we include in the electronics industry the following SITC Rev.2 categories: 716, 75, 76, 77, 87 and 88.

Denmark 22.4, Norway 19.6) it was 32.1 in West Germany, 44.1 in the Netherlands, 34.6 in the USA and 34.9 in Japan (OECD 1991).

The trade balance of the electronics industry went from slightly positive in 1975 to negative in the 1980s and remained negative in the 1990s; see table 4-1.[3] Data on the evolution of international competitiveness[4] is available for eight sub-sectors. See table 4-2. The data reflect Sweden's strength in communication technologies and, to a minor degree, in medical electronics. The weaknesses in consumer electronics, components, office machinery and computers are also apparent.

Of course, a negative trade balance in the electronics industry is not unique to Sweden; most countries appear to be in this position. As can be seen in table 4-3, all but three of the countries analyzed have a negative trade balance.[5] However, just as importantly, both Germany and Switzerland show a positive trade balance in electronics.[6] It is clearly possible, even for a small European country, to achieve a positive trade balance in electronics.

Table 4-1 Net Export Ratio* of the Swedish Electronics Industry, 1975-1992

Year	Net export ratio
1975	0.07
1980	0.03
1985	-0.01
1988	-0.07
1989	-0.09
1990	-0.07
1991	-0.06
1992	-0.05

* (X-M)/(X+M) (i.e. exports minus imports divided by total trade).

Source: Elaboration on SCB (1991, 1992, 1994)

3 It is not clear, however, to what extent all trade in power electronics and military electronics, two strong fields in the Swedish electronics industry, is captured by these statistics. Moreover, there are exports of electronics which are incorporated in mechanical engineering products, such as automobiles. Some of the electronics exports are therefore hidden. This applies, of course, also to imports.

4 See Ohlsson (1976) for a discussion of various indicators of international competitiveness.

5 See Appendix 4-1 for a discussion of the data used.

6 In table 4-3, we have included electrical machinery in the data, which may explain the discrepancy between tables 4-11 and 4-3 in the case of Sweden. This inclusion may seem questionable, but power electronics is today an important application area for electronics. Thus, if the power sector were to be excluded, we would underestimate the trade in electronics. In the case of West Germany in 1990, if we exclude SITC Rev.2 and 3, sectors 716, 773 and 778, which can, perhaps, be looked upon as clearer cases of electrical engineering, the trade balance is, however, still positive. For SITC Rev.3 the trade balance was 1.11 and for SITC Rev.2 it was 1.05.

Table 4-2 Net Export Ratios* for the Swedish Electronics Industry, 1975-1992

Year	Telecom-munication	Radiocom-munication	Consumer[a] electronics	Computers	Office machinery	Medical electronics	Industrial[b] electronics	Electronic components
1975	0.88	-0.06	-0.33	-0.11	-0.10	0.28	-0.20	-0.42
1980	0.84	0.21	-0.41	-0.09	-0.09	0.34	-0.09	-0.39
1985	0.73	0.14	-0.41	-0.14	-0.22	0.31	-0.07	-0.48
1988	0.63	0.22	-0.61	-0.27	-0.33	0.11	-0.11	-0.21
1989	0.58	0.22	-0.65	-0.28	-0.41	0.18	-0.12	-0.25
1990	0.62	0.21	-0.65	-0.31	-0.46	0.24	-0.10	-0.25
1991	0.58	0.27	-0.70	-0.33	-0.43	0.32	-0.04	-0.40
1992	0.58	0.34	-0.70	-0.38	-0.48	0.20	0.04	-0.40

* (x-m)/(x+m)
a Excl. microwave ovens
b Excl. robots

Source: SCB (1991, 1992, 1994).

Table 4-3 Trade Balance (Exports minus Imports) in the Electronics Industry* in OECD Countries, 1978-1992, selected years

	1978	1984	1990	1992
France	1.01	0.94	0.80	0.86
West Germany	1.68	1.31	1.13	na
Italy	1.07	0.91	0.76	0.76**
Japan	6.98	7.95	4.80	4.74**
Netherlands	0.98	0.90	0.81	0.83
Sweden	0.97	0.89	0.82	0.84**
Switzerland	1.97	1.45	1.31	1.34**
U.K.	1.09	0.78	0.86	0.89**
U.S.A.	1.14	0.76	0.79	0.74

* Defined as the following SITC Rev.2 categories: 716, 75, 76, 77, 87 and 88.
** 1991

Source: Elaboration on data kindly supplied by the United Nations Economic Commission for Europe.

Science and Technology Indicators

We will use three science and technology indicators: R&D expenditures, patents granted in the U.S.A., as well as bibliometric data.

Data on R&D expenditures (at the product group level)[7] in the electronics industry, see table 4-4, indicates that the allocation of resources to R&D, in real terms deflated by the salary increases of engineers,[8] grew by 5.7 percent per annum between 1983 and 1991.

This is not a high figure in an international perspective. In table 4-5, we have calculated the annual growth rate of R&D expenditures in the electronics industry (defined as electrical machinery, electronic equipment and components, instruments and office machinery & computers) in a number of OECD countries. Sweden ranks fairly low in this group; most countries have a higher growth rate and Sweden's position is above the average growth rate only due to the weak performance of the larger R&D spenders, the U.S. and the U.K.

---

7 This is in contrast to the branch level where the R&D is allocated to the ISIC group which dominates in the individual firm or plant.

8 I am grateful to Kristina Lindros at the Association of Engineers for supplying the data used to deflate the R&D figures.

Table 4-4 R&D Expenditures and Their Annual Growth Rate with Respect to Product Group Levels in the Swedish Electronics Industry*, 1983-1991, in 1983 Prices*

Year	R&D expenditure (billion SEK)	Annual growth rate
1991	4,652	
		6.9%
1989	4,069***	
		6.0%
1987	3,620	
		3.6%
1985	3,369	
		6.3%
1983	2,984	

* ISIC 3825, 383 and 385.
** The deflator used is the annual salary increases of all kinds of engineers.
*** The official data underestimate the expenditure by 1 billion SEK, according to officials at the Central Bureau of Statistics. That sum is included in the table.

Sources: 1991: Data received from the Central Bureau of Statistics. 1989: SCB U14 SM 9101. 1987: SCB U14 SM 8901. 1985: SCB U14 SM 8701. 1983: SCB U14 SM 8501.

In terms of patents granted in the U.S.A., the data in table 4-6 show that the Swedish share of all electronics patents is quite low (column 2) and that it declined in the period after 1980. However, this decline was not due to a decline in the number of Swedish electronics patents but to a rapid increase in all electronics patents granted in the U.S. to innovators.[9] Moreover, as indicated in column 4, Sweden became progressively less specialized in electronics in the whole period studied. Thus, the period shows a decline both in terms of Sweden's absolute share of U.S. patenting and the relative emphasis of electronics vis-à-vis other technological areas.

Finally, the share of published articles in electronics and computer science is much lower in Sweden than in the world as a whole. Moreover, Sweden's share in electronics actually declined between the period 1970-75 and 1985-1990. See table 4-7.

9 This increase in patenting in electronics may possibly be explained by an increased propensity among Japanese firms to patent in the U.S.. However, as was shown in table 4-5, the growth of R&D in electronics was much higher for Japan than for other large countries in the 1980s.

Table 4-5 Annual Cumulative Growth Rates of R&D in the Electronics Industry in OECD Countries, 1981-1989 (in constant 1985 U.S. dollars and purchasing power parities)

Country	Growth rate (%)
Spain	19.0
Japan	12.1
Finland	9.9
Italy	9.7
Norway	9.6
Germany	8.7
Belgium	8.5
Sweden	6.9
ALL COUNTRIES	6.4
France	6.0
USA	4.3
U.K.	1.9

* According to an official at the Swedish Central Bureau of Statistics, the official Swedish data underestimate the Swedish R&D in the electronics industry by approximately 1 billion SEK in 1989. A billion SEK has been added to the official data in this table, although in a deflated form.

Source: Elaboration on OECD (1991, 1993).

Conclusions

In conclusion, the Swedish electronics industry is

* relatively undersized [10] and Sweden has
* a negative trade balance in electronics.

These traits are shared with many other developed countries but there are countries other than Japan, including one small European country, which have a positive trade balance. The technology indicators reveal

* a less than impressive growth in R&D in the electronics industry;
* a declining share of U.S. patenting in electronics and in the relative importance of electronics in Swedish patenting; and
* relatively little emphasis on electronics and computer science in scientific publications.

[10] This does not, of course, necessarily mean that other industrial sectors, such as the paper and pulp industry, are 'oversized.'

Table 4-6 Sweden's Share of Electronics Patents Granted in the U.S. and Sweden's Revealed Technological Comparative Advantage in Electronics, 1963-1990

	(1) Number of electronics* patents	(2) Share of electronics patents	(3) Share of all patents	(4) RTCA**
1963-1968	455	0.86%	0.69%	0.80
1969-1974	777	1.11%	0.82%	0.74
1975-1980	729	1.30%	0.91%	0.70
1981-1986	673	1.16%	0.75%	0.65
1987-1990	497	0.96%	0.53%	0.55

* Instruments and controls, Photography and photocopy, Image and sound equipment, Calculators, computers and office equipment, Semiconductors, and Telecommunications.
** Revealed Technological Comparative Advantage, which is the share of Swedish patents in electronics divided by the Swedish share of all patents taken in the U.S.A. in each time period.

Table 4-7 Revealed Scientific Advantage within the Engineering Field in Sweden, 1970-1975 and 1985-1990

	1970-1975	1985-1990
Electrical engineering	0.72	0.82
Electronics and telecommunications	0.90	0.73
Computer science	0.71	0.75
Optics	0.98	0.91

Source: Persson 1991, based on online study of Compendex.

Thus, it appears that the Swedish technological system has not worked very well with respect to creating an internationally competitive and dynamic electronics industry.[11] Nor does it appear from the science and technology indicators that a major improvement in the performance of the electronics industry can be expected in the years to come.

This is quite disturbing for several reasons. First, the rapid technological and product development hinted at in the Introduction has had large and direct economic

---

[11] In contrast, it appears that the diffusion of electronics-based products is often rapid and extensive in Sweden. Sweden belongs to those countries which have the greatest 'density' in the use of a set of various products such as mobile telephones (Oskarsson and Sjöberg 1991), computers (Näringsdepartementet 1991) and a set of products used for automating the engineering industry (robots, numerically controlled machine tools, computer aided design, flexible manufacturing systems) (Carlsson and Jacobsson 1994). Other work (SIND 1988) suggests that the use of electronics in production processes in general is extensive in Sweden.

effects in that, globally speaking, the electronics and computer industry has for some time now had a higher growth rate than the average for the manufacturing sector and is today as large as the automobile industry (Näringsdepartementet 1991). In other words, the market growth for electronics products has been greater than for the manufacturing sector as a whole for many years (Edquist and McKelvey (1991, tables 18 and 19)).

Second, the absolute level of productivity is higher for electronics products than for the manufacturing sector. In the case of USA, Tyson (1992) shows that the difference is quite substantial. This means that the size and growth of the electronics industry has a direct effect on the productivity level of industry as a whole.[12]

Third, indirect and positive effects on the economy from having a strong electronics industry may also be identified, for example in terms of fostering a faster diffusion of electronics products (see Carlsson and Jacobsson 1994).

THREE ELEMENTS OF THE SWEDISH TECHNOLOGICAL SYSTEM FOR ELECTRONICS TECHNOLOGY

This section discusses how three central elements of the Swedish technological system for electronics technology may have influenced its somewhat dismal performance: the behavior of the higher educational sector, the character of the economic competence in Swedish industry, and government technology policy.[13]

Higher Education

We postulate that the number and quality of people with higher education in a specific technological field is an important determinant of the extent to which various actors perceive, acquire and exploit that technology. We further postulate that the size and quality of academic R&D in a specific technological field can have a strong influence on how an economy exploits the opportunities of a new technology. The response of the higher educational system to the rise and diffusion of new generic technologies is

---

[12] In the Swedish case, the absolute productivity (value added divided by the number of employees) in electronic products was only marginally higher in 1989 than in manufacturing as a whole. For ISIC 3825, 3832, 3831 and 385, the value added per employee was 410,000 SEK while for the manufacturing industry it was 398,000 SEK. However, in the period 1973-1989, the electronics industry, defined as above, increased its productivity much faster than the manufacturing industry. In current prices, the manufacturing sector increased its productivity by a factor of 5.6, while the electronics industry productivity grew by a factor of 6.6. If these differences in rates of change of productivity continue, the differences in absolute levels of productivity will, of course, widen (elaboration on SCB 1975 and 1991b).

[13] The specific knowledge/competence networks operating in this technological system will not be discussed in detail. For an analysis of the emergence of the strong Swedish position in mobile telephony, see Oskarsson and Sjöberg (1991).

therefore seen to be a critical factor in the adjustment process of a nation's industrial structure.

We will analyze the response of the Swedish academic system using two indicators: the rate at which the 'output' of electronics and computer science specialists increased in the period 1970-1989, and the number of new chairs instituted in electronics and computer science at the technical universities.[14]

Table 4-8 Number of Students Graduated and Enrolled in Electronics/Computer Science and in All Engineering in Swedish Universities of Technology, 1962/63-1988/89

	Graduated			Newly enrolled		
	(1) Engi-neering	(2) Elec-tronics*	(3) (2)/(1)	(4) Engi-neering	(5) Elec-tronics*	(6) (5(/(4)
1962/63	887	244	0.28	1,237	359	0.29
1971/72	1,577	334	0.22	3,110	761	0.24
1972/73	1,546	331	0.21	2,942	669	0.23
1973/74	1,857	418	0.23	3,043	641	0.21
1974/75	1,682	421	0.25	3,167	691	0.22
1975/76	1,749	403	0.23	3,192	651	0.20
1976/77	1,945	436	0.22	3,463	723	0.21
1977/78	1,984	397	0.20	3,114	755	0.24
1978/79	1,850	378	0.20	2,980	702	0.24
1979/80	1,880	403	0.21	3,058	755	0.25
1980/81	2,026	486	0.24	3,033	757	0.25
1981/82	2,002	478	0.24	3,091	697	0.23
1982/83	2,049	496	0.24	3,793	1,006	0.27
1983/84	2,232	570	0.26	3,781	1,081	0.29
1984/85	2,037	509	0.25	3,951	1,209	0.31
1985/86	2,184	545	0.25	4,173	1,345	0.32
1986/87	2,579	728	0.28	4,271	1,354	0.32
1987/88	2,527	721	0.29	4,050	1,330	0.33
1988/89	2,590	705	0.27	4,363	1,384	0.32

*Electrical/electronics and computer science engineers

Source: SCB (1990:10).

[14] These are indicators of change of the academic system which can be used in international comparisons and presumably precede changes in R&D, patents and bibliometric data.

Table 4-9 Number of B.Sc. and M.Sc. Graduates in Computer Science and Electrical/Electronic Engineering in the United States, 1971-1988 (in thousands)

	Computer Science*			Electrical/Electronics			
	B.Sc.	M.Sc.	Total	B.SC.	M.SC.	Total	TOTAL
1971	2.4	1.6	4.0	12.2	4.3	16.6	20.6
1975	5.0	2.3	7.3	10.2	3.5	13.7	21.0
1976	5.6	2.6	8.2	8.8	3.8	12.6	20.8
1977	6.4	2.8	9.2	9.9	3.8	13.7	22.9
1978	7.2	3.0	10.2	11.2	3.7	14.9	25.1
1979	8.7	3.1	11.8	12.4	3.6	16.0	27.8
1980	11.2	3.6	14.8	13.8	3.8	17.6	32.4
1981	15.1	4.2	19.3	14.9	3.9	18.8	38.1
1982	20.2	4.9	25.1	16.5	4.5	21.0	46.1
1983	24.5	5.3	29.8	18.0	4.5	22.5	52.3
1984	32.1	6.2	36.3	19.9	5.0	24.9	61.2
1985	36.9	7.1	44.0	21.7	5.1	26.8	70.8
1986	41.9	8.1	50.0	23.7	5.5	29.2	79.2
1987	39.7	8.5	48.2	24.6	6.2	30.8	79.0
1988	34.5	9.2	43.7	23.6	6.7	30.3	74.0

* Computer and information sciences

Source: U.S. Department of Education, National Center for Education Statistics, *Digest of Education Statistics*, 1988, 1989 and 1990.

One may distinguish between a reactive or a proactive science and education policy. A reactive policy derives the demand for research and for people with a higher education in, say electronics, from the present industry's demand. A proactive policy includes an element of 'technology push' and tries to supply in accordance with the demand from a future industrial structure. Moreover, as supply can, in part, determine demand (OECD 1992), a proactive policy can directly contribute to the restructuring of industry. We will argue in this section that the Swedish academic system has, on the whole, been reactive with respect to electronics technology. Indeed, the adjustment has been so slow and weak that it is questionable if the demand from even the present quite limited Swedish electronics industry has been met.

Adjustment of the Composition and Size of Output of Engineers. Table 4-8 gives the data for the number of graduated and newly enrolled engineering students in total and in electronics/computer science. A number of observations can be made from this table. First, the share of electronics engineers of the total number of graduated engineers was not larger in 1988/89 than in 1962 (28 percent) and indeed, this share actually declined in the 1970s.

Second, the share of newly enrolled students in electronics did not begin to increase until 1982/83 when it rose above 25 percent. This increase can largely be explained by the start of a Computer Science education in Gothenburg, Lund and Stockholm. In Linköping, this education started in 1975/76.

Third, it was only in the second half of the 1980s that the supply of electronics engineers increased in a substantial way.[15]

It would thus seem as if the whole of the 1970s was lost in terms of enrolling and graduating electronics and computer science students. It was only in the second half of the 1980s that a significant increase in degrees awarded took place. The question is, is this response poor in comparison with other countries?

Data from the United States suggest that the adjustment period in Sweden was unnecessarily long. In table 4-9 we can see time series for graduated engineers and computer scientists at the B.SC. and M.SC. levels in the United States.

The most noteworthy feature of this table is that there was already a substantial output of computer scientists at the end of the 1970s in the United States while this was not the case in Sweden.

The U.S. also responded faster with electrical/electronic engineers; the output of electronics engineers grew greatly in the first half of the 1980s while in Sweden it grew in the second half. The faster adjustment in the U.S. is particularly clear in table 4-10, where the number of graduated electronics engineers and computer scientists are set in relation to the size of the population, a crude measure of the importance of this field in

---

[15] Two Computer Science degrees at non-engineering universities have to be added to those given at the technical universities (shown in table 4-8). Data on newly enrolled students and degrees awarded in non-engineering universities are provided below.

Newly Enrolled Students and Degrees Awarded in Computer Science* in Non-Engineering Universities in Sweden

	Newly enrolled	Graduated
1977/78	122	-
1978/79	256	-
1979/80	252	15
1980/81	286	31
1981/82	320	79
1982/83	577	117
1983/84	583	149
1984/85	793	168
1985/86	841	227
1986/87	824	303
1987/88	834	356
1988/89	860	352
1989/90	864	354

* Systemvetenskapliga och Datavetenskapliga linjerna

Source: Information received from SCB.

the two countries. While the U.S. density was approximately twice that of Sweden's until 1977, it increased to nearly four times as many in 1985. Thus, the United States increased the output of electronics engineers and computer scientists at a much earlier stage than Sweden and (although the indicator is rather crude) produces far more graduates in electronics technology than Sweden. The U.S. system of higher education clearly reacted more vigorously than the Swedish one.

Table 4-10 Density* of Graduated Computer/Information Scientists and Electrical/Electronics Engineers in USA and Sweden, 1971-1989

	(1) USA	(2) SWEDEN**	(3) (1)/(2)
1971	0.91	0.42	2.17
1975	0.92	0.52	2.19
1976	0.92	0.50	1.84
1977	1.01	0.55	1.84
1978	1.11	0.50	2.22
1979	1.22	0.47	2.60
1980	1.43	0.52	2.75
1981	1.68	0.65	2.58
1982	2.03	0.70	2.90
1983	2.30	0.77	2.99
1984	2.70	0.90	3.00
1985	3.12	0.85	3.67
1986	3.49	0.97	3.63
1987	3.48	1.29	2.69
1988	3.26	1.35	2.41
1989	n.a.	1.32	n.a.

* Number of graduates divided by the population in 1980 x 10,000 (227 and 8 million, respectively).

** As is indicated in tables 8 and 9, the statistics are given for the academic year (e.g. 1972/3). Using the data in those tables to construct this table we have allocated the graduates to the later year, i.e. 1972/73 is changed to 1973.

Sources: See tables 4-8 and 4-9.

The contrast with Japan is just as great. In 1982 (the only year for which we have data) 19,428 B.SC. students graduated in electrical and electronics engineering while 1,671 received the M.SC. degree. The ratio between the Japanese and Swedish density for electrical/electronics engineers was as high as 3.0![16] Germany also had a significantly larger output of computer scientists than Sweden at the end of the 1980s (Näringsdepartementet 1991).

The significance of this slow Swedish response in computer science and electronic engineering education can be illustrated by a simple example.

In table 4-11, we have made an estimate of the ceiling for annual R&D growth in the period concerned for the electronics industry by estimating the maximum growth of electronics engineers who were 'available' for the R&D function in the electronics industry in the middle of the 1980s.

Table 4-11 An Estimate of the Ceiling to R&D Expansion in the Swedish Electronics Industry* in 1985/86

(1)	Total number of engineers	5,911
(2)	Total number of engineers in R&D and design work	3,680
(3)	Total number of electronics engineers in R&D and design work**	1,904
(4)	Maximum gross addition of electronics engineers to the R&D and design functions***	82
(5)	Line (4) divided by line (3)	4.3%

* Defined as ISIC 3825, 3831, 3832, 3845 and 385.

** We have assumed that the share of electronic engineers in the R&D and design function is the same as for all types of engineers in these five industries.

*** 537 (number of graduated electronics engineers) *times* 0.245 (which is the share of newly graduated electronic engineers that went to the five industries in the period 1981-1987) *times* 0.62 (which is the assumed share of electronics engineers that went into the R&D function).

Source: The Swedish Central Bureau of Statistics provided a special printout of the number of all engineers in various functions in the five industries. The data on the share of the newly graduated electronic engineers that went into the five industries derive from elaboration of data provided by the Swedish Association of Graduate Engineers.

The significant result is that the figure for the maximum addition of electronics engineers to the R&D function (82 engineers) is indeed very small.[17] Although some substitution of people with lower degrees or other backgrounds may have taken place and the sourcing of technology from the service sector probably expanded, the supply of new electronics engineers would appear to have set a fairly strict limit to the growth of R&D in existing firms and for the growth possibilities of the new technology-based firms, for instance those which from the early 1980s began to be spun off from the universities in quite large numbers (Olofsson and Wahlbin 1993). Indeed, against this

---

16 The population of Japan was 116 million according to the World Bank (1981). 73,593 persons graduated with a B.SC. in engineering while 7,363 graduated with a M.SC. (Fishwick 1988). For all types of engineers, the density amounted in that year to 2.76.

17 Using the SCB (1991) definition of the electronics industry raises the figure to 133.

background, the figures we saw in table 4-4 of the rather small increase in R&D in the electronics industry are not very surprising. A necessary condition for industrial renewal seems not to have been fulfilled.[18]

The Adjustment Pattern for Professors in Electronics and Computer Science at the Technical Universities. Perhaps the best indicator of the importance the state places on developing both general and, in particular, specific competence in an expansive technological field is the creation of new chairs. The number of new chairs is expected to influence the rate and quality of the expansion of both graduate and undergraduate education as well as research. Indirectly, we would expect the indicator to be linked to the rate of science-based spin-offs from the universities. For instance, it comes as no surprise that more than half of the spin-offs from the University of Linköping are within electronics and computer science (Olofsson and Wahlbin 1993), since this discipline was chosen as the specific profile of that University at its start at the end of the 1960s. Moreover, we would expect it to be positively related to the existence of various types of spill-over effects from academia to industry (Feldman 1994).

In table 4-12, we have listed all new chairs in six technical universities in electronics, computer science as well as parts of the field of physics. It is noteworthy that with the exception of Linköping Institute of Technology, there was very little emphasis on creating new chairs in electronics technology until the second half of the 1980s. Although we do not have access to international data for a comparative analysis, the response seems slow, given the increase in technological opportunities as early as the beginning of the 1970s.

Conclusions. In conclusion, both these indicators suggest a slow and weak response to the emergence and diffusion of electronics technology by the Swedish academic system. It was not really until the second half of the 1980s that significant changes took place, in terms of both number of students and new chairs. On the whole, the educational policy can be seen as reactive.

---

[18] Another way of calculating the influence of the low level of output of electronics engineers is the following. The share of the electronics industry (defined here as ISIC 3825, 3831, 3832 and 385) of the engineering industry's value added amounted to approximately 20 percent in both 1981 and 1989. In Germany, this share was 32 percent in 1988. Assume that we would have liked the Swedish industry to have achieved this share by 1989. That would have meant a growth in value added of 90 percent (instead of the actual 20 percent). How many more electronics engineers would have been needed? In the period 1981-1989, the electronics industry (using the SCB definition) increased its number of electronics engineers by approximately 1,700 (Jacobsson *et al.* 1989) but only managed to grow at the same rate as the engineering industry as a whole as indicated by a constant share in its value added (by 1989). If we assume a fixed ratio of changes in the stock of engineers to changes in value added, the electronics industry would therefore have had to recruit four and a half times the number of electronics engineers. This would amount to 7,800 which can be seen in contrast with the entire graduation of 4,800 in the period 1981/2-1988/89. In practice, the rather limited supply of electronics engineers coupled with a strong demand for these skills from other actors in the technological system, meant that very few of these 4,800 electronics engineers were 'available' to the electronics industry. Considering that only about 25 percent of all newly graduated electronics engineers went to the electronics industry (Jacobsson *et al.* 1989) in the period 1981 to 1987, only around 130 per year were 'left' for all functions in these industries.

Table 4-12 The Growth of New Chairs* in Electronics, Computer Science and Parts of the Physics Field at Swedish Universities of Technology, 1969-1991

	CTH	KTH	LTH	LiTH	LuTH	UUTH	TOTAL
Prior to 1969	12	10	5	-	-	1	28
1969-1974	-	3	2	11	-	-	16
1975-1979	2	-	1	1	1	-	5
1980-1984	3.5	1	-	3	1	2	10.5
1985-1991	3	5	4	9	4	1	26

CTH - Chalmers University of Technology
KTH - Royal Institute of Technology
LTH - Lund Institute of Technology
LiTH - Linköping Institute of Technology
LuTH - Luleå Institute of Technology
UUTH - Uppsala Institute of Technology

* Wherever possible, existing chairs that have been redesignated have been excluded. Details on the compilation of the data can be obtained from the author.

Sources: Data received from the various universities of technology and Ferm (1989).

The Character of Economic Competence in Swedish Industry

The Swedish economy is greatly dependent upon a few industries for its earnings of foreign exchange. This is not surprising; indeed, for a small country, access to a world wide market will tend to lead to a relatively high degree of 'specialization' with a consequently higher degree of product concentration of exports than that which is usually found in larger countries (Soete 1988). Indeed, much of the Swedish trade surplus stems from just a few industries: paper and pulp, wood products, ferrous metals, non-electrical machinery and automobiles. Central to these industries is metalworking and mechanical engineering technology.

Obviously, a small country with relatively narrow capabilities can gain substantially from trade.[19] At the same time, the product specialization may well be associated with a process where the traditional technological capabilities will be built upon and strengthened.

Such specialization is fine as long as the traditional capabilities can be used to create products and services that continue to give a high economic value in the world

---

[19] Still, for a small country, Sweden's range of capabilities is quite broad.

market. However, to the extent that technical change shifts the locus of industrial growth to largely different technological capabilities, a strong technological specialization might make it difficult to follow suit in the process of structural change. Put in other terms, the specialization pattern might be more or less rigidly 'locked' into traditional industrial activities (Soete 1988), at the expense of expansion in new areas.

The problems of a possible 'locked in' specialization pattern refers fundamentally to the behavior of firms with respect to their choice of business strategy. This strategy (its product/market choice) depends, in part, on the managerial capability in the firms. This includes the ability to identify, expand and exploit a set of technological and commercial opportunities: economic competence in Carlsson and Eliasson's (1994) terminology. This ability is unevenly distributed among firms. Moreover, it is reasonable to assume that the knowledge firms have regarding the opportunity set is not only highly differentiated but also quite limited. As Eliasson (1990) argues:

"...agents are not marginally uninformed as in classical search theory, but grossly ignorant, and differentially informed about small subsets of an immense state space. To become informed becomes an experimental learning process."

Firms therefore operate with different knowledge bases and under different assumptions concerning technology, markets, competitors, etc. The variations in the set of assumptions guiding management in their investment decisions, including their strategic choices, could be seen as arising from long historical processes, central to which are learning processes which are local and cumulative as well as path dependent (David 1988, Dosi *et al.* 1991). As Dosi *et al.* (1991) explain:[20]

"... a firm's previous investments and its repertoire of routines (its 'history') constrain its future behavior. This follows because learning tends to be local. That is, opportunities for successful new developments will be 'close in' to previous activities..."

A particularly important subset of management's experience and competence can be seen as being derived from the firm's technology base. The particular type of 'path' that the firm is set on (involving a particular technological trajectory) would be expected strongly to shape the technological and commercial opportunity set, as perceived by management. Allocative competence (what to produce and how to price it) and transactional competence (the make/buy/share decision) (Dosi *et al.* 1991) can therefore be expected to be intertwined with and dependent on technological competence.[21] The response capacity to technological changes can therefore be expected

---

20 Fransman (1990) introduces, in a similar way, the concept of 'bounded vision' which suggests that different kinds of organization have different limitations to their field of 'vision', that is (Fransman 1990, p.3),

"... their ability to perceive the importance of various areas of science and technology and to make serious attempts to assimilate and develop these areas...the resulting bounded vision implies that new technologies emerging from neighboring areas where the corporation does not have current activities are likely to take some time to penetrate the corporation's field of vision."

21 This is recognized Carlsson and Eliasson's (1994) definition of economic competence. Techno-economic competence may therefore be a more suitable term.

to vary between firms moving along very different technological paths where the perception of the opportunity set is bounded by the experience accumulated by management.

When new generic technologies become available, offering vast technological opportunities, the technological competence accumulated (shaped by long learning processes and constrained by a particular technological path) over many years by managers (and other actors in the surrounding technological systems) may limit the understanding of these opportunities and influence the reaction of firms to these opportunities.

In what follows we will explore how well available evidence is consistent with this view of 'how history matters.' We will first analyze the evolution of revealed technological comparative advantage for Sweden and argue that the historically very strong specialization in metals and mechanical engineering has been further strengthened at the expense of the newer field of electronics. We will subsequently suggest that this picture at the aggregate level is due not only to a lack of diversification of the existing firms into electronics but also to a lack of growth of smaller technology-based firms in electronics.

Revealed Technological Comparative Advantage - Metal/Mechanical Engineering Grows Stronger.

The evolution of the revealed technological comparative advantage of Sweden suggests that Swedish industry has been strongly guided by its history with respect to its approach to electronics technology. Using the Science Policy Research Unit data base on patenting in the U.S.A., we have aggregated the initial 33[22] classes into four categories of 'generic' technologies: chemicals, metal and mechanical engineering, electrical engineering, and electronics and computer engineering.

Table 4-13 shows that Sweden had quite a strong revealed comparative advantage in metal and mechanical engineering technology in the 1960s. This specialization became even more pronounced in the twenty years following.[23] At the same time, as was mentioned in section 2, Sweden's revealed comparative advantage in electronics technology was initially weak and, furthermore, gradually deteriorated in the period

---

[22] Seven classes were excluded as it was not possible to place them within a 'generic' technology label. Altogether, these classes represented 13 percent of Swedish patenting in the United States for the period 1981-1986.

[23] Other studies, using a different data base and a different classification procedure, show much the same picture. Noteworthy from these studies (Archibugi and Pianta 1992, Engelsman and van Raan 1990) is the very strong revealed advantage of Sweden in space vehicles and guided missiles as well as ordnance. In other words, Sweden is strong in defense and space electronics. Sweden patents a great deal also in paper, which is not surprising given the critical position of that industry in its industrial structure.

studied. Hence, taken together, Swedish industry increased its specialization in its traditional areas of competence in this period of time.[24]

Moreover, those areas in electronics technology where Swedish industry patents a lot are control systems and instruments, for both civilian and military applications. This category accounted for about half of the electronics technology patents in the period 1963-1990 (elaboration on data supplied to the Science Policy Research Unit by the U.S. Department of Commerce, Patent and Trademark Office).

Table 4-13 Revealed Technological Comparative Advantage[a)] in Swedish Industry, 1963-1990

Industry	1963-68	1969-74	1975-80	1981-86	1987-90
Metal and mechanical*	1.31	1.32	1.47	1.53	1.55
Electrical**	0.99	0.93	0.87	0.87	0.88
Electronic/computer***	0.80	0.74	0.70	0.65	0.55
Chemicals****	0.43	0.50	0.50	0.45	0.61

a) The share of Swedish patents in each of the four areas divided by the Swedish share of all patents taken in the U.S.A. in each time period.
* Metallurgical and other mineral processes; Apparatus for chemicals, food, glass, etc.; General non-electrical machinery; Non-electrical special, industrial equipment; Metallurgical and metal working equipment; Assembling and material handling apparatus; Nuclear reactors and systems; Power plants; Road vehicles and engines; Other transport equipment; Mining and wells machinery and processes; and Miscellaneous metal products.
** General electrical industrial apparatus, electrical devices and systems
*** Instruments and controls; Photography and photocopy, image and sound equipment; Calculators, computers and office equipment; Semiconductors; and Telecommunications.
**** Inorganic chemicals; Organic chemicals; Agricultural chemicals; Chemical processes; Hydrocarbons; Drugs and bio-affecting agents; Mineral oils and fuels, etc.; and Bleaching, dyeing and disinfecting.

Source: Elaboration on data supplied to the Science Policy Research Unit, by the U.S. Department of Commerce, Patent and Trademark Office.

In this area, the critical element in the innovative process is application knowledge, often in mechanically dominated areas (such as pulp and paper production), rather than leading edge knowledge in electronics technology.[25] Typically, Volvo's electronics patents are mostly in this category. Hence, a significant part of the Swedish patenting

---

[24] There are some indications that Swedish industry could have developed differently, i.e. that there were many lost opportunities. For instance, in the 1950s, Sweden had the most advanced computer, BESK, in the world, but it was decided not to continue to pursue that development. The Facit company, which was very large in mechanical calculators, had developed an electronic calculator earlier than the Japanese, but top management decided not to develop it (Ingberg 1993). Office automation systems were developed in the 1970s for demanding local customers. Finally, the rising share of Sweden in all electronics patents granted in the U.S. until 1980 also suggests some vitality in that period.

[25] A great deal of patenting occurs in the field of military electronics, too.

activity lies in 'mechatronics' (including medical electronics) which 'lies closer' to Sweden's traditional core areas than 'pure' electronics products.[26]

Patenting at the Firm Level. A study of patent activities (in the U.S.A.) at the firm level, in the period 1969-1988, provides additional insights. In table 4-14, three groups of firms can be identified. These firms accounted for nearly three-quarters of all electronics technology patents in the period 1969-1973 and more than two-thirds in 1984-88.

Table 4-14 Patenting Activity* in Electronics in Three Groups of Swedish Firms, 1969-73 and 1984-88

	Number of patents in electronics	
	1969-1973	1984-1988
Ericsson	113	106
ABB	100	75
SAAB	22	17
Bofors	23	19
Philips	11	10
Atlas Copco	3	2
Sandvik	1	1
Alfa Laval	1	3
Electrolux	13	5
SKF	9	6
Tetra Pak	1	2
SCA	0	0
Stora	0	1
Siemens	5	15
Volvo	2	17
Hasselblad	5	11
IBM	2	9
AGA	40	0

* Patents registered in the U.S.A.

Source: Elaboration on data supplied to the Science Policy Research Unit, by the U.S. Department of Commerce, Patent and Trademark Office.

[26] Moreover, it is reasonable to expect that a fair share of these patents were granted to firms which are not classified as electronics firms; their electronics activities are therefore 'hidden'.

First, there are those firms which already had a significant activity[27] in this technology field in the 1960s; Ericsson in radio and telecommunications (both civilian and military), ABB in e.g. process controls and in power electronics, SAAB in computers and military electronics, and Bofors in military electronics. These firms constitute the dominating patentees in electronics and are the main suppliers in three clusters built around tele- and radio communication, military and electric power generation/distribution. Many of the final users are technologically very competent buyers and have been involved in procurement programs where they have made the first and critical order for several important innovations supplied by the Swedish electronics industry. Some of these are high voltage direct current, thyristor controlled reactive power compensation (Engström 1992) and mobile telephony (Oskarsson and Sjöberg 1991). Typically then, the electronics firms have benefitted from a close relationship with technologically highly capable local customers, often state-owned, which often have carried not only technology but also visions and risk-reducing procurement into the relationship (See for example Edquist 1994, Fridlund 1994, Oskarsson and Sjöberg 1991). For example, Televerket (the Swedish PTT) was instrumental in forming a vision of the role of mobile telephony at a very early stage of that industry's development, thus encouraging Ericsson to invest in that industry before many of its potential competitors (Oskarsson and Sjöberg 1991).

Second, there is a set of firms belonging to the core of Swedish industry with little or no activity at both the beginning and the end of the period. Taken together, these firms had as much as seventy-nine percent of their patents in metal and mechanical engineering at the beginning as well as at the end of the period. Thus, many of the larger Swedish firms did not take part in the rapidly growing innovative activity in electronics technology but continued to build upon their traditional areas of strength, chiefly metals and mechanical engineering.

Third, only four firms show a significant increase in patenting in electronics technology: Siemens, Volvo, IBM and Hasselblad. AGA, on the other hand, had a substantial activity at the beginning of the period but no activity at all at the end.[28]

A fourth group (not in table 4-14) constitutes a long 'tail' made up of firms with mostly one, but also in a few cases up to four patents, in each of the time periods. In 1969-73, this group consisted of about 80 firms. This tail had expanded a great deal by 1984-88 and included a growing set of firms which had specialized in electronics.[29] Its

---

27 The limit is arbitrarily set at a minimum of six patents in the period 1969-1973.

28 The same applies to Karlstads Mekaniska Verkstad, Svenska Dataregister AB, and Aktiebolaget Atomenergi.

29 In this tail, one may identify two main groups. First, those firms which only or mainly patented in electronics (henceforth the electronics group) and second, those which have most of their patents in mechanical engineering (including metals) but also in electronics (henceforth the mechanical/electronics group). In 1969-73, the electronics group accounted for about 76 patents (in 56 firms) while the mechanical/electronics group accounted for 42 patents (in 26 firms). By the end of the period, the electronics group had increased significantly both in terms of number of firms and in number of patents; to 104 firms

share of all electronics patents increased at the same time from about 25 percent to about 31 percent of all patents in electronics.[30] This suggests that there is a growing technological activity in electronics which is undertaken quite apart from the larger R&D spenders in Sweden. The increasing importance of this 'undergrowth' is, of course, very interesting given the lack of activity of the larger mechanically based firms in diversifying into electronics.

It is, however, worrisome that only a few of the firms in the tail of 1969-73 had grown up to be large patentees at the end of the 1980s: Siemens, Hasselblad, Volvo, and IBM.[31] Siemens has a large and successful subsidiary in Sweden in medical electronics (Siemens Elema) to which the patents presumably belong. Similarly to the other firms in this group, Siemens enjoys the benefits of demanding customers locally, which in this case are the clinical departments in the hospitals (Hidefjäll 1995). Finally, Hasselblad (cameras) and Philips (consumer electronics) had a substantial patenting activity in both periods. Hasselblad[32] seems to be the only Swedish firm specializing in electronics which has grown to become a large patentee in this period. Volvo had a sizeable patenting activity in electronics technology in the 1980s but seems to have used its capabilities to support its traditional product areas rather than diversifying into product groups based mainly on the new technology. IBM-Sweden produces printers but the patents may well derive from having a special responsibility for production technology development within IBM.

The lack of firms with a growing and significant patenting activity raises the question about Sweden's ability to exploit the technological capabilities of the tail. After all, it would not have been unreasonable to expect more of these firms (the bulk of these appear to be small and medium sized) to have grown into firms with a more sizeable patenting activity. Will the same lack of growth turn out to be characteristic also of the present and enlarged tail? This may well be a major policy issue for the 1990s.

Conclusions. To conclude this section, the evidence is consistent with the hypothesis that a part of the explanation of the poor performance of the Swedish electronics industry could be sought in the particular path dependencies of Swedish industry which emerged decades ago and which have been strongly characterized by mechanical engineering technology. In this process, firms have largely chosen strategies which continue to build on the traditional areas of technological strength where the firms had built up strong international positions and have been highly

---

and over 121 patents. Thus, this group increased its share of all electronics patents from 15 percent in 1969-73 to 25 percent in 1984-88. An increasing share of Swedish patenting in electronics is therefore undertaken by firms in the tail, patenting only or mainly in electronics. The mechanical/electronics group, however, was reduced in both number of firms and number of patents (to 14 and 22, respectively).

30 I am grateful to Peter Lindvall for helping me with all the computer work for this part of my analysis.

31 In addition, Facit had six patents and Boliden five; both belonged to the tail in the early period.

32 For an interesting description of Hasselblad's venture into electronics and computer science, see Öster (1991).

successful.[33] Those few firms which managed to build up a substantial electronics activity share a common feature: they have all worked in close relationships with a technologically capable local customer, often state-owned firms.

We have also discovered that the undergrowth of firms in electronics is expanding but with the worrisome attribute that few seem to expand their patenting activities. We do not have a good answer as to why this is the case but we would expect that certain features of the technological system have a particularly important bearing on the behavior of these smaller firms. The performance of the system of higher education, discussed above, immediately comes to mind. Expansion in a situation characterized by a permanent shortage of qualified engineers is not easy. Other institutions, in particular the capital market and the tax laws, are of relevance, too. It is evident that the capital market has not functioned well with respect to the supply of risk capital (Ohlsson 1991, SOU 1992:95). This is more detrimental to small than to large firms, particularly technology-based ones (Storey and Westhead 1995).[34] Moreover, the tax laws of the 1970s and 1980s (with 52 percent tax on business profits and a 70-80 percent marginal income tax on dividends) produced not only a strong incentive to reinvest profits in existing lines of business but also exacerbated the capital market problem in that it led to a low mobility of capital between firms, e.g. to finance the expansion of small technology-based firms. Both these factors have biased the industrial development in favor of the larger firms which obviously tend to 'stick to their knitting.'

Government Technology Policy

The 1970s were characterized by a technology policy which was primarily concerned with the final user of electronics technology: health hazards, employment and skill effects, personal integrity aspects of information technology, etc., were the foci, and little government support was given to R&D in this technology (Glimell 1988). The Swedish Board for Technical Development (STU, later NUTEK)) thus had a very limited involvement in electronics technology. For instance, the support to Swedish research into semiconductors was a mere million SEK (approximately $200,000) per

---

33 Swedish industry tends to be rather conservative in its choice of strategy. In a study of the 100 most important Swedish innovations, Granstrand (1992) indicates that only a few were in product areas which were new to the firms. There is also fragmentary evidence suggesting that patents and small technology-based firms in new technological areas are often sold to foreign firms due to limited interest among Swedish firms.

34 Olofsson and Wahlbin (1993) note that academic spin-offs in the early 1980s tended to grow faster than earlier generations, a time which was characterized by a peak in the size of the venture capital market and the widening of the stock market to smaller firms.

year in the beginning of the 1970s, even though the microprocessor arrived on the scene in 1969.[35]

In the middle of the 1970s the sum had increased to 7 million SEK, but in the years 1972/73-1977/78, the share of information technology in STU's budget actually decreased (Glimell 1988). However, in spite of its small size, it was of decisive importance for the build-up of the Computer Science Department in Linköping (Friborg 1992) - computer science being a subject which had difficulties in getting funds through the 'normal' channels (STUinfo 1991).[36]

At the end of the 1970s - ten years after both the invention of the microprocessor and a Swedish Parliamentary Committee (1969) which emphasized the strategic importance of an advanced domestic semiconductor industry - STU began a long-term improvement of the Swedish resource base in electronics technology. In 1981/82, about 65 million SEK was allocated to improving the university sector's weak capabilities in this technological field (cf. section 3 above). This was followed by the National Microelectronics Program in 1983/84 which was co-financed by a set of institutions (STU, Swedish Telecommunication Authority, Defense Procurement Agency, etc.) and amounted to as much as 600-700 million SEK. A large part of the program (NMP4) focused on building up limited but advanced Swedish design and production capabilities of very large scale integrated circuits - an area which had been neglected ever since the 1960s. In this connection, STU put a great deal of effort into trying to get Swedish industry interested in its microelectronics policy (Ragnarsson 1992). However, Swedish industry, which had only a very marginal research activity in semiconductors (the total number of patents granted to Swedish firms in the United States in the period 1969-84 was only 74 or 0.42 percent all semiconductor patents granted), was not very interested in NMP.

A third program - IT4 - followed. In principle, this program was concerned primarily with system technology, but it also meant a continued involvement in component technology. The total budget amounted to 1,000 million SEK of which half was government funds. The interest on the part of industry for the IT4 program was greater - in part due to its orientation towards system technology, but also since STU and other actors had become better at pursuing such programs.

The Swedish competence base in electronics technology was improved by these programs (Technopolis 1992, Ragnarsson 1992), which were very large by Swedish standards. On the whole, however, the main effect of STU's technology policy was to compensate for the university sector's prior neglect of electronics technology. STU was therefore partly turned into an instrument for educational policy rather than one of industrial and technology policy. For example, STU has for some time financed as

---

35 In part, this can be explained by resistance from industry against an active technology policy, presumably for political reasons.

36 This is not surprising, given the above analysis of the adjustment pattern of the system of higher education.

much as two-thirds of the university sector's R&D in electronic component technology. Due in part to these efforts, it seems as if by the end of the 1980s, the university sector's competence base in electronics technology had reached an acceptable level, comparable to that in other advanced industrial countries. Moreover, industry was more interested in exploiting it (Ragnarsson 1992),[37] although a recent evaluation of the IT4 program indicates that academics had relatively little involvement in that program.

CONCLUSIONS

Sweden's electronics industry is small and experiences a negative trade balance. Swedish R&D in electronics does not grow very fast and the country's share of U.S. patenting in electronics as well as its revealed technological comparative advantage in electronics are declining. In addition, Sweden's scientific activities have been concentrated on other areas. The picture is thus bleak for the Swedish technological system for electronics, at least in terms of its support for the Swedish electronics industry.

We have tried to explain this poor performance by a set of factors. First, neither diversification into electronics by established firms nor the growth of smaller technology-based firms have acted as strong sources of renewal. The larger Swedish firms, with a few exceptions, have continued to specialize in their traditional, often mechanical, technological areas of strength for their product renewal. The exceptions are firms which already in the 1960s were technologically close to electronics and/or supplied electronics to the military (Ericsson, ABB and SAAB). While there is a growing 'tail' of smaller firms in electronics, it is worrisome that only a few firms from the tail of 1969-73 grew significantly in patenting activities in the period studied.

Second, in the few cases when firms have chosen to move into electronics, they have done so primarily for the purpose of improving traditional product areas; their search activities appear to have been close to their core areas, typically leading to innovations in controls and instruments.

Third, the academic sector has also performed in an unsatisfactory way in this technological field; it has been reactive (as opposed to proactive). The lack of 'output' of electronics and computer science specialists has reflected, and contributed to strengthening, industry's specialization in more traditional areas, or at least not made it easier for industry to tread new paths into the electronics technology field.

At the same time, academic research was neglected for many years. There are exceptions, however. For example, the academic sector played a role in the early phases of the very successful Ericsson venture into mobile telephony. A significant contribution was, for instance, made by a department at Chalmers University of Technology in

---

[37] A case in point is Ericsson Radar which has begun to develop strategic links to parts of Chalmers University of Technology.

developing a noise reduction algorithm (Oskarsson and Sjöberg 1991). The late awakening of the academic sector has, of course, a bearing on the performance of the universities in terms of 'output' of engineers, but it has presumably also (negatively) affected the frequency of spin-offs from the universities as well as various types of research spillovers.

The precise reason for the slow adjustment by the universities is not clear and warrants more attention. Tentatively, the reasons are probably to be found both within and outside the universities. As far as outside factors are concerned, there appears to have been a lack of pressure from industry to expand education into electronics and computer technology (Ingberg 1993) which, of course, does not come as a surprise since the industry tended to 'stick to its knitting.' Furthermore, the relevant state bodies appear to have reacted slowly to the demands from the universities. For instance, at the Royal Institute of Technology, a committee had pointed out the need for a new chair covering the field of computer science already in 1959. This chair was not granted until 1976 (Ferm 1989)![38]

Fourth, the technology policy of the state can, as the academic sector, either be oriented towards strengthening the existing firms and technological systems or towards fostering new technological systems. The latter can be done by various means, e.g. financial support of new start-ups, R&D subsidies, as well as fostering R&D with a special orientation at the universities.

It is clear that in the 1970s, the technology policy of the Government contributed to conserving the traditional specialization pattern of Swedish industry. There were some exceptions, however, such as the development of computer science in Linköping with money from STU. Since the early 1980s, the technology policy was reoriented through large programs which have aimed at radically increasing the capabilities and size of research in electronics at the universities (which, as noted above, were performing poorly in this area). At the same time, the state has made strong attempts to increase the capabilities and the awareness of electronics technology in industry, especially in the area of components. These programs have had a positive effect on the universities. Industry has benefited, too, at least from the IT4 program (Technopolis 1992).

Public procurement of new products incorporating new technology can be a powerful instrument to make firms tread new paths. This tool was used successfully in the 1970s and in the 1980s, although not as an explicit technology policy instrument, an example of which is the tele- and radio communications sector. However, the procurement instrument does not seem to have been coordinated with other technology policy instruments, thus limiting its impact. This is, of course, a very different picture from, say that of the United States, where the defense programs for a long time paid not

---

[38] Factors inside the university system matter, too. For example, the President of Linköping Institute of Technology argues strongly that the decision making process for allocating research funds is quite limited in its ability to implement radical reorientations (Erlander 1994).

only for development work by industry but also for a great deal of research in computer science at the universities.

Fifthly, the capital market and the tax laws have both contributed to the dominance of large firms in the Swedish industrial landscape, firms which tended to 'stick to their knitting.'

To conclude, our interpretation is that these factors, directly and indirectly through a set of feedback mechanisms, strengthened Sweden's orientation towards more traditional areas of specialization. Our study suggests that technical change is a cumulative, path dependent process which is associated with a highly stable pattern of specialization. The stability arises from an interaction of all elements in the technological system. This raises the difficult but important issue of how new technological systems emerge and become a powerful factor in the technological transformation of a nation. Obviously, in the Swedish case, a new and powerful technological system built around electronics and computer technology did not grow out of the older system built on metals and mechanical engineering.[39]

REFERENCES

Archibugi, Daniele and Mario Pianta, 1992. *The Technological Specialization of Advanced Countries, A Report to the EEC on International Science and Technology Activities.* Dordrecht, Boston and London: Kluwer Academic Publishers and The Commission of the European Communities.

Carlsson, Bo and Gunnar Eliasson, 1994. "The Nature and Importance of Economic Competence," *Industrial and Corporate Change*, 3 (1), pp. 687-711.

Carlsson, Bo and Staffan Jacobsson, 1994. "Technological Systems and Economic Policy: The Diffusion of Factory Automation in Sweden," *Research Policy*, 23 (3), pp. 235-248.

Datateknik. 1991. No. 13, 27-29.

David, Paul A., 1988. "Path-Dependency: Putting the Past into the Future," Stanford University, Institute for Mathematical Studies in the Social Sciences, Economic Series, Technical Report No. 553, November.

Dosi, Giovanni, David Teece and Sidney Winter, 1991. "Toward a Theory of Corporate Coherence", in Giovanni Dosi, R. Giannetti and P.A. Toninelli, (eds.), *Technology and the Enterprise in a Historical Perspective.* Oxford: Oxford University Press, 185-211.

Edquist, Charles and Maureen McKelvey, 1991. "The Diffusion of New Product Technologies and Productivity Growth in Swedish Industry," CCC Working Paper

---

39 A successful 'mechatronics' technological system did, however, emerge, as was shown in Carlsson and Jacobsson (1994).

No. 91-15, Center for Research in Management, University of California at Berkeley.

Edquist, Charles, 1994. "Offentlig teknikupphandling för behovstillfredsställelse och konkurrenskraft," Tema T Rapport 34, Linköpings Universitet.

Ehrnberg, Ellinor and Niklas Sjöberg, 1991. "Technological Transitions, Competition and Firm Performance," Paper presented at the E.A.R.I.E. 19th Annual Conference, 1992, Stuttgart.

Eliasson, Gunnar, 1990. *Commentary: Economies of Scale through Network Technologies and the Size of the State Space*, Booklet No. 306, Industriens Utredningsinstitut, Stockholm.

Engelsman, E.C and A.F.J. van Raan, 1990. "The Netherlands in Modern Technology: A Patent-Based Assessment," Research Report to the Ministry of Economic Affairs, Center for Science and Technology Studies, University of Leiden.

Engström, Gunnar, 1992. "Ledarkrav i en multiteknisk koncern som verkar i de tekniska frontlinjerna" (Leadership Requirements in a Multitechnical Company on the Technical Frontier), in Ove Granstrand (ed.), *Att leda företag i teknisk och ekonomisk omvandling,* Chalmers University of Technology, Gothenburg, 32-41.

Erlander, Sven, 1994. "Balans mellan forskningens aktörer" (Balance Among the Research Actors), *Ny Teknik*, 46, 3.

Feldman, Maryann, 1994. *The Geography of Innovation.* London: Kluwer Academic Publishers.

Ferm, E., 1989. "Elektroteknisk utbildning under 100 år. Utvecklingen av elektroteknikutbildningens innehåll vid KTH under 1900-talet" (Electrotechnical Education during 100 Years. The Development of the Content of Electrotechnical Education at the Royal Institute of Technology during the 20th Century), Stockholm Papers in History and Philosophy of Technology, Royal Institute of Technology Library, Stockholm.

Fransman, Martin, 1990. *The Market and Beyond.* Cambridge: Cambridge University Press.

Fridlund, M., 1994. "En specifik svensk virtuoskonst: Empiriska och teoretiska perspektiv på utvecklingsparet Asea-Vattenfalls historia" (A Specifically Swedish Virtuosity: Empirical and Theoretical Perspectives on the History of the Asea-State Power Board Development Duo), *Polhem Tidskrift för Teknikhistoria,* 12 (2), 106-131.

Fishwick, Wilfred, (ed.), 1988. "Structures of Technological Education and Contributing Social Factors," UNESCO Studies in Engineering Education. Paris: Unesco.

Friborg, Lennart, 1992. Personal communication.

Glimell, Hans, 1988. "Den svenska vägen till trängre kretsar" (The Swedish Road to Smaller Circuits), in Hans Glimmell (ed.), *Industriförnyelse i Norden. 80 talets programsatsningar på mikroelektronik.* Århus: Förlaget Samfundsekonomi och Planläggning.

Granstrand, Ove, 1984. "Technology Procurement as a Special Form of Buyer-Seller Interaction in Industrial Marketing," *CIM-Report No. R:1984:06*, Department of

Industrial Management and Economics, Chalmers University of Technology, Gothenburg.
Granstrand, Ove, 1992. Oral presentation at the inauguration of the School of Management of Technology at Chalmers University of Technology, January 31.
Granstrand, Ove and Staffan Jacobsson, 1991. "When Are Technological Changes Distruptive? - A Preliminary Analysis of Intervening Variables between Technological and Economic Changes," Paper presented at the Marstrand Symposium on Economics of Technology, Marstrand, August.
Hidefjäll, Patrik, 1995. "Patterns of Product Innovations in the Cardiac Pacemaker Industry" in David Bennet and Fred Steward (eds.), *Technological Innovation and Global Challenges. Proceedings of the European Conference on Management of Technology*. Aston Business School, 221-229.
Ingberg, Walter, 1993. Personal communication with Walter Ingberg, Ministry of Industry.
Jacobsson, Staffan, 1993. "The Length of the Learning Period. Evidence from the Korean Engineering Industry," *World Development*, 21 (3), 407-419.
Jacobsson, Staffan, Christer Oskarsson, and Lennart Bångens, 1989. "Industriell elektronik i Sverige - aktörsstruktur och tillväxtmönster" (Industrial Electronics in Sweden - Structure and Growth Patterns of Actors), *CIM-Report, R:1989:06*, Department of Industrial Management and Economics, Chalmers University of Technology, Gothenburg.
Jacobsson, Staffan and Christer Oskarsson, 1995. "Educational Statistics As an Indicator of Technological Activity," *Research Policy*, 24 (1), 127-136.
Näringsdepartementet, 1991. "IT 2000. Svensk informationsteknologis möjligheter under 1990-talet med förslag till riktlinjer för en IT-politik inom det näringspolitiska området" (IT 2000. The Possibilities for Swedish Information Technology in the 1990s and Recommendations for IT Policy), *DS 1991:3*, Stockholm: Allmänna förlaget.
Ohlsson, Lennart, 1976. *Svensk verkstadsindustris internationella specialisering* (The International Specialization of the Swedish Engineering Industry). Stockholm: Industriens Utredningsinstitut.
Ohlsson, Lennart, 1991. *Förnyelsepolitiken & Riskkapitalet* (Renewal Policy and Risk Capital). Stockholm: Industriförbundets Förlag AB.
OECD, 1991. *Industrial Structure Statistics*. Paris: OECD.
OECD, 1992. *Technology and the Economy. The Key Relationships*, Paris: OECD.
Olofsson Christer and Clas Wahlbin, 1993. "Teknikbaserade företag från högskolan" (Technology-Based Companies from the University), IMIT, Linköping.
Oskarsson, Christer and Niklas Sjöberg, 1991. "Bäst i Världen projektet: produktivitet inom mobiltelefonim" (Best in the World Project: Productivity in Mobile Telephony), mimeo, Department of Industrial Management and Economics, Chalmers University of Technology, Gothenburg.

Öster, Jerry, 1991. "Vår fortlevnad och våra tekniska satsningar" (Our Future and Our Technical Investments), in Ove Granstrand, (ed.), *Att leda företag i teknisk och ekonomisk omvandling*, Chalmers University of Technology, Gothenburg, 182-196.

Persson, Olle, 1991. "Forskning i bibliometrisk belysning" (Research in a Bibliometric Perspective), INUM, Umeå.

Ragnarsson, Sven-Ingmar, 1988. "Informationsteknologi" (Information Technology), mimeo, Swedish Board for Technical Development, November.

Ragnarsson, Sven-Ingmar, 1992. Interview with Sven-Ingmar Ragnarsson at NUTEK.

SCB, 1975. *Industri 1973*, Part 1, Stockholm: SCB.

SCB, 1985. *Statistical Report U14 SM 8501.*

SCB, 1987. Information received from SCB.

SCB, 1987a. *Statistical Report U14 SM 8701.*

SCB, 1989. *Statistical Report U14 SM 8901.*

SCB, 1990. "Civilingenjörsutbildningar genom tiderna i siffror" (Engineering Education History in Figures), Örebro: SCB.

SCB, 1991. "Elektronikindustrin i Sverige" (The Electronics Industry in Sweden). Örebro: SCB.

SCB, 1991a. *Statistical Report U14 SM 9191.*

SCB, 1991b. *Industri 1989*, Part 1, Stockholm: SCB.

SCB, 1992. "Data och elektronikindustrin i Sverige 1992" (The Computer and Electronics Industry in Sweden 1992). Örebro: SCB.

SCB, 1994. "Data och elektronikindustrin samt teleoperatörer i Sverige 1994" (The Computer and Electronics Industry and Telecommunications Operators in Sweden 1994). Örebro: SCB.

SIND, 1988. "Mikroelektronikanvändningen i industrins produkter och tillverkningsprocesser" (The Use of Microelectronics in Manufacturing Processes and Products), SIND PM 1988:1, Stockholm.

SOU, 1992:95. *Den svenska marknaden för projektkapital. Slutbetänkade av Industrifondsutredningen* (The Swedish Market for Project Capital. Final Report from the Industry Fund Committee). Stockholm: Allmänna förlaget.

Soete, Luc, 1988. "Technical Change and International Implications for Small Countries," in Christopher Freeman and Bengt-Åke Lundvall (eds.), *Small Countries Facing the Technological Revolution.* London: Francis Pinter, 98-110.

Storey, David J. and Paul Westhead, 1995. "Financing Technology-Based Firms in the United Kingdom." Paper presented to the Fifth International Conference on Economics of Innovation: New Perspectives in Technology Diffusion, Catholic University of the Sacred Heart, Piacenza, June 23-24.

STU info, 1991. "Vad har STU gjort för utvecklingen av informationstekniken och dess användning" (What Has STU Done for the Development of Information Technology and Its Application?), STUinfo (838-1991), Stockholm.

Technopolis, 1992. "Evaluation of the IT4 programme, Stockholm, May.

Tyson, Laura, 1992. *Who's Bashing Whom? Trade Conflict in High-Technology Industries.* Washington, D.C.: Institute for International Economics.

U.S. Department of Education, National Center for Education Statistics, Digest of Education Statistics, 1988, 1989, 1990.
U.S. Department of Commerce, Bureau of Census, Statistical Abstract of the United States, 1985 and 1990, Washington D.C.
The World Bank, 1981. *World Development Report 1981*. Washington, D.C.: The World Bank.

APPENDIX 4-1

We have calculated the trade balance in the electronics industry using three different sets of data. First, SITC Rev.2 at three digit level; second, SITC Rev.2 using more aggregate data, mainly two digit level; third, SITC Rev.3 which is only available from 1990, at three digit level.

The results are given in the table below where the first row refers to SITC Rev.2 at the three digit level. The second row is the same as in table 4-3 and refers mainly to the two digit level of SITC Rev.2. The third row refers to three digit level of SITC Rev.3.

It is noteworthy that with the exception of Switzerland, the results are much the same independently of the method used. In the case of Switzerland, we have added a fourth row, SITC Rev.3 at two digit level. The data at three digit level (both SITC Rev.2 and 3) excludes watches while the two digit level data includes watches. This probably accounts for the difference between the two and three digit levels.

Table A4-1 Trade Balance (X/M) in the Electronics Industry in OECD Countries, 1978-1992, Selected Years**

Country	1978	1984	1990	1992
France	1.01	0.94	0.80	0.85
	1.01	0.94	0.80	0.86
			0.77	0.83
F.R.G.	1.73	1.30	1.10	na
	1.68	1.31	1.13	na
			1.15	na
Italy	0.95	0.82	0.67	0.67*
	1.07	0.91	0.76	0.76*
			0.72	0.71*
Japan	7.06	8.16	5.05	5.00*
	6.98	7.95	4.80	4.74*
			5.07	5.02*
Netherlands	1.01	0.91	0.83	0.87
	0.98	0.90	0.81	0.83
			0.82	0.86
Sweden	0.97	0.88	0.83	0.85*
	0.97	0.89	0.82	0.84*
			0.83	0.86*
Switzerland	1.36	1.07	0.91	0.94*
	1.97	1.45	1.31	1.34*
			1.07	1.10
			1.29	1.39
U.K.	1.12	0.81	0.87	0.91*
	1.09	0.78	0.86	0.89*
			0.88	0.86
U.S.A.	1.19	0.79	0.83	0.78
	1.14	0.76	0.79	0.74
			0.82	0.78

* Refers to 1991.
** Row 1 is defined as the following SITC Rev. 2 categories: 716, 737, 751, 752, 759, 761, 762, 763, 764, 772, 773, 774, 776, 778, 871, 872, 873, 874, 881, 884, 898. Rows 2 and 4 are defined as the following SITC Rev.2 and Rev.3 categories: 716, 75, 76, 77, 87 and 88. Row 3 is defined as the following SITC Rev.3 categories: 716, 731, 733, 737, 751, 752, 759, 761, 762, 763, 764, 772, 773, 774, 776, 778, 871, 872, 873, 874, 881, 884, 898.

Source: Elaboration on data kindly supplied by the United Nations Economic Commission for Europe.

5

THE DEVELOPMENT OF BETA BLOCKERS AT ASTRA-HÄSSLE AND THE TECHNOLOGICAL SYSTEM OF THE SWEDISH PHARMACEUTICAL INDUSTRY*

Rikard Stankiewicz

INTRODUCTION

Technological competition arises from the differential ability of economic actors to *recognize* and *exploit* technological opportunities. Technological opportunities are a function of the knowledge and competence base. As knowledge and competence accumulate, new possibilities emerge and/or the probability of solving previously recognized problems changes. The process is relatively straightforward when the creation of knowledge occurs chiefly locally. Much of the past technological accumulation has been of that sort: it has taken place predominantly within a company, an economic district, or an industry. The striking feature of contemporary technologies, particularly so-called high technologies, is that their knowledge base lacks clearly defined social boundaries; it tends to be institutionally heterogeneous, multi-disciplinary, and geographically distributed. In such circumstances, the firms will find it difficult to fully internalize critical competencies. Instead, their internal activities will be strongly dependent on a largely exogenous knowledge base which undergoes rapid change.

---

* Parts of this paper have been presented earlier in Stankiewicz (1993). The analysis presented here is based on written sources which are identified in the text and on interviews with several scientists who played crucial roles in the development of beta blockers at Hässle and ICI. I would like to express my gratitude to these scientists and especially to Ivan Östholm of Hässle. Another invaluable source of information about the development of beta blockers has been Reinder Vos' book *Drugs Looking for Diseases*. While acknowledging the importance of all these sources I would like to stress that the responsibility for any errors in this paper is exclusively mine.

The dependence of firms on exogenous sources of knowledge places special demands on the structure and management of the firms themselves, as well as on the technological systems in which they are embedded. The aim of this chapter is to shed some light on the nature of these demands by examining in depth a particular instance of innovation in the pharmaceutical industry in the 1960s and 1970s: the development of a group of cardiovascular medicines known as beta blockers. We will describe how the relevant technological opportunities emerged as a result of the accumulation of knowledge, how they were recognized or failed to be recognized by the pharmaceutical firms involved, and how they were exploited by some of them. I will particularly focus on the Swedish firm Astra-Hässle and the two successful beta blockers it has developed: Aptin and Seloken.

The case of beta blockers was selected for several reasons: (i) it is particularly illustrative of the dynamics of pharmaceutical innovation in general; (ii) it is a uniquely well-documented case of parallel technological development in several companies and several countries, and (iii) it played a special role in the evolution of the Swedish pharmaceutical industry in the formative decades of the 1960s and 1970s (see the next section). By following the specific process through which beta blockers were discovered and developed, we will be able to examine the extremely complex interplay between the general evolution of the *knowledge base* of a technology, the *R&D and innovation strategies* at the corporate level, and the features of the larger *technological system* in which firms are embedded.

The chapter is structured as follows. The first section places the Aptin/Seloken case in the broader picture of the evolution of the Swedish pharmaceutical industry and of the Astra company in particular. The following section will give an account of the field of beta blockers. It describes Hässle's beta blocker program against the historical development of the field as a whole and compares it with similar programs in other companies. In section three, we will examine the factors at the corporate level which contributed to the success of Hässle. Section four broadens the perspective and discusses some of the key features of the technological system of the Swedish industry during its formative decades. Finally, section five discusses the ways in which that system has changed in recent years and the resulting challenges for Astra and other pharmaceutical firms.

HISTORICAL DEVELOPMENT OF THE SWEDISH PHARMACEUTICAL INDUSTRY

Recent Dramatic Success

The Swedish pharmaceutical industry has undergone a truly dramatic development in recent years. Within a mere 15 years it has been transformed from a relatively marginal component of Sweden's economy into one of its two main high-tech industries, the other one being telecommunications. Its average growth rate in the 1980s was in excess of 15 percent per annum and in the 1990s well over 20 percent (LIF 1992). The growth came predominantly from exports which today account for approximately 90 percent of the industry's total sales. In 1991, the foreign sales of medicines corresponded to nearly 3 percent of the country's total exports, thus surpassing or achieving parity with such classics of the Swedish industry as iron ore, pulp, wood products, and trucks (LIF 1992; LIF 1995). This is significant, considering that until 1980 Sweden was a net importer of medicines and that its international performance in pharmaceuticals was at best middling. By 1991, the country's trade surplus reached nearly 6 billion SEK (approx. $1 billion). Three years later it grew to over 12 billion SEK. Its total sales in 1994 exceeded 48 billion SEK.

Parallel with the improved performance there occurred an equally dramatic restructuring of the industry. At the end of the 1970s, the Swedish pharmaceutical industry consisted basically of 7 major companies (Astra, Kabi, Pharmacia, Ferring, Leo, Ferrosan, and ACO). Within a short time this number was reduced to 3: Astra, Pharmacia, and Ferring, the first two accounting for the lion's share of pharmaceutical production. Although there exist, along-side with the above mentioned three firms, a number of small ones, the latter have played a very minor role.

This extremely rapid evolution of the industry during the last 15 years has been as remarkable as it was unexpected. A government inquiry into the current state and future prospects of the industry in Sweden published in 1980 forecasted a future yearly growth rate of less than 8 percent (SOU 1980). It also expressed some anxiety about the allegedly sub-optimal size of the firms. How then is one to account for the transformation which took place in the 1980s and 1990s? Why did the accelerated growth occur? What factors explain its extent and timing?

At one level, the answer is simple enough. The commercial success was a result of superior innovative performance. The growth momentum was built up by a series of highly successful new drugs including heart medicines (Seloken, Plendil), asthma therapies (Bricanyl, Pulmicort), hormones (Genotropin), separation products (such as Healon), gastrointestinal drugs (Losec), and many other less dramatically successful but nevertheless profitable drugs.

But this leads only to the next question: how to account for this sudden wave of creativity? In order to answer it, we must first consider the timing of the events. The commercial success in the 1980s and 1990s built on R&D performed decades earlier. In fact, most of the break-through products listed above have their roots in R&D decisions taken in the 1960s and 1970s.

Before World War II

The Swedish industry was born in the first decades of the 20th century — a comparatively late date, at least relative to the major European countries and the United States. The Swedish companies, like many of their international counterparts, were predominantly off-shoots from pharmacies. Their main purpose was to supply the local market with standard medicinal substances. Domestic production of drugs discovered and developed abroad was a dominant business idea during the early decades. This orientation and the relative weakness of the chemical industry meant that at first there was little scope in Sweden for the emergence of large research-intensive companies. Indeed, prior to World War II there was very little R&D in the industry and only a few original products were developed — most of which enjoyed only a moderate success (cf. Norgren 1989).

The 'Golden Age:' 1940s and 1950s

Despite a certain vitalization of the industry during the war, the situation described above continued in the 1940s and 1950s. However, its precariousness was becoming increasingly apparent. The 1940s and 1950s are often referred to as the Golden Age of the pharmaceutical industry. A series of dramatically effective drugs including sulphonamides, antibiotics, hormones, and many others, were introduced on the market, establishing the credibility of an industrial approach to the discovery and development of new drugs. The 'industrial approach' was based on scaling up and managerially integrating the research techniques previously developed in the academic setting or in the chemical industry. As the power of that approach became apparent, the structure of the pharmaceutical industry underwent an important change: there emerged a small but rapidly expanding set of R&D-intensive firms which soon came to dominate the industry in terms of growth rates and profitability.

The follower strategy pursued by the Swedish firms did not allow them effectively to participate in the exploitation of technological opportunities during the Golden Age period. However, the managements of the companies were not oblivious to what was happening. They realized that unless their firms were transformed, they would not only fail to profit from the rapidly growing international market in patent-protected ethical drugs but would also lose their position as the dominant national suppliers. Thus, towards the end of the 1950s there was much soul-searching in the industry, resulting in a series of strategic decisions to strengthen the industrial R&D capability.

1960s and 1970s

Starting in the early 1960s, the Swedish companies began stepping up their R&D spending. By the end of the 1970s they reached the internationally respectable level of around 15 percent of turnover (SOU 1980). It took some time before the economic effects of these policies could be measured. In the 1950s and 1960s the industry's sales grew at a yearly rate of around 6 percent, increasing in the 1970s to around 8 percent (SOU 1980). The real sea change occurred only in the 1980s when the sales began to grow at an average rate of about 16 percent. In the 1990s, that figure increased to 27 percent (LIF 1995).

There can be no doubt that the commitment of the firms to high R&D spending has been a necessary condition for the commercial performance that followed. But it was hardly a sufficient one. Large R&D budgets are no guarantee of success. What matters is the quality of the programs and strategy which determine the productivity of the R&D investments. And the productivity of the programs set in motion by the Swedish firms in the 1960s and 1970s proved to be remarkably high. In fact, it contrasted sharply with the general decline in innovation experienced at the time by the international pharmaceutical industry (Gross 1983; Walker & Parrish 1986; DiMasi *et al.* 1994).

The case of beta-blocking drugs is particularly revealing of the processes and mechanisms behind this creative surge. It also throws an interesting light on the evolution of Astra, the most successful of the Swedish pharmaceutical firms.

Astra's profits and sales have been increasing rapidly in recent years, taking it from around the 40th place on the world list in the 1980s to the 17th in 1995. These recent successes have been due chiefly to two products: Losec, a drug against peptic ulcers — the development of which is described in the next chapter, and Pulmicort, an asthma medicine. However, the real breakthrough in the history of the company was the successful development of the beta blockers at Astra-Hässle. In fact, the beta-blocker Seloken continues to be the company's third largest product until today.

Until the 1960s, Astra could be considered a fairly successful small company with only one product of major importance on the international market: Xylocain, a local anaesthetic[1]. Although the company had invested heavily, and very successfully, in the development of the drug, Xylocain was not an Astra invention. It was discovered serendipitously by two academic scientists in Stockholm (Lindqvist & Sundling 1993). Beta blockers were a very different story: Aptin and Seloken were the first major drugs which originated in Astra's laboratories. This was a momentous event in the history of the company since it proved that it could successfully compete with other international R&D-intensive firms. It established an innovative culture at Hässle-Astra which subsequently produced other major innovations: the gastro-intestinal medicine Losec, presently the best-selling of Astra's products, and the cardiovascular drug Plendil.

The fact that these developments took place at Hässle makes the story even more fascinating. When acquired by Astra in 1942, the company was a very minor and unsophisticated pharmaceutical operation. It had no R&D to speak of. It had no experience with developing new products. And it was poor in practically every kind of resource (Östholm 1991; Holm 1992). Not surprisingly, therefore, few people in Astra saw Hässle as a major scientific or technological player. Yet this is what the company

---

[1] Astra was formed in 1913. It began conducting its own R&D on a very modest scale in the early 1930s. These activities were stepped up during the war when the company established its central research laboratory focusing chiefly on antibiotics and local anaesthetics.

became within less than two decades, starting in the late 1950s[2]. In the next two sections we will examine the processes through which that happened.

THE DEVELOPMENT OF BETA BLOCKERS — AN INTERNATIONAL PERSPECTIVE

What Are Beta Blockers?

The beta adrenoceptor-blocking drugs provide effective treatment for angina pectoris, arrhythmia, and high blood pressure. They are also used against palpitations and in the after-treatment of heart infarction. The drugs act by blocking the action of catecholamines — adrenaline and noradrenaline — in the heart, thus eliminating or moderating the deleterious effects of stress on the cardiovascular system.

Several different beta blockers have been developed since the early sixties. Although they are rather similar in their main effect on the heart, they differ in their precise mode of action and side effects. In particular, great efforts have been extended to make the drugs less 'aggressive,' i.e., less likely to cause excessive slow-down of the heart. Another important objective was to make them more selective, thus eliminating the negative side effects in asthma patients. This was achieved in the so-called 'second-generation' beta blockers.

Two companies have consistently led in the development of beta blockers. ICI was first to the market and, thanks to the success of Inderal, dominated the first generation of beta blockers in the 1960s. Hässle, which started its beta-blocker program almost at the same time as ICI, developed a successful first generation blocker, Aptin, and after that, the leading second-generation blocker, Seloken. Among other companies active in the field we may mention Ciba, Sandoz, Boehringer, Hoechst, Mead Johnson, and Squibb.

The Technical and Historical Background[3]

Adrenergic receptors mediate the action of catecholamines (adrenaline and noradrenaline) and related stimulants of the sympathetic system. Today we know that there are four types of such receptors: alpha 1, alpha 2, beta 1, and beta 2. Stimulation of the alpha receptors results in the constriction of the bronchi and blood vessels leading, among other things, to increased blood pressure. Stimulation of beta 1 receptors causes a rise in the rate and output of the heart. Stimulation of beta 2 receptors results in

---

[2] Interestingly, there are quite striking parallels between the development of Hässle and that of Draco, the other highly successful Astra R&D company which developed Bricanyl and Pulmicort (Larsson 1990).

[3] For a historical account see Sneader (1985) and Parnham & Bruinvel (1984).

dilatation of the bronchi. The knowledge of the adrenergic receptors and their function has evolved since the turn of the century. It constitutes an excellent example of the complex interactions between fundamental and applied research, science and industrial drug development.

Adrenaline was identified and described around the turn of the century. Its effects on various organs including the cardiovascular system were observed at an early stage. The functions and mechanisms of the action of adrenaline and other catecholamines were gradually elucidated during the following decades. The apparent biological importance of adrenaline has led to intensive exploration of its various analogues starting with noradrenaline. The accumulating knowledge of the structure and modes of action of these substances has been exploited by a number of companies, including Boehringer in Germany and Eli Lilly and Mead Johnson in the United States. One important field of application was that of asthma drugs, where isoprenaline has become widely used since the 1940s. Antispasmodic actions in other organs have also been investigated. A major problem in those fields was to find drugs which had no negative side effects on the heart and the central nervous system. Some cardiovascular applications were also developed at an early stage.

The first substances known to block the action of adrenaline were those derived from ergot extract. The first synthetic drugs of this type were created in the 1920s. In the 1930s a number of important analogues of these substances were synthesized and tested. Some of them exhibited interesting properties. Amphetamines, for example, were used for a time as anorectic drugs. Interestingly, while many of these drugs did appear to have adrenergic effects, none of them seemed to block the action of adrenaline and noradrenaline in the heart. This view of things prevailed until the late 1950s.

In the mid 1950s, Neil Moran, an American pharmacologist at the National Heart Institute, began suspecting that certain substances could indeed antagonize the action of adrenaline and noradrenaline in the heart muscle. These suspicions were strongly reinforced by the discovery at Eli Lilly of the dichloroisoproterenol (DCI) in 1956. The Lilly scientists made the discovery while testing analogues of isoprenaline in search of bronchodilators to be used in the treatment of asthma. DCI showed certain unexpected properties: it appeared to interfere with the action of catecholamines. Slater and Powell who investigated DCI realized that they had something important on their hands, but they did not quite recognize its full therapeutic significance. However, the publication of their results in 1957-58 alerted the scientific community. Moran and his collaborator Perkins acquired samples of DCI and subjected them to a series of tests. They saw immediately that the actions of DCI were consistent with the theory of adrenergic receptors published ten years earlier by an American pharmacologist of Swedish descent, R.P. Ahlquist. Moran and Perkins presented their results to the research community at a scientific meeting in 1957 and published in a journal in 1958.

Although there existed by the end of the 1950s a sizeable family of biologically and pharmacologically interesting substances 'descending' in one way or another from the catecholamines, the knowledge of the precise mechanism behind their effects was limited and fragmented. The discovery of DCI and its interpretation by Moran and

Perkins had a catalytic effect on the scientific and technological development. It brought about a convergence between the accumulating empirical knowledge about the pharmacological properties of the relevant groups of substances and the rapidly evolving knowledge of the mechanisms of nervous transmission in general and of the adrenergic receptors in particular.

The concept of receptor goes back to Paul Erlich's notion that, in order to be effective, a drug must be bound to a specific site on a cell. In 1905, John Langley proposed that catecholamines bind to specific receptors in various tissues. In the following years, comparatively little progress was achieved in the further development of that idea.

In 1948, however, Ahlquist published a paper based on his studies of the effects of catecholamines in various organs including the heart. He concluded that his data could be best interpreted assuming the existence of two different types of adrenergic receptors: alpha — the stimulation of which led to excitatory responses, and beta — which had inhibitory functions.

Although they were initially resisted by other scientists, Ahlquist's results pointed indirectly to the possibility of developing drugs which could have selective blocking effects on adrenergic receptors. However, no practical steps in that direction were taken until after the discovery of DCI mentioned above.

The understanding of the mechanisms through which catecholamines and their antagonists were acting has been crucial because it has helped to interpret the accumulating knowledge of the normal functioning as well as pathology of the cardiovascular system. Among other things, the cardiologists were increasingly aware of the importance of stress as a cause of heart disease. It became apparent that an effective way of dealing with several heart disorders was by controlling the influence of the sympathetic nervous system on the heart. The concept of blocking beta-adrenoceptors fitted very well in that context.

However, it must be emphasized that when the development of beta blockers started in the early 1960s, the knowledge of the biological mechanisms involved was far from complete. The receptors were still hypothetical entities, the chemical properties of which were unknown. Scientists were uncertain as to what types of receptors there were, what their precise function was, and by which methods they could be studied. The effects of drugs on organisms are usually quite complex and difficult to disentangle and understand. An emerging theory can be a useful heuristic guide and a source of inspiration in the search for useful drugs, but it rarely leads directly to the solution of a therapeutic problem. This was definitely true in the present case. Once the therapeutic significance of beta blockers became evident to the industry, extensive programs of experimentation aiming at the elimination of various knowledge loopholes had to be undertaken.

The knowledge thus produced led simultaneously to technological and scientific advances. In retrospect, we now know that this double-pronged approach proved more effective than the more conventional empirical strategies based on massive screening of substances against pharmacological models.

Industrial Beta Blocker Programs

The discovery of DCI triggered a series of events which within a few years led to the marketing of the first beta blocker. A dozen or so companies were drawn into this development. They fall into three main categories: (i) the would-be innovators who reacted to the DCI discovery at an early stage but failed to exploit it (Eli Lilly, Boehringer, and Mead Johnson); (b) the successful innovators (ICI, Hässle); and (iii) the followers (the remaining companies). Let us briefly describe the beta-blocker activities in these three groups of companies.

Serendipity Lost: Lilly, Mead Johnson, and Boehringer[4]

As already noted, the synthesis of DCI occurred in the context of Lilly's program aiming at new bronchodilatory agents. Lilly's Cardiovascular Research Group failed to appreciate the significance of the discovery. Nevertheless, the substance was tested on three critically ill patients. It turned out that DCI had a strong, negative side effect: it caused a major increase in heart activity. One of the patients died. This affected negatively Lilly's attitudes towards beta blockers and had a similar impact on the views of the American medical community at large[5].

Nevertheless, for a time, Lilly continued its research. Its scientists synthesized about a hundred analogues of DCI, hoping to find one with less intrinsic activity. Eventually the company applied for a patent for pronethalol, only to find out that ICI had already obtained one. Lilly's effort seemed to be half-hearted and was soon discontinued.

A similar development occurred at another American company, Mead Johnson, which in the mid-1950s was engaged in a program to develop antispasmodic drugs. In search of a new bronchodilator, a substance (sotalol) was synthesized in 1960 using isoprenaline as a lead. Since sotalol showed effects similar to those produced by DCI, extensive preclinical and clinical research was undertaken to further explore these findings.

In retrospect we know that sotalol was a precursor of a second-generation beta blocker. However, Mead Johnson failed to exploit its lead. Sotalol and related substances were tested as bronchodilators and as drugs against nasal congestion. Cardiovascular tests were made chiefly to assess the possible undesirable effects. The

---

4 The account of beta blocker activities at Lilly, Mead Johnson, Boehringer, and Ciba is based on Vos (1989), Shanks (1984), and Östholm (1991).

5 Actually, DCI was not the cause of the death. Nevertheless, the psychological impact was very strong. It illustrates the hazards of premature clinical experimentation.

clues which these tests produced did not receive sufficient attention. When Mead Johnson learned about the related work at ICI, the company decided to quit the field.

The developments at the German company Boehringer were somewhat more complex. As in the case of Mead Johnson, Boehringer's beta-blocker program started when its scientists learned about DCI. In one sense, the company was in an excellent position to exploit Lilly's discovery. It had a strong tradition in the field of heart and lung diseases and had at its disposal strong chemical and pharmacological expertise. DCI was a derivative of isoprenaline which Boehringer marketed since 1952 as a drug against asthma (Aleudrin). It was therefore natural for the company to take a closer look at DCI.

Parallel with that, Boehringer's scientists were working on new anorectic drugs, searching for ones without central stimulating effects. Since some of the substances under consideration were similar to DCI, they too were tested for adrenergic action. Several of them (particularly Kü446) turned out to be beta adrenoceptor blockers. However, just as in the case of Lilly and Mead Johnson, Boehringer failed to quickly recognize the cardiovascular therapeutic implications of what was found and treated the beta-blocking activity as yet another interesting pharmacological phenomenon. In fact, the work on DCI and Kü 446 was conducted against the background of a program aiming at an anti-asthma drug without the negative side effects of Aleudrin. As there was a new drug with just these properties about to be launched, the work on Kü 446 seemed to be of mainly academic interest.

Nevertheless, it was allowed to continue as a research project and led soon to the discovery of another promising substance, Lg 32. However, the testing of Lg 32 proceeded slowly, and when Boehringer finally got around to applying for a patent, it learned that Lg 32 was identical with propranolol, which was already patented by ICI. The irony was that Boehringer synthesized Lg 32 before ICI.

Although viewed as a sideline or exploratory activity, the project did generate several other promising substances. However, despite the fact that the Boehringer team came once more close to beating ICI when it synthesized the substance Kü592 in 1961,[6] the end results were meager. Kü592 was clearly interesting, but there were doubts about its possible negative side effects. The company did not seek patent protection until August, 1963. A that time, however, it was realized that ICI had already acquired patent rights for the substance. Eventually an agreement was reached with ICI allowing Boehringer to market Kü592 as Doberol in 1970, but only in Central and Eastern Europe.

---

6 This was 18 months before ICI carried out the same synthesis.

Serendipity Gained I: ICI[7]

The ICI beta-blocker program was the creation of James Black who came from the department of physiology at the Veterinary School in Glasgow where he worked on the role of the sympathetic nervous system in coronary diseases. He was strongly aware of the importance of stress as a cause of angina pectoris and of the role of catecholamines in these processes. In the course of his research, Black came across Ahlquist's paper. He gradually came to the conclusion that anti-adrenergic drugs could be used in treatment of coronary diseases. However, at the time there were no clues as to where and how the required substances could be identified.

Black's program on coronary heart diseases was initiated at ICI in 1958 and soon became focused on angina pectoris. The news of the discovery of DCI reached Black in 1959 and had a decisive impact on the program. Black's team investigated DCI and confirmed its beta adrenoceptor-blocking properties as well as its intrinsic activity. DCI then became a lead-substance from which pronethalol was developed in 1960. The clinical trials of pronethalol began in 1962. They revealed that the drug had certain adverse effects on patients. It was marketed under severe restrictions in 1963 but had to be withdrawn altogether in 1965 when its carcinogenic effects were revealed in long-term studies.

Fortunately, pronethalol was not the only substance investigated by the ICI team. Already in 1962, a promising successor to pronethalol — propranolol — was discovered. The substance became the focus of a new ICI program. The clinical tests commenced in 1964 and the drug was marketed as Inderal in the UK in 1965. It was well received by the medical profession and became an object of extensive clinical experimentation and research. In fact, it became the standard reference beta receptor drug.

However, propranolol was not entirely unproblematic. The definite knowledge of the mechanism of its action was still lacking. The drug turned out to be difficult to administer and was considered too 'aggressive.' These problems have been eventually mastered. However, at the time they caused ICI to focus on the development of a successor drug, practolol.

Interestingly, ICI was put on the track of practolol by the publication by Mead Johnson's scientists of their results concerning sotalol. The ICI scientists could soon show that practolol was a potent adrenoceptor blocker which had small effects on the central nervous system and no local anaesthetic action. The drug had little intrinsic activity which made it less aggressive than propranolol. Also, it was less likely to cause bronchospasm which was important from the point of view of patients suffering from asthma. In short, practolol was a very promising cardioselective beta blocker.

Practolol was marketed as Eraldina but had to be withdrawn from general use in 1975 after only 5 years. The reason was the serious side effects resulting from

---

7 Principal sources: Shanks (1984; 1989) and personal interviews at ICI.

prolonged oral administration. It is curious that, at the time, ICI did not make a stronger effort to save practolol or replace it by another similar drug. Instead, ICI concentrated on improving propranolol mainly by extending its field of application to hypertension. In part, the explanation of ICI's behavior can be sought in the deficiencies in the knowledge of receptors at the time. The ICI team did not understand the precise mechanism of practolol's action. Much of that ambiguity was resolved in 1967 when A. M. Lands, working at the Sterling Winthrop Institute, suggested that beta-receptors did not constitute a homogeneous class but rather consisted of two distinct types: beta 1 and beta 2. At the time, however, the ICI researchers appeared to pay little attention to Lands' theory. One wonders whether the departure from ICI of James Black had anything to do with it.

However, ICI eventually did develop a very successful cardioselective beta blocker. Its Tenormin became a leading product in the 1980s, and the main competitor to Hässle's Seloken.

Serendipity Gained II: Hässle[8]

The beta-blocker program at Hässle was the result of the decision in 1959 to radically restructure the company's R&D activities by focusing them on cardiovascular diseases. The research director of Hässle, Ivan Östholm, promoted extensive consultations between the Hässle team and several medical and academic advisers, including Lars Werkö — professor of cardiology at Gothenburg University, and Björn Folkow — professor of physiology.

On Werkö's advice, the first chosen target was arrhythmia. An existing drug, quinidine, was used as the point of departure. Its therapeutic usefulness was increased by administering it in the form of Durules, which Hässle developed at an earlier stage in collaboration with the faculty of pharmacy in Stockholm. However, the potential of quinidine was limited and there was a need for an entirely new drug.

In comparison with the other companies discussed above, Hässle virtually lacked the experience and resources necessary for the development of truly new drugs. Thus, Ivan Östholm faced the task of creating an entirely new R&D organization. In 1959, a department of chemical synthesis was set up and Arne Brändström and Hans Corrodi were recruited to lead it. A year later, a department of pharmacology was added. The know-how and the know-why about appropriate test models and techniques were sought through contacts with the academic community, both at home and abroad.

Initially, Hässle's team had no useful clues around which to reorient their new cardiovascular research program. Contacts with professor Folkow suggested that the stress hormones adrenaline and noradrenaline were involved in arrhythmia, but the

---

8 The present account of the developments at Hässle is based chiefly on Östholm (1991), Holm (1992) and on interviews with Ivan Östholm, Bengt Åblad, and Peder Berntsson.

knowledge of the precise biological mechanisms was scarce and there were no promising lead substances to focus on.

In 1961, the direction of the company's R&D changed radically when Arvid Carlsson, a newly appointed professor of pharmacology at Gothenburg University, became a consultant to Hässle. Carlsson, who specialized in the pharmacology of the central nervous system, was familiar with the general trends in the field of neuro-transmitters. He was in fact present at the symposium in Bethesda, Maryland, in 1958 when the discovery of DCI was discussed by Moran. Under Carlsson's influence, Hässle's program became redefined to aim at the development of beta adrenoreceptor-blocking agents with DCI as a lead substance.

Intensive work on beta blockers work commenced at Hässle in 1961. New methods for pharmacological testing were developed and more than 300 substances synthesized and screened. Some of them proved promising. However, in 1962 Hässle became aware that ICI was active in the field and had a head start. Moreover, it soon became apparent that many substances of interest had been previously synthesized and patented by other companies, mainly ICI or Boehringer. Hässle was close to abandoning the project.

Fortunately, the commitment of Hässle's team to beta drugs and Ivan Östholm's powers of persuasion saved the project. An important contributing factor was the influence of Arvid Carlsson who impressed on Östholm and other key persons that major therapeutic advances were not likely without new and deeper knowledge of biological mechanisms. Thus a novel type of research culture was implanted at Hässle. The ideal person to promote the growth and development of that culture was Bengt Åblad who in 1963 became the leader of the beta project. Åblad had an outstanding ability to combine sophisticated preclinical research with insights into medical needs. He also had a strongly innovative drive. His specialty was the animal and human pharmacology of the cardiovascular system. Before joining Hässle he worked in the department of pharmacology in Gothenburg where he acquired a Ph.D. in 1961. He also spent a short time with Moran in the United States.

The continued synthesis and screening program at Hässle resulted in 1964 in a promising substance: alprenolol. However, its future was immediately threatened. Some of the clinical tests gave alarming results, and once more the whole project was close to termination. Fortunately, a careful re-examination of the test results revealed that the substance was in fact both effective and safe. It was registered as Aptin in 1967 and successfully marketed in Sweden and several other countries as a drug against angina pectoris and certain forms of arrhythmia. In 1969, the use of Aptin was extended to include the treatment of high blood pressure (eventually the largest field of application of Aptin), and in 1974 to the field of prevention of heart infarction. During the 1970s, the annual sales of Aptin were around 100-150 million SEK.

Aptin has never achieved a market position equal to ICI's Inderal. Its main successes have been in Sweden, in the Nordic countries, and in Japan. The European continental market proved difficult. The plans to sell Aptin on the U.S. market had to be abandoned after the licensing arrangement with Warner Lambert proved unworkable. However, *the importance of Aptin went beyond its direct commercial value. It proved to Hässle's scientists that they could develop new drugs and that their approach was effective. Still more importantly, Aptin paved the way for the development of an extremely successful second-generation beta blocker, Seloken.*

It all started with certain serendipitous discoveries in 1963 and 1966 which suggested that beta receptors were not identical in different organs. Some isomers of Aptin appeared to be cardioselective. The problem was to find a substance which had both high receptor affinity and high cardioselectivity. ICI's practolol showed that this was possible.

Among the variants of Aptin synthesized in 1966, there was one which had a clearly stronger effect on the heart than on the vascular system and bronchi. Hässle scientists arrived independently at results pointing in the same direction as those produced by Lands. In 1967, a project was launched under the leadership of Bengt Åblad, and later Enar Carlsson, with the objective of finding a selective beta blocker. A program of synthesis and screening of about 400 substances was initiated. A number of promising leads were obtained. In 1969, two interesting cardioselective substances were synthesized, including H93/26 (metoprolol). Patent applications were filed. Both substances were submitted to clinical trials up to stage 2. From 1972, the work was focused on metoprolol.

An important step in the development of metoprolol was the realization by Åblad and Carlsson that the two kinds of beta blockers (beta 1 and beta 2) were found in different proportions in different organs. Beta 1 receptors dominated in the heart, beta 2 in bronchi and the vascular system. The implication was that only partial selectivity was possible. These results were obtained by Åblad and Carlsson in the course of basic research and illustrate how fundamental and applied work can cross-fertilize each other.

A successful launching of a new beta blocker required large resources. In order to overcome the resource bottlenecks, cooperation started with Ciba-Geigy. The latter company was to synthesize a large amount of metoprolol and carry out an international clinical test program. However, the cooperation proved unsatisfactory from Hässle's point of view. Ciba-Geigy was developing a beta blocker of its own, Trasicor, and was ambivalent to Hässle's product. The research collaboration was eventually discontinued.

Metoprolol was found effective against high blood pressure and angina pectoris. It was registered in Sweden in 1975, where it was marketed as Seloken. Continued clinical studies led to the broadening of its applicability and improved methods of administration. Seloken's success was also facilitated by the withdrawal from the market of ICI's Eraldina.

Thanks to the quality of documentation produced by Hässle, Seloken was soon introduced internationally. The drug was registered in the United States in 1978, where it was marketed by Ciba as Lopressor. Introduction in Japan took place in 1982, where it was marketed by Fujisawa. In 1982, Seloken became one of the most sold medicines in the world and Astra's largest product with yearly sales reaching 3 billion SEK.

The development costs of Seloken after its market introduction in 1975 were around 400 million SEK. New variants of the drug were introduced, including combination drugs Beloc comp, Treloc, and Belnif. Seloken ZOC was introduced in 1986 and has been successful in competing against ICI's Tenormin.

Attempts have also been made to develop a successor to Seloken. A new promising substance, H 104/08 (pafenolol), was found and tested but it proved to be carcinogenic and was abandoned. However, the search for a third-generation beta blocker continued under the leadership of Bengt Åblad.

The Followers[9]

Several other companies have added to the field of beta blockers at various stages.[10] In most cases, they have adopted a follower strategy imitating the existing products. This kind of approach does occasionally lead to valuable therapeutic advances, among others making it possible for doctors to individualize the treatment of patients. However, a majority of 'me-too' beta blockers varied rather little from the leading drugs and did not offer any serious challenge to them. Still, a me-too drug can become a serious contender if it is marketed by a company enjoying high reputation and a commanding, powerful marketing organization. In the case of beta blockers, Ciba-Geigy was such a company.

Ciba had a long tradition of R&D in the cardiovascular field. It also had experience with catecholamine analogues which it examined while searching for hypertensive drugs. However, the company's program on beta blockers seemed to have started only after ICI's work on pronethalol became known. A large number of pronethalol-like substances were synthesized and investigated, with no interesting results. When ICI acquired a patent for propranolol, the pattern was repeated: a series of similar substances were synthesized by Ciba. One of these was oxprenolol which was eventually marketed as Trasicor. The drug had a limited success in the skeptical German-speaking world but, because it was considered less 'aggressive,' became a serious competitor to ICI's propranolol on the British market.

As already noted, parallel with its work on Trasicor, Ciba-Geigy collaborated with Hässle on the development of Seloken. Despite that, Ciba-Geigy underestimated the advantage of the cardioselective second-generation beta blockers. In the event, it was Sandoz that became the most successful follower company, without however threatening the dominant position of ICI and Hässle.

The Main Features of the Development of Beta Blockers

Let us now briefly summarize the main features of the development of adrenergic beta blockers.

---

9 Based on Vos (1989), Östholm (1991), and Holm (1992).

10 In the early 1970s there were some 50 beta-blocker patents filed by different companies.

The discovery of DCI triggered an intense R&D activity leading, within a comparatively short time, to the development of a series of highly innovative drugs. This was possible because by the early 1960s most of the knowledge required to set up meaningful industrial R&D programs was in existence. Yet, as we have seen, the actual process of discovery and development of beta blockers was by no means easy and free-for-all. *The ability to bring together the relevant pieces of knowledge and to fuse them into successful R&D programs was, in fact, quite rare.* It required a capacity to absorb and utilize insights from on-going research in several disciplines and the ability to overcome many residual ambiguities and knowledge gaps. Last but not least, it demanded the ability to overcome considerable resistance in the medical communities, industrial R&D bureaucracies, and the academic world. Thus, it is not entirely surprising that only five companies initiated beta-blocker programs in the early 1960s and that only two of them were truly successful.

Indeed, the story of beta blockers was quite dramatic. Like all genuine innovation processes, it was characterized by many twists and turns. Opportunities were seized and lost, projects were initiated, redirected, scaled down, or abandoned. A handful of individuals played a great role in the unfolding of events. But the ultimate success of these individuals depended critically on their ability to develop and inspire large teams of competent and dedicated specialists.

Also important were the alliances forged by them with other key individuals and groups within the managements of their companies as well as in academia and the medical community. All these features come into sharp focus when one examines the critical incidents in the history of beta blockers. Two kinds of such incidents have been particularly important: (i) the recognition and exploitation of *serendipities* in the R&D process and (ii) the handling of *crises* repeatedly arising in that process.

The role of chance in drug development has been and continues to be large. Important discoveries are constantly being made as by-products of investigations and activities directed at other unrelated objectives. Clinical observations of side effects of existing drugs, unexpected results of preclinical tests in industrial labs, as well as observations made in the course of basic research can all provide clues to new medicines. Serendipity occurs when the significance of such clues is perceived and correctly interpreted. In some cases, this is quite easy; more often, however, the significance of a serendipitous discovery is deeply hidden in a jumble of confusing signals and becomes apparent only gradually. In either case, what is needed is the 'prepared mind.' The discovery of DCI, for example, meant different things to different individuals and groups.

The superior ability to evaluate information, whether systematically sought or accidentally generated, has been a consistent feature of the successful teams. These teams were also in a better position to handle crises which arose when (i) the progress of a project became suddenly blocked by the discovery of unexpected side effects, (ii) it was realized that a substance one invested heavily in was actually ineffective, or (iii) it was realized that a certain approach had been preempted by competitors. The first of these factors, the observation or fear of serious negative side effects, has been particularly critical, considering the fact that throughout the sixties, the entire pharmaceutical industry was reeling from the shock of the thalidomide catastrophe. Under these circumstances, the mere suspicion of serious side effects could undermine a project. It is difficult to deal effectively with the problem of negative side effects unless one has a reasonably complete knowledge of the specific mechanisms through which the drug exercises its primary and secondary effects. Such knowledge is necessary for the design and interpretation of preclinical and clinical tests. In its absence, projects are likely to be abandoned prematurely or pursued at a sub-optimal level. Clearly, the ability to utilize serendipity and to handle crises is largely a function of the knowledge base and the research strategy of the teams involved.

In his detailed study, Reinder Vos (1989) interprets the development of beta blockers in terms of a theory according to which the discovery and effective development of new drugs depends critically on clinical research and insights based on therapeutic practice.[11] He points, among other things, to the fact that the views held by the German and American cardiologists had pervasive negative effects on the course of beta-blocker programs in the American and German companies. The opposite was true of Sweden and Britain.

I fully agree with Vos about the importance of the therapeutic practice and clinical observation in the development of drugs. That importance is particularly great when it comes to finding new indications for existing drugs. Such extensions are often of great therapeutic and economic significance. Equally, the information about side effects may provide useful clues to the development of entirely new drugs. However, Vos' analysis appears to be somewhat one-sided. The success of ICI and Hässle depended on more than the access to first-rate clinical research necessary as it was. *Rather, it depended on the ability of these companies to link up the clinical and chemical competence into a coherent whole by relying on basic biological and pharmacological knowledge.*

Drug discovery and development is a complex process demanding cooperation of many disciplines ranging from fundamental biological and chemical fields to various clinical specialties. It also draws on the accumulated experience of the medical profession. Companies attempt to incorporate those competencies in their R&D organization and other functions, including leadership. However, *the composition of the knowledge base and the precise relations between its elements change over time*. This change involves cumulative developments as well as discontinuities. The older components of the knowledge base mature and become widely available to the industry as a whole. Newer elements are added and, in some cases, define what one might call the *critical competence*, i.e., competence which endows a company with a distinct competitive advantage in a certain field of drug development.

---

[11] Vos believes that the apparent decline in the 1960s and 1970s of the productivity of pharmaceutical R&D was a result of the weakening of the link between drug development and medical practice.

Until the 1960s, or thereabouts, the chemical approach dominated much of drug development.[12] This was partly due to the fact that many of the leading firms were off-shoots from the chemical industry. In these companies and in their emulators, chemistry was the critical competence shaping the R&D philosophy, organization, and the corporate power structure. This dominant R&D culture had one important consequence. While extolling the combination of chemical manipulation and pharmacological screening as the royal road to drug discovery, it created a certain aloofness between the industry on the one hand and the bio-medical community on the other. As a result, the industry had a limited capacity to absorb and effectively utilize the rapidly growing body of bio-medical knowledge.[13] The story of beta-blockers illustrates this point. ICI and Hässle were superior to their competitors in their ability to bridge the gap between the chemical and clinical aspects of drug development. This was achieved by focusing on the pharmacological and biological mechanisms of disease and drug action. The approach allowed the scientists at the two companies to tap a larger knowledge base and gave them an edge over the competitors who followed more conventional R&D strategies. Boehringer, Eli Lilly, and Mead Johnson have all had extensive experience with substances closely related to beta blockers. On repeated occasions, they were ahead of both ICI and Hässle in their synthesis programs. But they were slow in recognizing the biological and therapeutic significance of their results.

Thus, in a sense, the story of beta blockers is one of the competition between different approaches to drug development. Still the question remains: why have some of the companies chosen the right approach and applied it effectively, while others have not?[14] In the following section we shall try to answer this question with regard to Hässle.

WHY HAS HÄSSLE'S PROGRAM SUCCEEDED?

Why has Hässle succeeded? How was it possible for a small company with very limited resources to do so well in competition with large firms? How could an R&D team with

---

12 The does not mean that chemical approach was <u>the</u> paradigm of pharmaceutical technology. In fact, many of the most important achievements of the Golden Age occurred in the pronounced biomedical and biotechnological fields such as hormones and, especially, antibiotics. In so far as it is possible to talk of paradigms in this context, the pharmaceutical industry has always been 'multi-paradigmatic' with the different approaches to drug discovery and development both vigorously rivaling and complementing each other.

13 For a discussion see Bartholini (1983) and Carlsson (1983).

14 By this I do not wish to imply that there is but one right approach to drug development. What is appropriate differs considerably from one field of pharmaceuticals to another and over time. The same results can be achieved by different routes, and good luck definitely plays a major role. However, given a certain state of knowledge in a specific field, some R&D strategies are likely to be more efficient than others.

practically no experience of developing genuinely new drugs select and effectively execute a highly innovative drug development strategy?

The answers to these questions necessarily involve many different considerations. Some of them reflect factors external to the company, others are specific to Hässle. Here I will concentrate on (i) the particular style of innovation management and research culture at Hässle, (ii) the ways in which the company managed to exploit the academic infrastructure, and (iii) the relation between Hässle's R&D program and Astra's corporate leadership. In the next section we will look at a wider picture and consider the systemic ramifications of Hässle's success.

The Beginner's Advantage: The Emergence of Hässle's R&D Strategy

The contemporary pharmaceutical industry is characterized by a dual structure. It consists, on the one hand, of R&D-intensive companies predominantly engaged in product competition and, on the other, generics producers engaging in price competition. The division is a result of the R&D dynamics of drug development as well as of certain economic and institutional factors, including intellectual property rights and pricing policies.

The above pattern has emerged gradually during the post-war decades. By the end of the 1950s it became increasingly apparent that a strong investment in R&D was a prerequisite for economic success in the industry. It was against that background that Astra decided to step up its R&D activity. However, the specific implications of the decision were a matter of interpretation. Shuld one have a central ized or decentralized organization? Should the emphasis be on radical innovations or on an incremental strategy? Should one diversify or specialize? What existing models of R&D organization and strategy should be emulated? None of these questions had a clear answer in the 1950s. Nor do they have such answers today. They reflect eternal dilemmas rather than problems allowing final solutions.

Let us take the question of radical versus incremental innovation. One striking feature of the pharmaceutical industry is that almost all the companies (including the biggest ones) derive most of their revenues from a *few highly successful drugs*. Such drugs tend to offer a clear-cut advantage over others in a specific therapeutic field. Companies which wish to grow and/or stay highly profitable must produce such 'blockbusters' at regular intervals. Not all succeed. Big established firms can survive periods of 'drought' by pursuing expensive defensive strategies extending the life-times of successful drugs, developing 'me-too drugs,' or by buying technology in the form of licenses or through acquisitions. Such actions are necessary components of most companies' technology strategy. However, an over-reliance on them, while risky even in the case of the largest companies, is still more dangerous in the case of small and medium-sized firms.

Yet the alternatives are by no means obvious. The discovery of radically new drugs (based on previously unknown/unused principles) is widely regarded as a chance event of low probability. This was, at any rate, the dominant view in the 1950s and 1960s. Against that background, the decision that Hässle, lacking relevant experience and resources, should engage in a major program to develop new drugs must have appeared daring to say the least. A more modest approach oriented towards incremental technical improvement seemed more appropriate. However, Hässle and its research director, Ivan Östholm, refused to be 'reasonable,' fearing that the conventional approach would conserve Hässle's weakness and eventually threaten its survival. By adopting a radically new approach, they were able to turn some of the company's weaknesses into advantages.

The fact that Hässle was relatively unencumbered by an established tradition made it easier to create a genuinely new research organization and to develop innovative approaches. This, of course, did not happen automatically. Rather, it was a result of experimentation with various alternative R&D strategies. One approach was to search for new drug candidates among various substances extracted from plants. Clues from traditional medicinal practices were sought and investigated. The results, however, were discouraging. Another, more conventional approach, based on chemical manipulation, was followed in search of new anaesthetic drugs. Interesting lead substances were generated, but the project was eventually abandoned to leave room for the cardiovascular program.

The latter underwent, as we have seen, an evolution from a comparatively traditional approach focusing on an existing drug to one involving search for new pharmacological principles based on the knowledge of underlying biological mechanisms. That approach, which was proposed by Hässle's consultant, professor Arvid Carlsson,[15] found an enthusiastic supporter in the company's research director Ivan Östholm and, eventually, crystallized into an explicit research philosophy.

Hässle's Knowledge Base

All companies strive to adjust their competence profiles to the changing knowledge base. However, these adjustments are usually made with a certain *time lag*. Such lags open *opportunities for newcomers* to enter and/or advance their position in the industry.

In the mid-1950s, some intensive discussions were under way within Astra concerning the company's general R&D strategy. The upshot was the decision to expand the volume of R&D and strengthen the medical science competence in the company. The latter policy was to a large extent due to the influence of consultants from within the Swedish academic community. Bio-medicine and, particularly, clinical research have been traditionally strong in Sweden. Thus it was natural for the Swedish pharmaceutical industry to rely on the comparative advantages of the Swedish R&D

---

15 For an account of the approach, see Carlsson (1983).

system. This was reflected in the case of Astra in the policy to locate its daughter companies (Hässle and Draco) in the neighborhood of major universities in Sweden.[16]

Although embraced by the company's top leadership, the policy of emphasizing biomedical orientation was by no means universally endorsed within Astra's R&D organization and management which were committed to the traditional chemical paradigm. However, the policy resonated well with the hopes and aspirations of the fledgling R&D organization at Hässle. This, as we have seen, was largely due to the strong influence exercised by a few key academic consultants.

This openness to academic advice was in part a result of the weakness of Hässle's research organization. The only way in which the company could build up its competence rapidly was by turning to where such competence was plentiful, i.e., the universities. That, however, was by no means a simple matter as the attitudes of the academic scientists to the pharmaceutical industry were at the time reserved, if not outright negative. In order to solicit the cooperation of academia, Hässle had to develop an entirely new relationship based on subtle give-and-take on both sides.

The manner in which that was done is instructive. The following features stand out: (i) there was *a deliberate policy* of seeking contacts with the universities;[17] (ii) both *formal and informal links* were established; (iii) the importance of *mutual respect, tolerance, and personal friendship* were strongly emphasized; (iv) t*he academic advisers were influential on all levels of decision making* (from setting up of the general strategy to the choice of research methods); (v) *multiple links with key research environments* were established including *consulting, personnel mobility, and collaborative research. As a result, the research culture established at Hässle contained many elements emanating from the university. This made continuous communication and collaboration easier.*

The close relations to academia provided Hässle with important product ideas, problem-solving capability, highly qualified manpower, and a contact surface with the best international science. The high standard and large volume of clinical research in Sweden was important for Hässle in the course of testing and documenting the drugs. However, it was important in yet another manner. The research-orientation of a large fraction of the Swedish medical profession made the Swedish doctors, on the whole, less conservative and better equipped to evaluate and use new drugs than were many of their foreign colleagues.

Managing the Creative Process

A common drawback of 'science-driven' technology development programs is that they easily become separated from technological and commercial realities. The challenge is

---

16 Pharmacia followed the same approach by linking up with Uppsala university.

17 Cf. Östholm (1983).

to develop teams which are competent simultaneously in three areas: science, technology, and application. The success depends on (i) *the right composition of the team*, (ii) *the pattern of communication and interaction within it*; and (iii) *the leadership*. Hässle's beta-blocker team grew gradually, and the company's R&D director was capable of recruiting several competent and highly motivated scientists. That was not an easy achievement, given the lack of research tradition in the company. Östholm's success must be attributed in large measure to his personal leadership qualities. It may also have depended on the fact that, a pharmacist by training, he was not strongly identified with any single scientific discipline. Thus his role was that of a team-builder, a network-creator rather than an expert-specialist.

Indeed, an essential feature of Hässle's team was that it was *well balanced in terms of chemical, clinical, and biological/pharmacological competence*. Furthermore, due to the small size of the team, it was relatively easy to establish close communication between different specialties. There was *high social integration* (personal relations, friendship, shared leisure) and important *professional overlaps*. Thus, the clinical scientists were heavily involved in preclinical research, and there were chemists with well developed biological insight and imagination. Many of these scientists had close personal links to one or several academic research environments, and/or innovative clinical environments. All these features were essential to create *a science-informed but clinically-anchored research program*. This was indeed different from what went on in several of the other companies working on beta blockers, where the teams were bureaucratically fragmented and where there often existed high social barriers between different professional groups.

The ability to combine the academic-analytical attitude with the industrial-entrepreneurial was embodied in the director of the beta-blocker program, Bengt Åblad. Here we see a striking parallel to the situation at ICI, where a similar role was played by James Black. Both Black and Åblad had an outstanding ability to spot therapeutic pearls in scientific oysters. It is rather difficult to see how Hässle and ICI could have succeeded without them.

Generally speaking, the problem of *recruiting and retaining truly creative scientists* in an industrial setting is one of the critical issues in the pharmaceutical firms wishing to succeed in the field of 'drug discovery.' Hässle and its research director have been very successful in that respect. Another factor contributing to the effectiveness of Hässle's team were, paradoxically, *resource restrictions*. Hässle could not afford massive synthesis and screening programs. 'Working smart' was a necessity. This might have legitimized unconventional approaches originating in the academic research programs.

Objectives and Priorities

A company can afford only a limited number of product development programs. The wish to spread risks and/or to avoid conflicts within the organization may induce excessive diversification. Generally, it is better to run a few well-staffed and well-financed programs than many under-critical ones. But such a policy forces some very tough and potentially disruptive decisions.

The success of Hässle depended to a large extent on the fact that beta blockers were given high priority. Several of the other companies active in the field failed to assess correctly the potential of beta blockers and to make the transition from 'skunk works' to 'major project' in time. They may have had 'good reasons' for de-emphasizing beta blockers. The point is, however, that *in a 'hot' field you have to make up your mind and choose between pursuing the project aggressively and giving it up altogether.*

However, the ability to determine priorities is only one aspect of goal setting. The way in which the objectives are formulated and communicated in the R&D organization is at least as important. Here conflicts can easily develop between the culture of R&D departments and that of management. The risks of such conflicts are especially great in large, bureaucratic organizations. Managements tend to cast commercial objectives in terms of *probabilities* and *quantitative* economic indicators of market share, profits, etc. The problem is that the long-term R&D objectives cannot be so formulated.

Hässle has been able to avoid these conflicts. Its leadership emphasized the necessity of setting the objectives in terms of *qualitative therapeutic goals*, mindful that an over-emphasis on quantitative management will tend to favor development activities extending/defending the existing products and technologies at the expense of radical innovation.

However, this did not mean that the company followed a relaxed and overly liberal policy. On the contrary, the severe resource restrictions forced it repeatedly to make tough priority decisions. The policy was to concentrate resources on a few promising projects.

The Interface between R&D and Corporate Management

The choice of priorities, and decision-making generally, is rarely a result of application of some abstract formula; rather, it reflects *complex negotiations and power-play within the company.*

Let us now briefly consider the way in which Hässle's R&D activities and decision processes were embedded in the larger framework of Astra.

The R&D staff proposes, the corporate leadership disposes. The latter functions as a surrogate market for the former. The relationship between the two is complex and often problematic. Several of the beta-programs have faltered because of the negative or lukewarm attitude of top management. In some cases, the reason was that the research teams lacked the necessary autonomy. In others, they lacked the scientific and technical authority necessary to persuade the skeptical corporate leadership.

There can be little doubt that the decentralized structure of Astra has been important for the success of the beta-blocker program. The policy of allowing daughter companies to compete among themselves by shaping their own R&D portfolios and selecting their own strategies has prevented the bureaucratization and over-management which has been characteristic of many large pharmaceutical firms. In particular, the research directors enjoyed considerable autonomy and influence. Such an approach was particularly useful at a time when genuinely novel approaches to drug development were being introduced and tested. In a tightly integrated hierarchical organization, the scope for genuine innovation is usually severely constrained: the established R&D structures and the middle management tend to take a conservative attitude and often manage to prevent or slow down change.

Hässle's programs and general policy have been questioned and criticized within Astra on repeated occasions. This was indeed a natural consequence of the competitive decentralization policy within Astra. The fact that Hässle has been, in most cases, able to defend its position, depended on several factors. The energy and the persuasive powers of the company's R&D director, Ivan Östholm, were extremely important. Östholm enjoyed good relations to Astra's managing director, Arne Wegerfelt, who intuitively trusted Hässle's team and its approach. Also important was the support which Hässle's team had in the academic community. Finally, we should mention the fact that in the 1960s and 1970s, Astra's board was chaired by Jacob Wallenberg. The spirit of industrial enterprise and commitment to long-term investment in R&D which pervaded Wallenberg companies has helped to create a climate which was congenial to innovative entrepreneurship exemplified by the beta-blocker program at Hässle. These criss-crossing alliances and influences helped to counterbalance the opposition from the established R&D organization at Astra.

However, none of these factors would have saved Hässle's beta-blocker program unless its viability could be demonstrated within a relatively short time. The development of Aptin gave Hässle's team the necessary *credibility* and established its *negotiating position* within Astra. Furthermore, the success of the entire beta-blocker program helped to create a corporate culture which bred further successes, particularly Losec, the most successful drug in the history of the Swedish industry, and one of the most sold drugs in the world (see next chapter).

BETA BLOCKERS AND THE TECHNOLOGICAL SYSTEM OF THE SWEDISH PHARMACEUTICAL INDUSTRY

We have defined *technological systems* as "knowledge and competence networks" supporting the development, distribution and utilization of technology in established or emerging fields of economic activity (Carlsson & Stankiewicz 1991). They consist of networks of firms, R&D infrastructures, educational institutions, and policy making bodies. Technological systems can be seen as mediating structures between the knowledge base of a sector and the firms active in it. Their quality influences the degree to which the firms can exploit the emergent technological opportunities. Hence, the importance of technological systems increases with the complexity and heterogeneity of the knowledge base of innovation. The field of pharmaceuticals is an excellent illustration of this tendency.

The technological system of pharmaceuticals consists of three main elements: the firms, the public (primarily academic) R&D system, and the health care system. The functioning of the system is influenced by the properties of the economic and regulatory regimes (e.g., pricing regulations, health care financing, intellectual property rights, and safety regulation) under which it operates.

The case study of beta blockers presented above illustrates the ways in which the properties of a technological system influence a firm's ability to perceive and exploit technological opportunities. Without doubt, Hässle benefited from the following properties of the technological system of the Swedish pharmaceuticals:

The Existence of a Well-Developed Public Biomedical R&D System

The Swedish R&D system has a tradition of strength in both basic and applied areas of the bio-medical field (Archibugi and Pianta 1992; Persson 1991). Biochemistry, pharmacology, and various clinical specialties have been particularly well developed.[18] These were the critical disciplines in pharmaceutical research in the 1960s and 1970s. Indeed, companies such as Merck, Pfizer, Smith Kline Beecham, Glaxo, Wellcome, ICI, and Astra which have become the industry's stars since the 1970s have all been based in the countries particularly strong in the biomedical sciences.[19]

The strong biomedical research helped the Swedish companies both directly and indirectly. The Swedish scientists enjoyed good international reputation and were well connected to the international scientific networks through which crucial scientific information was transmitted. The case of beta blockers which clearly shows the importance of these factors was by no means an isolated one. In his detailed study of the role of universities in the development of the Swedish pharmaceutical industry, Lennart Norgren (1989) found that out of 25 major product groups developed by the Swedish companies, 17 had their roots in the academic environment. Clearly we are dealing with a strongly systematic factor rather than an isolated lucky coincidence.

---

18 This was particularly true of clinical pharmacology where Sweden, along with the United States, had a leading position (Werkö 1986).

19 Archibugi and Pianta (1992) found that the Anglo-saxon and Scandinavian countries had a comparative advantage in the biomedical and biological sciences as opposed to the continental European countries such as Germany and Switzerland which had the advantage in physics and chemistry. Further details are given in Pianta and Archibugi (1991). Quay (1988), looking at the pattern of internationalization of the pharmaceutical R&D activities, found that the countries attracting foreign R&D operations were primarily the United States and Britain, followed by Canada and the Scandinavian countries.

The Properties of the Swedish Health Care System

In the post-war years the Swedish health care system has undergone a phase of rapid growth and modernization. The health care expenditure grew more rapidly than the economy, reaching 9.4% of GNP in 1980 (LIF 1995). That made Sweden one of the most health care-intensive countries in the world. The system that emerged was highly competent and innovation-oriented. This innovation friendliness was partly a result of the growing resources and partly of the system's structure which was characterized by a strong emphasis on hospitals rather than on small clinics and private practices. As pointed out by Vos (1989), due to the large technical resources at their disposal and the access to specialized competence, the doctors working in hospitals are generally more willing to experiment with new therapies. The receptiveness of the Swedish health care system to Aptin has certainly played a major role in the success of Hässle's beta-blocker program.

The Strong Linkage between the Health Care System and the Universities

An important feature of the Swedish R&D infrastructure has been the high degree of interpenetration between the academic system and the health care system. Roughly 40 percent of biomedical research in Sweden is conducted in the academic hospitals (SCB 1995). The latter have frequently played an important role in the innovation process by helping to target industrial R&D projects and through active participation in the development and evaluation of new therapies. The strong emphasis on research in the education of Swedish doctors has been another factor facilitating interactions between the health care system, the universities, and the industry. Medical practitioners with significant research experience can be expected to be more receptive to innovations and more willing to take part in their scientific evaluation than those without such experience.

Together with the favorable regulatory and pricing systems (Jönsson 1994), the above features of the technological system of Swedish pharmaceuticals definitely facilitated the remarkable development of Astra and other pharmaceutical firms in Sweden.

The Dynamics of the Technological System of Pharmaceuticals in Sweden

It should not, however, be assumed that there was anything preordained about the process. The system elements listed above did not by themselves guarantee success; something more was needed to bring them together into an effectively functioning whole. In fact, it could be claimed that Astra has not so much benefited from an existing technological system as helped to create one.

First of all, the knowledge and competence network which gave Hässle its competitive advantage over the other companies was by no means implicit in the institutional pattern of the R&D system at the time. Nor was it a result of a conscious design by policy makers keen on promoting industrial innovation. The motives for the building up of the bio-medical research complex in Sweden derived instead from a much more general policy of strengthening and developing the health care system. The generally positive attitude towards enhancement of welfare translated in the area of science policy into generous support of biomedical research. As one would expect in a highly professionalized field, the academic institutions played a dominant role in that process, thus contributing to the balance between basic and applied aspects of research. However, this also meant that the critical innovative capabilities were embedded in an institutional setting largely indifferent, if not hostile, to commercial or industrial endeavors. As we have seen, Hässle had to work hard to overcome the skepticism towards collaboration with industry which existed at the time in the scientific and medical communities (Östholm 1991).

The absence, at the government level, of a coherent industrial and technology policy for pharmaceuticals had several causes. Firstly, the existing system appeared to work well, so there was little incentive for major policy initiatives. Secondly, there were institutional obstacles to the emergence of such policy. Pharmaceuticals were generally regarded as a part of the national health care system rather than as an industrial sector in its own right (Werkö 1986). There was also an atmosphere of tension and even hostility between the industry and the government caused by recurring nationalization debates which culminated in the 1970s. Thirdly, the field of pharmaceuticals fitted poorly into the general pattern of technology policy which predominated in Sweden in the 1960s and 1970s. That policy was strongly oriented towards the support of small and medium-sized firms with limited R&D resources. The general assumption was that the pharmaceutical firms could fend for themselves. Finally, the characteristic features of pharmaceutical R&D, such as its high specificity, strong dependence on basic research, and intense product orientation, made the application of the typical tools of R&D policy (i.e., support to collective or collaborative applied research) difficult.

In these circumstances, the task of integrating and mobilizing the resources of the technological system became the responsibility of the individual firms calling for considerable foresight and organizational innovativeness. It is important to realize that the emphasis on cooperation with academic research which helped Hässle to achieve its success was by no means universal among the contemporary pharmaceutical companies. Although this sort of collaboration was a conspicuous feature of the pharmaceutical industry during the first half of this century, it decreased markedly in the 1950s and 1960s. The decline could be partially attributed to the decreased dependence of academic scientists on firms for financial support (Swann 1988). But at least equally it was due to the increasing introversion of the pharmaceutical companies which, having built up large internal resources, lost the immediate motive for close cooperation with the universities. Thus, there opened up a sort of 'cultural gap' between a large part of the industry and the rest of the R&D system. This gap was not due to some dramatic 'paradigm shift' or other technological discontinuity. It grew gradually, reflecting a natural tendency in firms to minimize the high costs of technology and competence transfer between institutionally dissimilar environments, to facilitate technology appropriation, and to maintain organizational coherence. Thus David Schwarzman (1976), basing himself on data from Schnee (1973), concluded that the academic system had ceased to be a major contributor to drug discovery. More than 80 percent of all new drugs between 1963 and 1970 were said to originate in the corporate labs, mainly of the largest firms. Only 9 percent originated in the universities. Between 1935 and 1949 these figures were 52 and 34, respectively.

Ironically, however, the 1960s were the very time when the productivity of the industrial R&D in pharmaceuticals (measured in terms of new drug approvals) was drastically declining (Walker & Parrish 1986; DiMasi *et al.* 1995). This negative tendency lasted until the late 1970s. To a significant degree it may have been caused by the toughening of safety regulations following the thalidomide catastrophe in 1961. However, this is hardly a complete explanation. In a technological field in which the knowledge base is heterogeneous and dynamic, corporate introversion and conservatism carry high risks. The growing divergence between the industrial and academic parts of the technological system of pharmaceuticals has certainly been among the main causes of the declining R&D productivity in the 1960s and 1970s. It was not just the diminishing frequency of innovation that began to worry the industry but also its declining quality as manifested in the shift from radical innovation to incremental and defensive ones (Gross 1983).

As we have seen, Hässle was among those companies which resisted the prevailing trends. Östholm and his colleagues helped to re-invent the industrial-academic partnership in pharmaceuticals long before the rise of genetic engineering made the need for such partnership obvious to everyone. When in the 1980s the industrial-academic alliance became popular once again, the way in which Astra-Hässle managed the university interface came to be viewed as something of a model (Östholm 1983).

DISCUSSION: THE CHANGING DYNAMICS OF PHARMACEUTICAL INNOVATION

I have argued above that the remarkable success of Astra in recent years can be attributed to its superior ability to exploit the technological opportunities which the rapid growth of biomedical knowledge created in the 1960s and 1970s. That ability was partly inherent in the corporate culture and strategy of the company and partly in the properties of the technological system in which it was embedded. The latter worked well because, relative to the needs of existing industry, it could generate plentiful and well-balanced knowledge base and, due to its high connectivity, create the conditions for the effective utilization of that base. Another important factor was the favorable innovation climate created by the rapid growth and modernization of the health care system. Astra-Hässle, Astra-Draco, as well as some of the other Swedish firms (particularly Kabi and Pharmacia) were capable of effectively exploiting these opportunities.

Since the 1980s, both the Swedish and the international pharmaceutical industry have undergone several significant changes regarding their knowledge base, corporate dynamics, overall structure, as well as the economic and regulatory regimes. The question is how well the technological system of the Swedish industry has adapted to these changes. It is impossible to give here a complete answer to this question. However, I will try briefly to outline the most important developments and challenges which need to be taken into account.

The Changing Knowledge Base and R&D Dynamics of Pharmaceuticals

A permanent feature of pharmaceutical R&D throughout its history has been its growing complexity, sophistication and costs. These trends have been accelerated rather than attenuated in recent decades. The rising safety and efficacy standards require strengthening of the pre-market selection processes. This means more R&D, both clinical and preclinical. Contrary to expectations, the technological and scientific advances in molecular biology, biotechnology and information technologies do not appear to make pharmaceutical innovation cheaper. The emergence of new technological opportunities opens up new fields of competition, demands expansion of corporate competencies, and forces capital investment in new instrumentation. The companies are also compelled to devote increasing resources to monitoring of the rapidly evolving knowledge base. In the United States the total costs per approved NCE (New Chemical Entity) doubled in constant dollars between 1963 and 1993 (DiMasi 1995). The fastest growing have been the preclinical costs which now account for more than half of the capitalized drug discovery and development costs (DiMasi *et al.* 1991).

The speed and complexity of these developments require a high level of flexibility and responsiveness from the R&D infrastructures. The pressures on the latter are intensified by the sharp increase in the demand for research personnel created by the expanding industrial R&D. These problems are particularly severe in smaller countries.

In the earlier decades there existed in Sweden a kind of 'excess capacity' in the R&D system which the pharmaceutical industry could utilize in the process of rapidly building up its own R&D organizations. The presence of plentiful academic competence in the critical fields of biochemistry, pharmacology, and clinical research — in combination with the flexibility of the companies whose research programs were not yet fully formed — meant that the matching of supply and demand for knowledge could be relatively easily achieved. These conditions no longer obtain.

In the 1980s there was a marked slowdown in the general level of public spending on R&D. Between 1983 and 1993 the academic research in Sweden grew by a mere 12 percent. During the same period the total industrial R&D in the country grew by 50 percent (SCB 1996). The disproportions are even more glaring when we look at the pharmaceutical field. While the academic medical research grew slower than any other field, the research performed by the pharmaceutical industry almost tripled (*ibid.*). The slow growth of the research system jeopardizes its adaptiveness and innovativeness. The problem is not only whether or not the system can take up new research specialism but also, and perhaps primarily, whether it can allow them to *expand quantitatively at a sufficient rate*. The latter is of great importance in the fields of R&D with large potential economic impact.

The response of the Swedish and other R&D systems to the advent of genetic engineering illustrates the difficulties involved. There can be no doubt that the Swedish universities were very well positioned to play a major role in the new biotechnologies. But as in most other European countries their ability to rapidly expand the activity in the relevant areas was limited (Stankiewicz 1986). Thus there developed competence bottlenecks impelling the Swedish firms to look for sources of competence and technology abroad (Johansson 1990).[20]

The threat of competence bottlenecks is by no means limited to biotechnology. The very rapid growth of the industrial R&D in the last decades creates recruitment needs which the Swedish academic system has increasing difficulties to meet. A recent survey of the demand for Ph.D.-level R&D personnel in the Swedish pharmaceutical firms indicates that serious manpower shortages are developing (LIF 1994). The total demand for Ph.D.s from 1994 to 1999 is estimated to be 523 - only 15 percent of which will be replacements. This demand corresponds to one-third of all Ph.D.s who will be produced in the relevant disciplines in Sweden during the period. In certain fields (pharmacology, galenics, and medical statistics to mention a few) the demand will exceed total supply.

Clearly, the Swedish pharmaceutical industry is beginning to outgrow the capacity of its national technological system. It is symptomatic that between 1985 and 1994 Astra's research personnel grew twice as fast abroad as in Sweden.[21] Increasing internationalization appears inevitable and places new demands on the national R&D policy.

Changing Corporate Dynamics

Astra today is a very different company from what it was 30 years ago. Its size, resources and geo-economic extension have grown enormously. Its corporate culture has evolved. What consequences do these changes have for the company's innovative performance?

The transformations which Astra underwent are by no means unique. The history of the pharmaceutical industry is one of progressive consolidation and corporate

---

20 The most dramatic instance of that has been the collaboration between the Swedish Kabi and Genentech on the development of genetically engineered human growth hormone, which under the names of Somatonorm and Genotropine have become the company's main cash generators (McKelvey 1994).

21 Source: Bolags Fakta 95-3; CD-rom.

growth. The latter has occurred both as a result of commercial success and through repeated mergers and acquisitions. In fact, since the emergence of R&D-intensive companies in the first half of this century, corporate growth has been one of the preeminent strategic objectives of most pharmaceuticals firms.

And yet the impact of corporate size on innovative performance has been a highly controversial issue. Large companies should, in principle, be more able to absorb technical and financial risks and therefore to pursue more aggressive R&D policies. They should also be able to accumulate larger and more diversified internal knowledge bases allowing them to maintain wider contact surfaces with the surrounding firms and R&D infrastructures. That should help them better to exploit the arising spillover effects. Some evidence supporting that view has recently been presented by Henderson (1994) and Henderson & Cockburn (1994). Also DiMasi *et al.* (1995) present data indicating that the cost effectiveness of pharmaceutical R&D appears to increase with firm size, particularly as regards preclinical research. The reasons for that are seen in the large firms' ability to spread fixed costs, pool risks, and utilize internal spillovers.

Contrary statistical evidence has, however, been presented recently by Graves and Langowitz (1993). Indeed, skepticism (or at least agnosticism) as to the blessing of corporate growth are widespread among the observers of the pharmaceutical scene. The examples of large companies suffering extended periods of decline are numerous.[22] In fact, all highly successful firms face a particular set of risks arising out of the very structure of innovation in pharmaceuticals. Spectacular growth frequently depends on one or a few extraordinarily successful products (Grabowski & Vernon 1994). The windfall profits continue until patent protection expires or a successful competitor drug enters the market. Before that happens, the company has to produce a new set of important innovations or face a painful contraction. In this situation some companies fall victims to what has been called the "negative spiral of success" (Spilker 1994). The spiral involves the following sequence of events. Faced with huge cash flow, the company engages in a massive expansion of its R&D including both heavy physical investment and rapid hiring of staff. This haste, in combination with overconfidence and complacency, results in an incoherent and unproductive R&D portfolio. When the company's patents expire, and no new block busters have been produced, severe cost and image problems develop. The company begins to plummet on the ranking list.[23] Crisis management ensues, leading to the sale of the company or its merger with another firm.

The above description is, of course, oversimplified. Corporate growth brings with it a variety of other, more subtle changes which can negatively influence innovativeness. As companies grow there occur important shifts in the balance between corporate subcultures associated with different key functions: R&D, marketing,

---

22 Bayer, Hoechst, Roche, Warner Lambert, and Upjohn are among the more spectacular examples.

23 For example, Upjohn ranked 17th in the 1970s and only 25th in the 1990s. During the same period Bayer dropped from 2nd to 15th place.

production, and finance. With size and centralization there often comes a tendency to move from the entrepreneurial to more 'managerial' form of leadership and decision making. The substantive technical goals become secondary to corporate commercial/financial objectives. The entrepreneurial daredevil commitment to one or a few projects is replaced by a risk containment mentality of portfolio management. The long-term research is excessively 'academised.' There is also a tendency towards routinizing and scaling up the R&D strategies which have been successful in the past, creating a sense of control and rationality which often proves illusory. Finally we should mention a propensity to increasing introversion in the company's attitude to its scientific and technical environment.[24] In a situation when the growth of the knowledge base of an industry is rapid and occurs largely outside of the company, these sorts of developments can have serious consequences.

The slide down the negative spiral can be arrested as exemplified by such companies as Merck, Pfizer, and more recently Roche. There are several different approaches which maturing companies can adopt (cf. Spilker 1994). One could perhaps speak of two main types of such strategies: "generative" and "harvesting."[25] The *generative strategy* is based on the premise that the company can rejuvenate itself and maintain innovativeness through creativity. The recipe is to invest heavily in the in-house long-term research, seeking major therapeutical breakthroughs. Such breakthroughs can establish long-term advantage which creates the necessary time and resources needed to develop new major products (the positive spiral of success). To this end it is necessary to promote internal diversity/decentralization and intrapreneurship, while all the time maintaining close links to the public R&D system, mainly universities. Merck is a company which exemplifies such a strategy.

The *harvesting strategy* is predicated on the company's ability effectively to exploit current opportunities through pursuit of an aggressive acquisition strategy combined with a powerful research organization which allows the company to win the innovative 'end game.' Crucial to success is a highly focused, commercially driven R&D program backed up with powerful marketing resources. Here large size is a major prerequisite of success. Glaxo can be seen as an example of this strategy.[26]

The ideal, of course, is to combine the attributes of both strategies, and indeed most firms position themselves somewhere between the extremes. However, the behavior of companies is determined by more than their internal dynamic. It also reflects the changing economic and regulatory regimes under which they operate and the overall structure of the industry.

---

24 This need not necessarily mean that the quantity of interactions decreases. It is the quality that matters. How rapidly are the *new* links established with the emergent research fields? At which organizational levels are they established and which forms do they take? Indeed, unless it is constantly reshaped, the external network of a company can become a source of conservative bias rather than innovation.

25 This applies to all high tech fields, not just pharmaceuticals.

26 A journalistic account of the differences between Merck and Glaxo has been given by Lynn (1992).

New Economic Regime

Pharmaceuticals is one of the most regulated fields of industry as regards safety, marketing practices, and price-setting. Together with other economic institutions and the prevailing market conditions, these regulations define the 'economic regime' under which the industry operates. Changes in any of the elements of that regime can influence the rate and direction of innovation in the field as well as the competitive situation of firms.

Since the 1960s, the international pharmaceutical industry has been exposed to increasing regulatory pressures in the area of safety, resulting among other things in escalating R&D costs and decreased effective patent protection time. For the most part, however, the impact of these developments has been relatively easy to absorb, given the fact that the economic regimes under which the industry has operated in most Western countries have tended to reward innovation. The health insurance systems and various forms of subsidies have favored new medicines. That, together with the rapid growth of health care expenditure, has acted as a stimulus to the development of the industry. Indeed, the profitability of pharmaceuticals has tended to be above the industrial average (OTA 1993).

Recently these economic regimes have come under review by governments seeking to cut the growth of the costs of medication.[27] The causes and implications of that growth are much debated. The general trend, however, is to favor the use of generic drugs, change the structure of subsidies and health insurance systems, and strengthen the hand of consumers (mainly institutional).

The Swedish industry is, of course, affected by these trends. In recent years, the Swedish health care system has undergone a process of contraction. The country's health care expenditure decreased from 9.4 % of GNP in 1980 to 7.25 in 1993 (LIF 1995). This should be compared with 14.1 for the USA, 9.9 for Switzerland, 9.8 for France, 8.6 for Germany and 8.5 for Italy. Also, the per capita expenditure on medication has been much smaller in Sweden than in other comparable countries (*ibid.*). In spite of that, new measures to cut the costs of pharmaceuticals are being proposed by an official committee on the financing and organization of the Swedish health care system (SOU 1995).[28]

These trends threaten to weaken the technological system of the pharmaceutical industry in several ways:

---

27 For a review of international developments in this area, see Shulman and Lasagna (1996).

28 In fact, the prices of medicines in Sweden have increased less than inflation, suggesting that the price setting mechanisms have been functioning reasonably well (Jönsson 1994; LIF 1995). The growth in the total cost of medications appears to be due mainly to the increased volume and greater use of new drugs. Much of that could lead to decreases in other costs of health care.

• Less generous pricing and reimbursement policies, and a systematic bias in favor of greater use of generic medicines could decrease the firms' earnings from new innovative drugs. This, happening at the time when the costs of development of new drugs are escalating, could significantly decrease the incentives of companies to engage in risky innovative R&D.

• Even though the dependence of the Swedish firms on the domestic market is limited (only about 10 % of the production is sold in Sweden), the market continues to be important for launching new products.

• There is an obvious risk that decreased spending on health care may hurt clinical research, resulting in less research on new medicines in the Swedish hospitals and, as a consequence, in weakening of the vital link between the health care system and the industry.

The Structure of the International Industry

The expansion of the knowledge base, the changing corporate dynamics, and the new economic and regulatory regimes are transforming the structure of the industry in ways which are yet difficult to ascertain. Two trends seem, however, to emerge: (i) the drive towards increasing consolidation and globalization of the industry; and (ii) the growing complexity and heterogeneity of the industry and its technological system.

In recent years there has been a series of spectacular company mergers in the international pharmaceutical industry, crowned with Glaxo's acquisition of Wellcome. The Swedish firms are in no way immune to the general trend. The process of consolidation in Sweden has in fact been quite dramatic. While in the 1960s and 1970s the Swedish industry consisted of several small or medium-size companies striving to close the gap with the foreign competitors, today it is dominated by two very large firms ranking 9th (Pharmacia-Upjohn) and 17th (Astra), respectively, in the world.[29]

This consolidation is not primarily driven by the belief that bigger companies are necessarily more creative. Rather, the mergers and acquisitions occur because of (i) the defensive value of size (overcoming temporary product drought, containing financial and technical risks, protection against hostile take-overs, etc.)[30] and because (ii) the great complexity of the innovation process in pharmaceuticals requires large sets of

---

[29] This process of growth and consolidation has followed two different patterns. Pharmacia and Kabi absorbed, step by step, several smaller companies in Sweden, then merged with each other and, subsequently, went international by first acquiring Farmitalia Carol Erba and then merging with Upjohn. Astra, on the other hand, has grown organically, occasionally absorbing small companies or establishing alliances with other firms.

[30] The recent acquisition of Wellcome by Glaxo is an example. Glaxo's own massive investment in R&D has not resulted in proportionate results. It is hoped that Wellcome's research portfolio will help Glaxo to overcome the new product drought. Upjohn's merger with Pharmacia seems to have a similar background.

complementary assets. Even if highly creative, the smaller companies find it increasingly difficult to bring their products to the market: the barriers to entry, both technological and financial, are simply too high. This explains the apparent failure of the new biotech firms to displace the established pharmaceutical companies, as well as the dire straits in which many of the older small and medium-sized companies find themselves.

What we are seeing, then, is a series of experiments seeking a new formula for corporate success. Many different combinations are being tried out. Some of the mergers and acquisitions have been vertical, others horizontal. In some cases there have been clear complementarities, in others size alone seems to have been the objective. The relative success of these experiments has also varied considerably.[31]

The drive towards consolidation raises some important questions about the nature of innovation in pharmaceuticals. Is the industry a clear instance of the Schumpeter Mark II model characterized by dominance of large R&D-intensive corporations, or is the present wave of mergers a mere prelude to a profound restructuring of the industry following the Schumpeter Mark I model, i.e., one led by new entrepreneurial firms? Some observers (such as Kenny 1986) believe that the upsurge of new biotech firms indicates that the latter is the case. However, the evidence so far does not fit either of the models very well. I would argue that the current developments should be seen as one facet of the changing pattern of technological division of labor in the industry.

The industrial structure which emerged during the Golden Age was based on a pronounced dualism of, on one hand, the R&D-intensive companies (engaging in product competition) and the generics producers and distributors (engaging in price competition) on the other. This dualism is now being partially abandoned due to the following combination of factors:

(i) The new economic regime induces more and more companies to follow the double track policy of competing in both market segments. Thus the sharp distinction between the *R&D*-oriented and the *generics and distribution*-oriented companies is becoming increasingly blurred. In order to survive under the new regime, many large firms are reconceptualizing their role as that of systems integrators rather than narrow product specialists.[32]

(ii) At the same time, the increasing dependence of pharmaceuticals on basic research has created a space for R&D companies many of which are academic spin-offs and/or operate close to universities. Although most of such companies aim at becoming full-fledged independent pharmaceutical firms, very few actually do. Instead, most of them perform the function of developing technologies, the final exploitation of which takes place in the large established firms or in an alliance with them.

---

31 As examples of less successful mergers one could mention Ciba - Geigy and Eastman Kodak - Winthrop. The failures may be due to the absence of real complementarities. They may also be the result of incompatible cultures.

32 One striking example of this is the acquisition by Merck of the distribution company Medco.

It seems, in fact, that we are witnessing a certain degree of *vertical disintegration along the R&D chain*, with the drug discovery and technology development companies becoming a distinct segment of the pharmaceutic industry (Whittaker and Bower 1994).[33]

The R&D firms have a number of advantages as incubators of pharmaceutical innovations (Stankiewicz 1994). (i) They are capable of functioning as bridges to academia — a matter of fundamental importance in a strongly science-based field. (ii) They create a mechanism for channeling additional funds (among others, venture capital) into risky long-term technical development. (iii) By making technological opportunities visible to many alternative investors/exploiters, they create a kind of technological market in which ideas and resources can be better matched. It is likely that a growing proportion of truly innovative drugs (such as Hässle's beta blockers once were) will in the future be discovered and initially developed by this segment of the industry rather than by the international giants whose role might increasingly be that of downstream developers, systemic innovators and global distributors of therapeutic products. This tendency is reflected among other things in the growing proportion of biotechnology-related drugs among the NCEs introduced in recent years (Bienz-Tadmor *et al.* 1992; Grabowski and Vernon 1994; DiMasi *et al.* 1994). The great increase in networking and collaboration between the large pharmaceutical companies and R&D firms (Whittaker and Bower 1994) points in the same direction.[34]

This, of course, is not to say that the large companies will give up altogether on the upstream elements of the drug discovery and development process. Many engage in massive expansion of their in-house preclinical research, either through internal growth or through acquisitions.[35] They are also expanding their direct links to the universities. However, there is much that suggests that the pharmaceutical R&D firms will play a vital role and that the ability to cultivate good relations with these firms will be a precondition of effectively exploiting the industry's expanding knowledge base.[36]

---

33 In fact, a similar phenomenon seems to occur in the area of clinical testing which is increasingly being performed by specialized consultancies (Cookson 1994).

34 Whittaker and Bower have found that between 1977 and 1987 the largest pharmaceutical companies increased the number of external R&D alliances six-fold, on average. Most of these linkages are to biotech firms.

35 For example: Roche's purchase of Genetech in 1990, Ciba's acquisition of Chiron in 1994, and Sandoz' acquisition of Systemics. Occasionally, older companies with excellent innovative record but limited ability to survive on their own are acquired by the leading firms. Johnson & Johnson's acquisition of the Belgian firm Jansen and Roche's of Syntex belong to this category.

36 One important reservation would be this. The new R&D technologies, such as automated chemical synthesis and screening, could evolve in a direction which creates great economies of scale at all stages of the research and development process, including drug discovery. This could restrict the role of small R&D firms greatly. However, if the price of these new technologies declines, the reverse is likely to happen, and the access to insights and ideas will again become decisive.

Astra's Response

One of the general conclusions emerging from this study is the decisive importance of the corporate culture and technology strategy in actualizing the potential of technological systems. A rich technological environment and an innovation-friendly economic regime are all important prerequisites of successful innovation. But they can be utilized only in the presence of effective *technological entrepreneurship*. The case of beta blockers gave us an insight into the nature and prerequisites of such entrepreneurship in a science-based field. The decentralized structure of Astra in the 1960s and 1970s gave its R&D companies considerable autonomy. Hässle was initially quite small. It pursued a limited range of objectives which were formulated as tangible medical goals. It sought large therapeutic advance as the way of achieving commercial success and was very pragmatic in its choice of means to reach these objectives. It had lean, flexible organizational structure and was very extroverted in its relations with the scientific and medical environments. All these features were highly conducive to the spirit of entrepreneurship and innovation, not unlike that found in the best of R&D firms today. The question is how that spirit can be preserved in a significantly changed organization operating in a radically transformed environment.

Astra continues to be an ascending company. Many of its products, including Losec (see next chapter), are still in a relatively early stage of their life-cycle and exhibit high growth rates.[37] The company also has a documented ability to expand application and prolong the life-cycle of its major products (as in the case of Xylocain and Seloken)[38]. Thus, in the short and medium term, Astra is very strong. What about the longer term?

As we noted earlier, the company has demonstrated the ability to repeat success. However, we noted also that most of these successes had their roots in a corporate culture and technological system of the 1960s and 1970s. At this point it is extremely hard to say whether the 1980s and 1990s will prove to have been equally creative. Although the company continues to introduce new drugs at regular intervals, none of the introductions since Losec and Plendil (both in 1988) have created much excitement. This need not be significant since it generally takes some years before the true potential of a drug becomes apparent.[39] The prospects of products still in the pipeline are even harder to assess. Astra now has some 25 drugs in various stages of clinical evaluation (Anon. 1995). However, the constantly increasing dependence on Losec[40] which now

---

37 Other examples: Plendil and Imdur.

38 In local anaesthetics - a classical niche area - Astra has a dominant position, and Xylocain constitutes the company's fifth largest product. In the cardiovascular field, Seloken, the company's third largest product (after Losec and Pulmicort), continues to grow and increase its market share.

39 Losec was initially thought of as a niche product complementary to rather than challenging Tagamet and Zantac.

40 The patents protecting Losec in its major markets will expire between 2001 and 2004. They have already been challenged by Takeda, Esai and most recently Byk Gulden.

accounts for more than 40 percent of Astra's sales (and keeps growing) is a problem. The company clearly needs better to balance its product range (Anon. 1995).

Successful companies do not drastically change their structures or strategies. Astra's development in the 1980s and 1990s is characterized by continuity. The basic policy of locating R&D in its autonomous research companies Hässle, Draco, Pain Control, and Arcos has been retained.[41] The company continues to broaden its contact surface with the universities, as revealed in the pattern of its research collaboration (Okubo *et al.* 1995). The Swedish universities, particularly Karolinska Institute and the University of Lund, remain the company's primary partners. The policy of supporting professorships and appointing prominent academics to important managerial positions continues. Mindful of the limitations of the Swedish R&D system, Astra is widening its contact surface with the knowledge base in other countries. Over the years, the company has set up several research units in the U.K., U.S., Canada, India, Australia, and Hungary. These function as satellites to the research companies in Sweden.[42] However, recently a new research company has been set up in the United Kingdom after the acquisition by Astra of Fisons' R&D capability. Astra has also stepped up its direct collaboration with foreign universities and begun to link up with foreign R&D firms (see below). All of this happens against the background of rapid growth of the company's total investment in R&D.

Judging by these developments, Astra seems to be in a good position to avoid the 'negative spiral of success,' or at least its more obvious manifestations. But, as argued earlier, the threats may be more subtle. Although the structure of the company appears to be largely unchanged, its main innovation engines, the research companies, have altered character. They have become quite large and, for natural reasons, devote much of their effort to the defence and extension of their territory.[43] The balance between different subcultures within the research companies and Astra at large is also likely to have changed, and there seems to be a higher degree of central control in all respects. These changes can alter the ecology of innovation and entrepreneurship characterizing the company at its earlier stages.

As noted earlier, there are two main ways of meeting these threats. One is to adopt the 'harvesting' strategy, staking the future of the company on the ability to quickly absorb and exploit external resources through mergers & acquisitions combined with aggressive marketing and powerful in-house research capability specializing in the end-game phase of drug development. Large size and financial strength are the main

---

[41] The companies are specialized as follows: Hässle — cardiovascular and gastrointestinal diseases, Draco — respiratory and inflammatory diseases, Astra Pain Control — anaesthetics, and Astra Arcos — CNS and infectious diseases.

[42] Hässle and Arcos, for example, have links to research units in Boston, U.S., and Astra Pain Control to one in Montreal, Canada. Arcos also has links in London, U.K.

[43] Astra Hässle now employs around 1,000 persons, of whom half are academically trained and 200 hold Ph.D. degrees.

prerequisites of that approach. The other way is to follow a 'generative' strategy requiring identification and development of early clues into radically new drugs and aiming at technological leadership in key areas. This second type of strategy requires the sort of entrepreneurship which made Aptin/Seloken, and later Losec, possible.

So far, Astra has pursued mainly the second of these strategies. The question is how successful it has been. Ivan Östholm (1995), now retired, voiced some concerns about the decline of technological entrepreneurship in large firms. He suggests that greater rewards and incentives for entrepreneurs could help to rejuvenate such firms. This may well be so, but other measured are also required.

Since the new technological opportunities, especially the important ones, are likely to emerge at the fringes of the corporate competence base and will often require behavior and strategies which run against the grain of the dominant corporate culture, the management of the interface between the company and its technological system acquires paramount importance:

(i) The relationships with the universities and the health care system continue to be crucial. But they become increasingly difficult to manage. The very scale of industrial R&D today poses a potential threat to the autonomy and diversity of academic research. It is essential that the corporations continue to follow and strengthen an enlightened and generous policy recognizing its long-term advantages.

(ii) The same is true of the relationships between the existing companies and the new R&D firms. The latter are a vital part of the emergent structure of the pharmaceutical industry and potentially the most effective mechanism for linking it to the evolving knowledge base. It is in these firms that the scientific and technical inventiveness and entrepreneurship are most likely to be found today. This has been recognized by the Swedish firms which in recent years formed several alliances involving new biotech ventures, both abroad and in Sweden. Kabi, for instance, was the first important client of Genentech (McKelvey 1994). Initially, however, Astra was rather restrictive in its relations to the R&D firms. The main link was to the Swedish academic spin-off firm Symbicom in which Astra acquired a 25 percent interest in 1987. Recently, however, there seems to have occurred a major change in policy. A short time ago, Astra acquired the remaining shares of Symbicom and incorporated it into Hässle. It also entered cooperative agreements with several U.S. R&D firms (T-Cell Sciences, Genome Therapeutics Corporation, Centaur Pharmaceuticals, and Cytotherapeuticals) and with Canada's Allelix Biopharmaceuticals.[44]

The relationships between the R&D firms and the big mature companies are complex and quite sensitive. Those of the latter which learn how best to cultivate the

---

[44] Indeed, the generally increasing reliance on external sources of ideas can be seen in the origins of the drugs now in clinical evaluation. Of 25 such drugs, three come from Fisons, one from Takeda, one from Dainippon, and two from U.S. R&D firms. An additional three products are based on patent-free substances which originated in companies other than Astra. This makes 10 externally generated drugs out of 25 in the clinical part of the R&D pipeline (Anon. 1995). Note that the three patent-free drugs are all asthma medicines which Astra-Draco uses in its Turbuhaler device.

independent or semi-independent R&D firms are likely to have a competitive advantage. It seems reasonable to assume that the companies pursuing mainly 'generative' strategies are more likely to form congenial relations with the R&D spin-offs. Just as in the case of academic-industrial relations, the recipe for success is likely to include fairness, mutuality, respect, personal mobility, geographic proximity, long-term orientation, and money. It will also require appropriate policies at the level of the technological system.

Policy Implications

As we have seen, one of the main features of the technological system of pharmaceuticals has been its rapid internationalization. Companies such as Astra strive to plug into the best R&D systems wherever these are found. There can be little doubt that that process will continue. Does this imply that the importance of the home country decreases, or even becomes negligible? Not necessarily.

The case of beta blockers illustrates the importance of both the extended international network and of the local one. Hässle would not have succeeded without either of them. A strong home base is a prerequisite of effective access to the global knowledge pool. There is nothing in the recent development in pharmaceutical R&D that points in a different direction. The companies will gravitate towards the locations where the agglomerated resources create the best environment for successful innovation. Such an environment has been created in post-war Sweden for reasons which had little to do with industrial considerations. Ironically, the risk now is that that environment will erode because of the *continued* neglect of industrial considerations. If this is to be avoided, three policy issues have to be addressed: (i) the size and quality of the R&D infrastructure; (ii) the support to R&D start-ups and other academic-industrial interfaces; and (iii) the potential impact of the health care reforms on clinical research and other aspects of the technological system of pharmaceuticals.

1. Given the rapidity and scope of change in the knowledge base of pharmaceuticals and the constraints on the R&D system imposed by the country's size, the neglect of industrial considerations in the shaping of Sweden's R&D policy in biomedicine is no longer tenable. However, as the case of biotechnology demonstrates, little has so far been achieved in this area. The efforts to set up a comprehensive national program have failed repeatedly, largely because of the institutional and cultural heterogeneity of the biotechnical and biomedical sector (Stankiewicz 1987)[45]. NUTEK (and its predecessor STU) — the main agency supporting research efforts in industrial

---

[45] This is by no means an exclusively Swedish problem. Similar tensions and difficulties have arisen, for example, in Britain (cf. Balmer and Sharp 1993).

technology — has never played a major role in pharmaceuticals,[46] and its recent efforts to take leadership in biotechnology have not been wholly successful. The medical research council, the other potential candidate for the role, has in recent years been the scene of major conflicts which at least in part could be attributed to the tensions generated in a traditional academic system by the advent of 'big science' and strategically/industrially oriented research.

However, a series of recent institutional innovations could create a better alignment between industry and academia. The most promising among these has been the establishment of a major separate Foundation for Strategic Research. The Foundation aims at rapidly building up the scientific and technical competencies in new areas deemed of strategic interest to the Swedish industry. Biomedical research is one of its main priority areas.

2. The heavy dependence of the pharmaceutical industry on the universities can create problems of its own. New ways have to be constantly sought for interfacing the two without negative consequences for either. In particular, a new attitude has to be developed as regards the technological spin-off (from both universities and industry). The somewhat simplistic view that these companies are Schumpeterian innovators who will quickly mature into large firms eventually replacing the old ones has been a source of unrealistic expectations and misguided policies (Stankiewicz 1994). It has to be modified and redefined in the context of the emerging structure of the pharmaceutical industry.

3. The regulatory regimes (safety, marketing, pricing/reimbursement, and intellectual property) under which the Swedish industry operated in the post-war decades have by and large been conducive to innovation. The new health care policies now under development should be carefully scrutinized with respect to their potential effect on the industry's ability and willingness to innovate.[47] It is crucial that new drugs leading to significant therapeutic advances are adequately rewarded. Such innovations are likely to pay for themselves by reducing costs in other parts of the health care system. It is also very important that the health care delivery organization (doctors, hospitals) continues to be receptive to therapeutic innovations and willing to participate in their development by engaging in clinical research and experimental evaluation.

The failure of policy in any of these areas is likely to have negative consequences for the firms directly involved and, to an even greater extent, for the ability of Sweden as a country to capitalize on its historically accumulated scientific and technological assets in the field of biomedicine. A well developed and functioning technological system of pharmaceuticals is a key component of a larger biomedical and biotechnolog-

---

[46] This is not to say that the organization was unimportant. It did play an active role in specific areas, including support in the 1980s to genetic engineering research, support to Hässle scientists in their work on Losec, financing of several new firms such as Medivir, and so forth.

[47] In this context, it is remarkable that the industry is not represented in the Committee on the Financing and Organization of the Swedish Health Care System (HSU 2000).

ical complex. It is possible that large companies such as Astra or Pharmacia-Upjohn can compensate for the weaknesses in the Swedish technological system by shifting their operations and interactions to other countries.[48] A high degree of internationalization is in fact both necessary and beneficial. However, at a certain point, the gradual shift of alliances may reach a critical point at which the technological system itself begins to slide down the negative spiral: the decreased dependence of firms on the Swedish R&D infrastructure could diminish the strength, coherence and connectivity of that infrastructure, creating even greater incentives for 'globalization' of industrial R&D, which would further weaken the infrastructure.

REFERENCES

Anonymous, 1995. "Efter Losec" (After Losec), *Affärsvärlden*, 22 November.

Archibugi, Daniele and Mario Pianta, 1992. *The Technological Specialization of Advanced Countries.* Dordrecht: Kluwer Academic Publishers and The Commission of the European Communities.

Balmer, Brian and Margaret Sharp, 1993. "The Battle for Biotechnology: Scientific and Technological Paradigms and the Management of Biotechnology in Britain in the 1980s," *Research Policy*, 22 (4), 463-478.

Bartholini, Giuseppe, 1983. "Organization of Industrial Drug Research," in F. Gross, (ed.), 123-140.

Bienz-Tadmore, Brigitta, Patricia A. DiCerbo, Gilead Tadmor, and Lois Lasagna, 1992. "Biopharmaceuticals and Conventional Drugs: Clinical Success Rates," *Bio/-Technology*, 10, 521-25.

Carlsson, Arvid, 1983. "The Role of Basic Biomedical Research in New Drug Development," in F. Gross (ed.), 35-42.

Carlsson, Bo and Rikard Stankiewicz, 1991. "On the Nature, Function and Composition of Technological Systems," *Journal of Evolutionary Economics*, 1 (2), 93-118.

Cookson, Clive, 1994. "Contract Research. Big Changes in Attitudes," *Financial Times*, March 23.

DiMasi, Joseph A., 1995. "Trends in Drug Development. Costs, Times and Risks," *Drug Information Journal*, 29, 375-384.

DiMasi, Joseph A., Henry G. Grabowski and John Vernon, 1995. "R&D Costs, Innovative Output and Firm Size in the Phramaceutical Industry," *International Journal of the Economics of Business*, 2 (2), 201-219.

---

[48] This is most obvious in the case of Pharmacia - Upjohn, but it is also increasingly true of Astra. During the past ten years Astra's research personnel abroad grew twice as fast as that at home, increasing from 16.5 percent in 1985 to 30.8 in 1994 (Source: Bolags Facta. CD-rom). The case of Hoffman La Roche provides an excellent example of how far the trend may go. Its recent expansion in the United States (due among other things to the purchase of Genentech and Syntex) leads one to wonder whether Switzerland is still its home base.

DiMasi, Joseph A., Ronald W. Hansen, Henry G. Grabowski and Louis Lasagna, 1991. "Costs of Innovation in the Pharmaceutic Industry." *Journal of Health Economics*, 10, 107-42.

DiMasi, Joseph, Mark A. Seibring, and Louis Lasagna, 1994. "New Drug Development in the United States from 1963 to 1992." *Clinical Pharmacology and Therapeutics*, 55, 606-622.

Grabowski, Henry G. and John Vernon, 1994. "Innovation and Structural Change in Pharmaceuticals and Biotechnology," *Industrial and Corporate Change*, 3 (2), 435-449.

Graves, Samuel B. and Nan S. Langowitz, 1993. "Innovative Productivity and Returns to Scale in the Pharmaceutic Industry," *Strategic Management Journal*, 14, 593-605.

Gross, Franz (ed.), 1983. *Decision Making in Drug Research*. New York: Raven Press.

Henderson, Rebecca, 1994. "The Evolution of Integrative Capability: Innovation in Cardiovascular Drug Discovery," *Industrial and Corporate Change*, 3 (3), 607-630.

Henderson, Rebecca and Iain Cockburn, 1994. "Scale, Scope and Spillovers: The Determinants of Research Productivity in Drug Discovery," draft manuscript, MIT, Sloan School of Management, Cambridge, MA.

Holm, Erik, 1992. *Hässle 1904-1988*. Mölndal: Astra Hässle.

Johansson, Katarina, 1990. *Introduction of Gene Technology in the Scandinavian Pharmaceutical Industry*. SBF-report nr 4. Stiftelsen Bioteknisk Forskning. Lund.

Jönsson, Bengt, 1994. "Pricing and Reimbursement of Pharmaceuticals in Sweden," *PharmacoEconomics*, 6, 51-60.

Kenny, Martin, 1986. "Schumpeterian Innovation and Entrepreneurs in Capitalism: A Case of the U.S. Biotechnology Industry," *Research Policy*, 15, 21-31.

Larsson, Hans Albin, 1990. *DRACO - Astras Foskningsföretag i Lund* (DRACO - Astra's Research Company in Lund). Lund: DRACO AB.

LIF (Läkemedelsindustriföreningen, The Swedish Pharmaceutical Industry Association), 1992. *Svensk läkemedelsindustri. Nationell spjutspets med global räckvidd* (The Swedish Pharmaceutical Industry. National Arrow-Head with Global Reach). Stockholm: Läkemedelsindustriföreningen.

LIF, 1994. *Forskarförsörjningen. Flaskhals för svensk läkemedelsindustri* (Supply of research personnel. The bottleneck of the Swedish pharmaceutical industry). Stockholm: Läkemedelsindustriföreningen.

LIF, 1995. *Fakta 95. Läkemedelsmarknaden och sjukvården* (The Drug Market and Health Care). Stockholm: Läkemedelsindustriföreningen.

Lindquist, Kjell and Sven Sundling, 1993. *Xylocain - en uppfinning, ett drama, en industri* (Xylocain - An Invention, A Drama, An Industry). Södertälje: Astra.

Lynn, Matthew, 1992. *The Billion Dollar Battle. Merck v. Glaxo*. London: Mandarin.

McKelvey, Maureen, 1994. *Evolutionary Innovation: Early Industrial Use of Genetic Engineering*. Linköping Studies in Arts and Sciences, No. 107. Linköping.

Okubo, Yoshiko, Staffan Hultén, and Jean-François Miquel, 1995. "Networking as a Firm Strategy in Scientific Production," conference paper presented at the 1995 Meeting of the R&D Dynamics Network in Stockholm.

Norgren, Lennart, 1989. *Kunskapsöverföring från universitet till företag* (Knowledge Transfer from Universities to Companies). Stockholm: Publica.

OTA, 1993. *Pharmaceutical R&D: Costs, Risks and Rewards.* Washington, DC: Office of Technology Assessment.

Pianta, Mario and Daniele Archibugi, 1991. "The Analysis of National Scientific Specialization with Bibliometric Indicators," unpublished manuscript.

Parnham, Michael J. and J. Bruinvel (eds.), 1984. *Discoveries in Pharmacology*, vol. 2. Amsterdam: Elsevier Science Publishers BV.

Persson, Olle, 1991. *Forskning i bibliometrisk belysning* (Research in Bibliometric Light). Umeå: INUM.

SCB (Statistiska centralbyrån, The National Swedish Central Bureau of Statistics), 1995. *Forskningsstatistik* (Research Statistics). Statistiska meddelanden U 13 SM 9502. Stockholm: SCB.

SCB, 1996. *Vetenskaps- och teknikindikatorer för Sverige 1996* (Science and Technology Indicators for Sweden, 1996). Stockholm: SCB.

Schwartzman, David, 1976. *Innovation in the Pharmaceutical Industry.* Baltimore and London: Johns Hopkins University Press.

Schnee, Jerome, 1973. "Innovation in the Pharmaceutical Industry, mimeo., Columbia University, Graduate School of Business, New York.

Shanks, Robin G., 1984. 'The Discovery of Beta Adrenergic Blocking Drugs," *Trends Pharm. Sci.*, 5, 405-409.

Shulman, Sheila R. and Louis Lasagna (eds.), 1996. *Cost Containment, Healthcare Reform and Pharmaceutical Innovation. Taking Stock 1995.* Supplement 2 to *PharmacoEconomics*, Vol 10.

Sneader, Walter, 1985. *Drug Discovery: the Evolution of Modern Medicines.* Chichester: John Wiley & Sons.

Spilker, Bert, 1994. *Multinational Pharmaceutical Companies. Principles and Practices.* New York: Raven Press.

SOU, 1995. *Reform på recept.* Delbetänkande av HSU 2000. (Reform on Prescription. Proposals of the Committee on on the Financing and Organization of the Healthcare System). SOU 1995:122. Stockholm.

SOU, 1980. *Svensk läkemedelsindustri* (The Swedish Pharmaceutical Industry). Industridepartementet. SOU 1980:33. Stockholm.

Stankiewicz, Rikard, 1986. "Utvecklingsdynamik inom genteknisk FoU i USA, Japan och Europa" (Genetic Engineering - R&D Dynamics in the USA, Japan and Europe), in Guno Haskå and Rikard Stankiewicz, (eds.), *Strategier för Bioteknisk Forskning och Utveckling* (Strategies for Biotechnology Research and Development). Lund: Stiftelsen Bioteknisk Forskning and FPI, 5-16.

Stankiewicz, Rikard, 1987. "Bioteknik i årets forskningspolitiska proposition" (Biotechnology in the 1987 Science Policy Bill), in Jon Sigurdson and Rikard

Stankiewicz (eds.), *Sveriges forskning. En kommentar till den forskningspolitiska propositionen 1987* (Research in Sweden.A Commentary on the Swedish Science Policy Bill of 1987). Lund: FPI, 47-60.

Stankiewicz, Rikard, 1993. "The Development of Beta-Blockers at Hässle-Astra," paper presented at a workshop on "Interdisciplinary approaches to innovation based on the creation, diffusion and exploitation of organizational knowledge" at NISTEP, Science and Technology Agency, Tokyo, March.

Stankiewicz, Rikard, 1994. "Spin-off Companies from Universities," *Science and Public Policy*, 21 (2), 99-107.

Swann, John P., 1988. *Academic Scientists and the Pharmaceutical Industry. Cooperative Research in Twentieth Century America.* Baltimore and London: Johns Hopkins University Press.

Vos, Reinder, 1989. *Drugs Looking for Diseases*, doctoral dissertation, University of Groningen.

Walker, Stuart R. and John A. Parrish, 1986. "Innovation and New Drug Development," in Bryan C. Walker and Stuart R. Walker (eds.), *Trends and Changes in Drug Research and Development*. London: Pharmaceutical Society of Great Britain, 1-27.

Werkö, Lars, 1986. *Medicin för miljarder. Har läkemedelsindustrin någon framtid?* (Medicines for Billions. Has the Pharmaceutical Industry a Future?). Stockholm: Timbro förlag.

Whittaker, Erica and D. Jane Bower, 1994. "A Shift to External Alliances for Product Development in the Pharmaceutical Industry," *R&D Management*, 24 (3), 249-260.

Östholm, Ivan, 1983. "Cooperation with University Departments - A Swedish Model," in Franz Gross (ed.), 227-230.

Östholm, Ivan, 1991. *Från örtavkok.... till läkemedel. Framgångar och bakslag i medicinernas värld under 50 år* (From Herbal Remedies to Modern Medicines. 50 Years of Successes and Failures in Pharmaceuticals). Stockholm: Apotekarsocieténs förlag.

Östholm, Ivan, 1995. "Innovatörerna stoppas" (Innovators Are Stopped), *Dagens Nyheter*, 24 August.

6

THE PHARMACEUTICAL AND BIOTECHNOLOGICAL COMPETENCE BLOC AND THE DEVELOPMENT OF LOSEC*

Gunnar Eliasson and Åsa Eliasson

INTRODUCTION

The wealthy industrial economies have long pinned their hopes for future industrial riches on high-technology or science-based industry. Knowledge would make it possible for firms in those countries to produce high-value products and thus be able to pay high hourly wages. Simple skill-based or manual labor-based production could be handed over to the learning NIC countries and the emerging East European producers. Such prospects place education and industrial competence development in focus. More than before, governments feel a responsibility to improve the future through industrial policy.

The first high-tech policy excitement was electronics (later called IT). It arrived in the early post-war years, grew rapidly and is now the dominant technological force in global industrial production. And it continues to grow; only it is not science-based and is not the result of industrial policy. It originated predominantly in private industry research laboratories and entrepreneurship in the United States, and development there was fast (Eliasson 1995a, 1996a). European industrial policy makers, on the other hand, encouraged investments in technologies of the past. The IT train came and went in Europe, which was left behind without a competitive computer industry.

Biotech is the next hope for a science-based industry. This industry did not exist in today's modern form until 1975. It was founded on a series of methodological developments in the U.S. around that time, creating the potential for genetic engineer-

---

* Earlier versions of this chapter have been discussed at seminars at the Department of Industrial Economics and Management at the Royal Institute of Technology (KTH) and at ERMES, Université Panthéon Assas (Paris 2). A shorter version of this paper, focused on the biotech competence bloc, has been published as "The Biotechnological Competence Bloc," *Revue d'Economie Industrielle*, 78 (4), 1996.

ing. In biotech, Europe appears to have a high-quality research base. Thus, provided other supporting elements of a *biotechnological competence bloc or technological system*[1] are not lacking, such a bloc should develop. One ambition of this paper is to identify the complementary elements of this bloc and assess the potential for positive development. So far, however, this development appears to have taken place only in the United States. Since 1975, venture capital on the order of magnitude of $20 billion has been poured into the U.S. biotech industry, and about 1,000 new firms have been established. The expectation that fueled such large investments was that by "creating a more efficient path to new drugs, biotech entrepreneurs would take over the pharmaceutical business" (*Business Week*, Sept. 26, 1994, p. 68). This time we also have a genuine exception to the general rule: a truly science-based industry. U.S. biotech industry is globally dominant and clustered around a few advanced university centers in particular: the San Diego, San Francisco Bay, and Boston-Cambridge-Worcester areas. To this should be added the belt of pharmaceutical companies in New York, New Jersey and Maryland, which also has significant biotech activity, even though it is dominated by older U.S. pharmaceutical giants. There is very little to be seen in Europe and Japan so far. The train has arrived at the station. Will it leave again without Europe?

THE PROBLEM

Biotechnology differs fundamentally from other industrial technologies in the origin of its competence base and in its diffusion. *First*, biotechnology originated in academia and therefore corresponds to the ideal picture of a science-based industry pushed in academic literature (see e.g. Nelson 1986; Stankiewicz 1986; Jaffe 1989). The academic research lab is at the core. Technically there are no principal differences between the academic and the industrial biotechnological laboratories.

Second, production in the biotech sector, notably in the pharmaceutical industry, is mainly laboratory work, and most of the rest is marketing. When completed, clinically tested and authorized, the actual manufacturing cost of a drug is relatively insignificant.

Third, new discoveries of products or drugs are almost always the result of a combination of different clusters of scientific know-how: in chemistry, molecular biology, cell biology physiology and toxicology, pharmacology, drug delivery, medicine and health care, diagnostics or treatment techniques and, not least, clinical documentation and laboratory technology. As a consequence, innovations and industrial

---

[1] For convenience, and in order to avoid the awkwardness of 'the technological system for biotechnology,' the term 'competence bloc' is used interchangeably with 'technological system' in this chapter. In this context, there is no need for a distinction. In other contexts, there is. See Eliasson (1995a, 1996b) where the concept of a 'competence bloc' was first introduced, and Eliasson and Eliasson (1996) where it was more fully developed.

applications occur predominantly in a *technological system or competence bloc* where a sufficiently diverse environment exists and where new and often unforeseen synergy effects can be captured.

Fourth, and not mentioned in many of the studies referred to, drug or new product discovery is *experimental*. While econometric studies link measured successes to measurable determinants, we learn nothing of failed experiments. There may be many failures, and perhaps most of the determinants of success are present also where the experiments have failed. Hence, failed experiments are a significant part of the total development costs in industry.

This chapter is organized as follows. We outline the principal discovery processes involved in biotech innovative activity, the formation of competence blocs, and the diffusion of related technology. This is both to demonstrate the role of infrastructure in the form of knowledge capital and the need to take economic analysis down to the micro level in order to understand what is going on. It is of particular importance for our purpose to understand and model how economic and biotechnological competence have merged in competitive markets behind the commercially successful innovations and, if possible, why and under what circumstances business mistakes have occurred. Since biotech industry, more than any other industry, is the offspring of academic research, it is of particular interest to determine whether the origin of capital spending (private or public) makes any difference. It is also of interest to clarify the market selection mechanisms behind successes and failures.

There is first a theoretical presentation of the principles involved. We then draw on existing empirical literature to narrow the range of possible interpretations. Finally, we present a case study to illustrate how close to premature termination was the world's perhaps most successful drug project, Astra's peptic ulcer drug Losec. If the project had been prematurely closed, would somebody else have discovered the drug or something equally good? When and where? What can we learn from this in organizing research?

The case study illustrates nicely both the cross-disciplinary nature of commercially oriented biotech research, notably to develop pharmaceutical substances, and the need to focus research efforts across disciplinary boundaries and methods. The critical importance of criteria for selection of projects comes out clearly. Since criteria differ from discipline to discipline, depending on the particular concerns, such choices may become academically disruptive. Thus, for instance, the almost completed Losec substance was at least twice close to being thrown out, once because of a statistically faulty experimental design and once because of faulty interpretation ('diagnosis') by American pathologists who mistakenly thought that discovered growths on rats were cancer tumors (see further section 7). Of course, as important as choice of criteria is the selection of the team of managers and specialists and, above all, project leaders, that ultimately determines the success of a development project.

The purpose of industrial research is to develop new products that please customers and to design new process methods that make possible more cost-efficient production. The *possibilities* are *technically defined*. But there are *economic restrictions* that reduce the economically viable *opportunity set*. Such development is resource-using, and the costs for the development of such new products or production methods cannot be calculated with certainty. Contrary to the calculation of manufacturing costs in repetitive production which is subject to probability laws, the range of uncertainty surrounding innovative R&D ventures is immense and not estimable. Hence, critical choices rather than rational calculation determine outcomes. Such choices are subject to two types of error:

(1) losers are kept on for too long,

(2) winners are rejected.

The competence to make informed and rational decisions of the two kinds resides at various levels in the firm and varies in relative intensity between large and small firms. Managers in big corporations with large financial resources are excellent at terminating bad projects early but also lack the creative understanding of explorative research needed to recognize new but odd winners. They normally have to work through expert reference teams which may be even less creative than the business people (see Losec case below). Above all, in order to manage R&D successfully, top management has to include somebody with experience from R&D management, not necessarily the scientific knowledge but experience from working with and managing the kind of people that populate R&D laboratories. Few of the big Swedish companies, for instance, appear to have such persons in their top management teams. Thus, even though resources are amply available, they may not be innovatively allocated. This is to be expected. Since a fairly small number of people dominates decision making at the top, the incompetence in recognizing and evaluating the economic consequences of new technical choices increases, the closer to the top the decision is taken.

In small firms, on the other hand, the varied backgrounds of a large number of independent people influence a large number of decisions, in contrast to the more single-minded policies of a few large corporations. In this context, the varied competence input of venture capitalists should not be forgotten. Venture capitalists often serve as discoverers of new but odd products that the standardized selection systems of credit institutions and big business have failed to recognize (Eliasson 1996c). As a consequence, *a more varied competence is involved in the processing of decisions in a large number of small firms compared to the case where an equally large volume of resources is processed in the hierarchies of one large firm.* This increases the probability of committing errors of type 1, which can mean disaster for a small firm, but reduces the incidence of giant errors of type 2, barring the unlikely case when a single-minded large investor happens to stumble upon the correct innovative decision. *Ex post*, this is fine. *Ex ante*, however, such arguments carry no weight. This means that *the optimal organization to exploit the competence base of an industry is to have many small players and many venture capitalists trying it out in the market.* Then the success projects get a fair chance by meeting many competent resource providers. This is a workable organizational proposition, provided there are no scale economies in R&D, which is rarely the case (see later). And if there are, there may be organizational solutions on a small scale as well (see further the case of KaroBio). In modern industry, the management competence to establish sufficiently porous filters to optimize rather than minimize development costs should be the overriding concern. The costs of carrying mistaken research a little bit further are small compared to the business costs of never finding winners. Existing economic evaluation methods are, however, biased in favor of rejecting winners (see further the Losec case).

THE HYPOTHESIS

Considerable controversy has surrounded the science-technology-industry links. While the political debate often assumes a direct, positive relationship between higher education and research, on the one hand, and productivity growth on the other, those who have studied these relationships are not so sure (Eliasson 1986a, 1993, 1994; Rosenberg 1994). It is rather typical that successful industrial countries like Japan and Germany do not particularly encourage the more advanced forms of theoretical academic inquiry but rather encourage applied technical research. Furthermore, academia has historically been a hostile environment to industry, and applied research has much too often been looked at as a low-brow form of academic inquiry. There are exceptions, however. The German chemical industry is said to have a university foundation (Locke 1989, p. 62), and the German Technical Institutes (Technische Hochschulen) that developed independently of, and outside, the university tradition are generally considered to have contributed to German economic development during the 19th century. Some of the basic inventions in information technology (not most, and few in later years, however) originated in the laboratories of technical universities and/or semi-public research institutions such as Bell Labs. Audretsch and Stephan (1996) and Audretsch (1995) argue that the puzzle of small and new firms innovating even in the absence of substantial investments in R&D can be explained by networks created by scientists in linking firms with knowledge created at universities. Instead of becoming routinized, as Schumpeter (1942) predicted, at least in biotechnology, a considerable chasm has appeared between the entities creating new knowledge and those appropriating it. The new IT industries, however, mostly originated in corporate development labs (Eliasson 1996a).

Some changes may now be under way. More academically oriented research close to large firms may be turning academic research in a more entrepreneurial direction, and while IT has one foot in academia and one in industry, the new microbiology, biochemistry, biotechnical, and pharmaceutical fields certainly breed on research coming out of science labs in university environments in the United States. Zucker, Darby, and Brewer (1994) argue that the more than seven hundred active biotechnology firms established in the U.S. in the last two decades are based on new developments in basic science. The timing, location and birth of biotech companies, they find, are primarily explained by intellectual capital, notably the local number of highly productive 'star' scientists. The establishment and growth of new biotech firms are closely related to the presence of active star scientists and their collaborators. The stars do not have to be entrepreneurs themselves, but the intellectual capital that grows around them appears to be the key input in the high-tech new entry process. Life has, however, been tough for these firms, and the market is crowded with competitors (*Business Week*, Sept. 26, 1994, pp. 66 ff).

Zucker, Darby and Brewer (1994) also conclude that status is not necessarily associated with an academic institution. Scientists in private industry laboratories, they argue, write as many scientific articles as do university scientists, even though some of the scientists we have interviewed expressed skepticism about this claim. Henderson and Cockburn (1993) also found that productivity effects in industrial (pharmaceutical) R&D improved significantly if participation in research programs could be run within the context of large and more diversified R&D efforts. Opportunities then increased to exploit knowledge across programs within the firm, and internal and external spillovers could be more efficiently absorbed. Such cross-disciplinary use of spillovers is more typical of industry labs than of academic labs, where fragmentation prevails.

The biotech industry is based on biological discoveries and systematization which occurred in academia. Fast commercial applications mean that commercialized research in this field is still scientifically interesting and embodies new, rather than routine, applications of already known principles. This also means that scientific learning in this area has been transferred through participation in market competition (also see Eliasson 1996b).

As a consequence, this paper is particularly concerned not only with the ways science, technology and economics have been merged in the biotech industrial competence bloc, but also with the scientific contribution (the policy issue) to this competence bloc (the following sections). For instance, how have market considerations in the biotech industry influenced the orientation of biological research? To what extent has decentralized biotech competence among venture capitalists promoted the funding of this new intractable industry in the United States compared to Europe? To what extent has biological science been influenced by the necessary across-specialties orientation of industrial R&D, and to what extent has the extreme specialization in biotech research (Eliasson 1994) contributed to and/or held back the development of the industry? Briefly, what kind of intellectual disruptions occur in a laboratory oriented towards developing products with pre-specified properties when the laboratory is staffed with specialists (once) devoted to promoting new knowledge creation?

This mostly qualitative analysis aims at assessing the capacity of a university environment, a region, or an economy to develop a competence bloc spontaneously. Entrepreneurship and the existence of venture capitalists will play an important role in the analysis.

WHAT IS BIOTECH?

"Biotechnology is the integrated use of biochemistry, microbiology and chemical engineering for achieving the technological application of the capacities of microbes and cultured tissue cells" (Rehm 1982, quoting the definition of biotech of the European

Federation of Biotechnology). This definition covers a whole range of industrial activities from food processing – including wine making, dairy production, chemical feedstock production and bioenergy production – to the production of fine chemicals and even metals. Medical technology is excluded by this definition, and gene technology is mentioned in passing as part of the biotech field, when such methods are applied in biotechnology.

At that time (around 1980), however, a biotechnological revolution was well under way in the U.S., soon to give an entirely new content to the above definition and placing gene technology at the center.

An alternative definition which is used in the current paper is the following: *Biotechnology is the art of manipulating biological subsystems for the purpose of discovering a new product or making existing products better, using particular methods, techniques, or processes.* The industry thus has a technical process-based definition, not a product definition. The biological subsystem is taken out of its total systems context and placed in a 'test tube' or in some experimentally controlled environment. It is 'improved' and then put back into its natural context where it hopefully behaves as in the test tube. (This in itself constitutes a problem, since most biological subsystems are integrated parts of the whole, and there are many, perhaps a thousand, such systems.) Currently, the biotech field is thought of as consisting of three or four sub-areas: (1) recombinant DNA (rDNA) technology, (2) the use of antibodies, including phage-display, and (3) protein engineering. Sometimes also (4) fermentation and volume production of (generic) biological substances, such as the raw substance for penicillin production, are defined as a separate biotech area. These are all empirical classifications denoting fields where methods have already been developed to perform such partial systems manipulations. The first field, rDNA, was opened up in 1973 by Boyer and Cohen who published the first test tube manipulations of DNA that could be used in biological systems. The first step towards manipulating DNA, the basis of life, to our needs had been taken.

Plants that can stand cold climates and cows that milk more which were earlier created through cross-breeding can today be achieved through recombinant DNA technology. Classical cross breeding is afflicted by frequent negative side effects, since also bad qualities can be reinforced. Today, relevant genes may be identified and modified through rDNA technology (gene manipulation) such that only the new good qualities are (in principle) inherited. Tomatoes that take longer to rot after picking have been manipulated that way.

The second field was opened up around 1975 by Köhler and Milstein who developed the hybridoma technique to construct monoclonal antibodies (they received the Nobel prize in 1984). Our immune system uses antibodies to recognize foreign molecules in the body very precisely, and to attach to, signal, and even destroy the foreign molecules. Thus, to use monoclonal antibodies is a way of mimicking the role of antibodies in the immune system. This technique is used both for clinical diagnosis of diseases and for identifying specific proteins. The technique was developed after the principles of the immune system had been better understood. Phage-display is a new method where phages (the viruses of bacteria) are used to produce, for example, large amounts of antibodies quickly.

The third field - protein engineering - was developed during the 1980s. Many researchers and institutes were involved. Protein engineering is focused on the end product - proteins - in living systems. Proteins - the effector molecules in cells - are manipulated such that they can become better for our purposes. Technically, this field has benefited from the development of X-ray crystallography and computers. This method makes it possible to see the 3-dimensional structure of the protein which makes it easier to make intelligent guesses as to what to change in order to improve the protein. Examples are the dirt-degrading enzymes in detergents. Protein engineers fixed the enzymes to work effectively also in the (for them unnatural) washing machine environment.

The pharmaceutical industry uses biotech technology intensively from all four areas. However, part of the pharmaceutical industry does not and therefore technically does not belong to the biotech industry. The pharmaceutical industry produces products designed to interact with, or influence, biological processes in human beings and animals. This means that biological knowledge enters into the development of most drugs, even though the drugs themselves are completely 'chemical.' The clinical testing of pharmaceutical products for negative side effects, which is the most resource-using part of pharmaceutical production, certainly involves biotechnology. As a consequence, a biotech *competence bloc or technological system* and its spillover range require a definition that is sufficiently broad to include all technologies needed to develop, produce, distribute, and use the product. The competence bloc, in fact, has to be even more broadly defined. The value of a new drug is determined not only by the absence of side effects but also by its potential to cure, or reduce other health costs, for instance hospital care. The productivity effects on health care production are in some cases enormous. On the average, a one dollar increase in drug expenditure reduces costs for hospital care in the U.S. by $3.65, and prolongs life (Lichtenberg 1996). The technological system behind the pharmaceutical industry, hence, has to include not only pharmaceutical firms but also pharmaceutical delivery techniques and medical care (as shown also in chapter 5).

Similarly, biotech is a much wider technology than that and operates far outside the pharmaceutical product range, for instance detergents, chemical stabilizers and preservatives.

The empirical nature of this classification, being dependent on particular manipulation techniques that constantly change, apparently means that the taxonomy currently used will not be stable for long.

THE COMPETENCE BLOC AS A TECHNOLOGICAL SYSTEM

Erik Dahmén (1950) coined the notion of a 'development bloc' signifying a cluster or a network of integrated physical production and distribution activities that create strong economic synergies at some higher level of aggregation and particularly over time. The automobile would be a typical example around which large industrial complexes have developed, enormous investments in road building and auxiliary facilities have taken place, and from which significantly different ways of life (residential areas, urbanization, etc.) have emerged. Carlsson and Stankiewicz (1991) have developed the concept further, resulting in the notion of a technological system in which technical knowledge plays a critical role in creating productivity at some higher level of aggregation, for instance around factory automation. The recent spillover literature (summarized in chapter 8 below) discusses the same technical and economic phenomenon in terms of strong, positive externalities associated with particularly advanced (read R&D-intensive) industries that spill technology to related industries. The macroeconomic effects induced through spillovers appear to be very large (see chapters 8 and 9 in this volume). A technological system or competence bloc encompasses the total infrastructure needed to *create (innovation), recognize (venture capital provision), diffuse (spillovers)* and successfully *exploit (receiver competence) new ideas* in clusters of firms. It has to be defined sufficiently broadly, so that all possible important synergies across different competence areas are captured. This is much more than the public goods externality properties of spillovers observed by Henderson (1974) and used as the foundation of the so-called new growth theory. The technological system is concerned primarily with the dominant intangible human-embodied competence associated with a particular economic activity.

A technological system is highly diversified and organized to make cross-fertilization over many technologies possible. Participating actors both *contribute* to technological spillovers and *use* the spillovers emanating from other actors. The system has to operate as an attractor that – once the innovative seeds have taken root – will spontaneously develop into a synergistic bloc. All benefit, and there is a mutual interest in participating in the competence exchange both ways. This is synonymous with saying that entrepreneurs have to be present within the system. Entrepreneurship is needed to recognize and to take advantage of the technological opportunities offered. Particular combinations of competencies create a business success. Technological diversity and the potential for cross-disciplinary (across specialties) synergies characterize a viable technological system. A particularly differentiated system may create many successes and thus operate as a synergistic spillover source.

To have a business potential, a competence bloc requires a minimum of more or less related competencies embodied in:

- educated and talented people,
- *technical specialists* and "*bridge builders*" who integrate technical specialties,
- *entrepreneurs* who integrate technical competencies under the restriction of economic criteria,
- *venture* capitalists,
- corporate management and financial specialists.

To carry a new product to industrial production, all these competencies must also be integrated with

- marketing knowledge and
- manufacturing skills.

To this has to be added an environment that makes expectations of reasonable or better rewards for those who succeed credible. There appears to be a long way to go before such environments have been created, notably in Europe.

The *experimental nature* (see Eliasson 1987, 1990) of innovative activity makes the intersection of technical or scientific development, entrepreneurship, and venture capital supply particularly important. Many experiments are needed to produce a success, but that success also has to be *recognized* as a business project. The venture capitalist plays a critical role here and the intellectual capacity to recognize successful business experiments is an important element in a viable technological system (Eliasson 1996c). Standard administrative tests rarely discover new solutions since they have been designed to recognize only known phenomena. Discovery therefore requires particular creative competencies of recognition, and in this task, competent venture capitalists play a critical role.

It is highly unlikely that the necessary ingredients of a technological system are in place from the beginning. A viable technological system or competence bloc develops spontaneously through attracting complementary competencies (technology) that locate there because of the existence of complementary technologies. The complexity - to warrant the name - of such competence blocs is enormous and incomprehensible *ex ante*. Technological systems thus cannot be planned ahead. They develop spontaneously in the market, and the primary role of policy is to support institutions that contribute to their rapid formation such as an educated population, the right cultural and legal institutions that shape attitudes, competent and wealthy venture capitalists of sufficient diversity, etc. See further chapter 11.

This list shows that a technological system is *both technical and economic*. Some of the knowledge originates in science, some in the business and financing community, some in pure talent and human effort. Some is culturally founded. A society that looks down on the desire to accumulate private wealth as unethical, for instance, is likely to achieve little in accumulating industrial competence and wealth.

It is to be expected that a large part of the competence mass of such a bloc resides in the firms of particular industries. What happens, for instance, if existing firms cannot take advantage of ingenious but odd ideas and if entrepreneurship is lacking?

The competence bloc develops endogenously in the market, supported by its institutions, notably those that support innovative activity, recognition of opportunities, the diffusion of new technology, and its implementation in firms.

1) *Innovative* activity occurs in research laboratories and in firms and the results (variety and new technology) have to be 2) *recognized* as opportunities. Venture capitalists play a critical role here, and are instrumental in the 3) *diffusion* of new technology through the mobility of people and through entrepreneurship (see Eliasson 1996a). The introduction of new technology in firms requires 4) *local receiver competence* (Eliasson 1986a, 1990) embodied in educated, experienced and creative people.

We want to underline the Schumpeterian innovation side of a technological system, emphasizing *recognition* of new successful business experiments (discovery) rather than the systematic building of knowledge. *If you filter research results by known scientific criteria, you lose the radically new results.* Innovative experiments require creative recognition that goes beyond established filters to be discovered. When you do not know what you are looking for, you cannot ask the appropriate questions. Thus, in biotech, when you screen with known methods, you only find substances with known effects (Östholm 1995, p. 62). Innovation, hence, is a matter of formulating and testing new experiments and the ability to discover unique solutions that would not have been detected by standard criteria. In such an experimental process, *venture capitalists* play a critical role. The competent venture capitalist takes initiatives. This is not typical of Sweden. It is rare in Europe. An administrative system, however, staffed with competent people, is normally incapable of inducing, discovering, and selecting unique innovations. It therefore lacks the active elements of a technological system. The marketing people do not recognize the product. Financing people cannot calculate and, therefore, kill projects prematurely. Other talents are needed. The market (read a diversified venture market) is the most efficient filter when it comes to optimizing the use of the existing competence base and avoiding throwing out winners, since it mobilizes a maximum of competence. Finally, in understanding the origin of a business success, the entire supporting competence bloc has to be understood. Let us see what it looks like in biotech industry.

THE CONTENT OF THE BIOPHARMACEUTICAL COMPETENCE BLOC

The development of a drug illustrates the complexity of the biotechnological pharmaceutical (the biopharmaceutical) technological system or competence bloc.

The Discovery Process

Drug discovery is a complicated, drawn-out, and often frustrating process. In the 1950s and 1960s, pharmaceutical research began in a rather random fashion. People in the laboratories had ideas of what they were looking for, of course, but normally substances were not synthesized in a very systematic fashion. The substances synthesized were then filtered through prepared screening programs designed to indicate known positive effects. Substances that passed this test (filter) were called *lead compounds*. At that time, something like 5,000 substances or so were needed to arrive at one lead compound. Today we are talking about hundreds of thousands of chemical substances to be filtered through modern computer-based filters to arrive, eventually, at one successful substance.

This new method was adopted already in 1961 at Hässle (see Gross 1983, p. 237) and used in discovering Seloken, and partly also in discovering Losec (see below).

Nowadays, synthesizing scientists start from prior general scientific knowledge about biological mechanisms in order to narrow down the range of substances that are likely to function as expected, and before the costly testing procedure begins.

The modern discovery process begins, however, long before that. First of all, a *medical need* (the market) has to be identified. Once the market niche has been identified, medical, chemical and biological competence has to work together to *identify* clusters of substances and discover the *substance* (the *candidate drug*) to be tested with a high probability of being able to cure the targeted illness. There may be several biological solutions (see Losec case: neutralize acid, stop producing acid, kill the bacteria). As already mentioned, today this may result in a library of hundreds of thousands of possible chemical substances to be synthesized and then tested, one at a time. From this library maybe only one will remain as potentially interesting.

Particularly difficult, important, and still lacking in the prior scientific screening is knowledge about the human biological systems' properties that determine 'down-stream' systems' responses to particular, partial manipulations of the system, such as the *administering of a medicine*. Earlier, negative side effects were sometimes detected late and far into clinical testing, after millions of dollars had been spent. This was the case with Entero-Vioform, Thalidomide and Zemilidin (of Astra)[2]. Today, such late detection is rare. Molecular modeling and modern methods such as crystallography may be and have been useful in leading faster than before to lead compounds, and on the basis of a library of fewer substances than before. As mentioned, biological knowledge, biotechnology, and new computer-based screening methods are increasingly combined to make pre-sorting faster and more efficient (see the KaroBio case below).

The lead product thus found is then improved upon, and better substances emerge as *candidate drugs* in the *drug discovery phase*. Such candidate drugs are then subjected to *safety studies* on test animals. While modern molecular modeling methods may be powerful in discovering known positive and negative effects, they have, so far, been of little value in filtering out substances with negative side effects, simply because all negative effects that have to be discovered prior to clinical use are not known ahead of time.

After safety studies on animals, *Phase I studies* begin on healthy human beings. The purpose now is to see whether the test persons tolerate the highest doses believed to be needed to achieve the desired effects.

Phase II studies initiate what has come to be called the *clinical testing program* on real ('sick') patients to ascertain the expected effects on the targeted illnesses. If results are positive, *Phase III studies* begin on large patient groups to learn what happens under

---

2 Zemilidin (Östholm 1995, p. 106 ff.) was introduced as an anti-depressant in 1969. After having been administered clinically to some 200,000 patients it was discovered that a few patients developed a rare muscle disease. Since Zelmilidin was very effective as an anti-depressant, it was debated within Astra and among the medical profession whether the substance should, nevertheless, be used. Astra, however, did not dare to keep it on, and Zemilidin was withdrawn from the market in 1983. A slightly modified formula is currently marketed by Eli Lilly as *Prozac*.

normal clinical conditions on a varied patient population. This involves thousands of patients and very large investments. Only after this phase is the substance cleared for routine use in health care.

There is also a *Phase IV study program (drug surveillance)* of the substance in clinical use to identify *interaction effects* when used together with other substances and rare but possible negative effects not discovered in clinical testing.

Most of this elaborate testing was developed by the pharmaceutical industry itself to manage the risks involved, but a large part exists because of government regulation, notably after the Thalidomide incident. The enormous costs involved make the organization of this testing both a matter of industry concern and of policy efficiency.

At all stages of developing a new substance, laboratory technology is instrumental. Thus, Pharmacia laboratory instruments, particularly for separation techniques, are said to have supported an early development of biotechnology in Sweden. In fact, to capture all synergies discussed in this section, the technological system should be defined to include the entire *health care* sector, that is, hospital care, pharmaceutical and biotechnological industries, as well as the medical and biotechnological laboratory instrument industry.

KaroBio - An Intermediary Player

KaroBio is a biopharmaceutical company, located at the research park (Novum) at Huddinge Academic hospital outside Stockholm. KaroBio focuses on human diseases that involve nuclear receptors, such as breast cancer, inflammatory diseases, and cardiac arrhythmia. The company was founded in 1987 by two experts on nuclear receptors with some connections to the investment community. They decided to focus on nuclear receptor research. To get a positive cash flow into the company, they initiated several additional short-term projects. After some years of trouble, however, they realized that these short-term projects generated costs rather than cashflow. KaroBio now concentrates on nuclear (hormone) receptors only – its unique biotech specialty – and operates as an *intermediary between academia and the pharmaceutical industry*. This intermediary role is particularly interesting and illustrates how the functions of a large firm can be decentralized through the market, making R&D and innovative activity more efficient. KaroBio looks for and *discovers* business opportunities in the academic research environment. By contributing complementary knowledge, entrepreneurship, and venture capital, it puts business life into the biotechnical knowledge base residing in academia. KaroBio, however, stops short of the large scale and enormously costly routine clinical testing of a promising drug. This phase is best taken over by a traditional, large pharmaceutical company which often is bad at innovative activity and discovery but good at repetitive, routine testing using known, established methods. Routine testing and large-scale screening are also steps in the process of developing a drug, today normally beyond the resources of a small company. *KaroBio*, thus, *is an organizational solution that lowers the risk of committing errors of both type I and II.*

This intermediary function, however, demands a very broad competence base that cannot be developed within one small company. KaroBio has, therefore, formed *partnerships* with either the university or other biotech/pharmaceutical companies. KaroBio directs the partners' research and uses their knowledge for its own commercial goals. KaroBio then carries the research up to the phase of clinical tests.

Nowadays it is almost impossible for one company to develop an innovative product internally from scratch. The very large companies have the resources but are usually conservative and too slow in changing direction. Hence, they are not good at early recognition of new promising ideas that may lead to new types of products. Small and new firms may be innovative but normally lack both resources and competence to take the discovered substance through all steps of testing, introduction, and marketing to industrial-scale production. As the biotech field becomes larger and increasingly dynamic, it becomes important for companies to focus on what they are good at. This makes them even more conservative in dealing with ideas and technologies not related to their strategic focus.

This means that in a dynamic and constantly changing market the successful company often develops the potential for its own destruction, if it cannot find an *organizational solution to deal with focus and variety simultaneously*. Few companies have succeeded with that ambition. The new 'organizational' solution is for the large firms to leave innovative activity to the small idea-rich firms and enter at later stages to acquire ready solutions or entire firms and do what they do best: carry the new product to industrial scale production and distribution. KaroBio is an early operator in this intermediate field. Success here, of course, depends on the existence of many competing buyers of the innovative firms that raise the price level of these firms and of good scientists and entrepreneurs sufficiently to increase the supply of innovations to levels compatible with a rapid development of the industry (Eliasson 1986b). This incentive structure is an important element of the competence bloc.

The large traditional pharmaceutical companies are good at clinical testing of given substances for known biological effects. It was earlier common to detect negative side effects late and far into clinical testing, after millions of dollars had been spent. Here academic biological knowledge, biotechnology, and new, computer-based screening methods make pre-sorting faster and more efficient. KaroBio has developed a specialized business niche in substance screening, using essentially three highly advanced scientific methods:

- computer-aided drug design and crystal structure analysis,
- tissue profiling, and
- combinatorial chemistry.

These methods are used to filter out substances from the library with a higher probability than other substances of having the desired effects.[3] Such lead compounds

[3] Different strategies apply depending upon the nature of prior knowledge. If the structure of the receptor is known from crystal structure analysis, for example, molecules based on computer analysis are constructed. Computer programs can then determine whether the molecules fit into the 'lock.' Binding studies then follow,

can often be identified with modern computer-based screening methods because scientists know what they are looking for. Negative side effects are more difficult to identify because unknown side effects cannot be identified by such methods.

This pre-sorting can reduce the number of substances from 100,000 to, say, between 10 and 20 to be tested on animals. One, maybe two, promising substances are then tested on humans. This screening know-how of KaroBio still requires specialization on particular proteins to be commercially viable. KaroBio has particular knowledge in hormone receptors (special proteins that bind signaling substances) and also has acquired knowledge about techniques for those receptors. The company has therefore decided to focus on four particular receptor classes: thyroid, estrogen, glucocorticoid, and orphan receptors. The receptors and the signaling substances that bind to them work like a lock and key. (If you know exactly through 3-dimensional structure analysis how the receptor (the *lock*) looks, you can design small molecules (*keys*) that bind to them and block the action of the receptor. KaroBio uses very advanced protein engineering, X-ray crystallography, and computer-aided modeling techniques to figure out what the receptor looks like and what signaling substances (keys) should be used and combinatorial chemistry to improve on compounds. Tissue profiling and binding studies are used for continued screening and testing.) This is how pre-sorting is used to economize on the discovery process by reducing by several orders of magnitude the number of substances to be synthesized and tested. When a desired compound has been identified, toxicological studies on animals for negative side effects begin. Clinical testing is enormously costly; high-level medical competence (normally based in academic teaching hospitals) and closeness to the same hospital-based specialties are required.

In addition to this, pharmaceutical competence enters when the most effective substance delivery system is to be determined: intravenous injection, nasal spray, or oral intake, for example. Negative side effects can be associated with some delivery methods. It has happened that a potent drug has had to be abandoned because no effective delivery method without side effects could be found.

The Biotech Competence Bloc - What Do We Know?

The biotech competence bloc can also be identified by a classification of industrial activities, the people working there, and the nature of their research. The biotech industry has been innovatively presented in a series of National Bureau of Economic Research (NBER) working papers by Lynne Zucker, Michael Darby and colleagues.

---

using tissue profiling to identify positive and negative effects.

If little prior knowledge exists, the strategy may be to buy a 'randomized' library of some 250,000 substances to select, through binding studies, a set of some 20 compounds with the desired effects. Promising substances are improved through combinatorial chemistry to reduce the number of compounds which are then tested by tissue profiling. Tissue profiling and binding studies are examples of 'artificial' animal testing. Before these sophisticated techniques existed, the whole library would have had to be tested on animals which, of course, would not have been possible.

They have found that the strong regional clustering of the biotech industry to the Boston, San Francisco Bay, and San Diego areas is strongly linked to the presence of intellectual capital embodied in academic biotech 'stars' in the universities in the same geographic regions (Zucker, Darby, and Armstrong 1994; Zucker and Darby 1995b). While Europe may have scientific and academic knowledge, the U.S. still has somewhat more than half of the world's 'stars' in biotech (Zucker and Darby 1995b), and since (see below) the rest of the world appears to lack several of the other necessary contributing infrastructure factors, such as an entrepreneurial mentality and bold and competent venture capitalists, the complete dominance of U.S. biotech industry appears to be explained. This outcome can be contrasted against the consensus emerging out of a conference in 1982 (see Giersch 1982) that Japan was well on its way to becoming the superstar actor in biotechnology, based on its political determination and forceful political administration to become so.

Active biotech stars, according to Zucker, Darby and Brewer (1994), are defined as persons associated with a large number of genetic sequence discoveries reported in 4,315 articles in major journals, co-authored with 7,718 other scientists, named 'collaborators'; they identified a total of 337 such stars in the U.S. Great universities are likely to grow and recruit such scientists. These stars can be statistically linked (as the principal determinant) to the location and growth of the industry.

The creation of biotech competence blocs and the growth of biotech firms do not follow 'star formation' under all circumstances. Incentives to move scientific discoveries to commercial applications have to be in place, which means that rents have to be naturally appropriable by the discoverer. This occurs when the intellectual capital is tacit (the authors do not use that term, but rather talk about 'natural excludability') and cannot be communicated, except through 'hands-on experience,' or can be kept vested in a closely knit group of collaborators who share the rent (Zucker, Darby, Brewer and Peng 1995). This organization of closely knit rent-sharing 'trusts,' the boundaries of which 'act as information envelopes,' of course poses an academic dilemma. Should academic results be published (if possible) with such detail that the experiment can be repeated and checked by other teams, or do the business results ('products') serve as sufficient 'scientific documentation?'

With tacit knowledge dominating the competence capital, the primary route to sourcing new knowledge must be through the hiring of new personnel with the requisite knowledge (Eliasson 1994, 1996a; Zucker and Darby 1995a). For a commercial break-through to occur, however, the scarce knowledge of new research techniques and the idea of how to apply these techniques in innovative ways have to be combined, notably across specialties. To achieve that, university scientists and firm scientists have to collaborate to build bridges across specialties, and such collaboration in fact increases the publishing of scientific articles among both the university and the firm scientists (Zucker and Darby 1995b).

Since incumbent firms are slow to develop ties with university stars, the latter try to start new biotechnology firms to commercialize their knowledge (*ibid.*, p. 2). The significant involvement of star scientists in cooperation with firms in commercializing scientific knowledge (i.e., entrepreneurship), virtually absent in France, Germany, and Canada (*ibid.*, p. 15), appears to explain the U.S. lead in commercial biotechnology.

University policies that "limit professors' ability to contract freely with, and establish ownership positions in firms may protect the norms of disinterested science but at the cost of limiting technology transfers" (Zucker, Darby and Armstrong 1994, p. 3) and growth of new industry. California has been particularly open in this respect. As a consequence, policy makers have a role to play in opening up the competence base residing in the academic world for commercial exploitation.

Economic activity in the biotech industry, and especially in pharmaceutical firms, is almost all R&D and marketing. Productivity and organizational efficiency therefore refer to those activities, both of which are subject to scale economies. However, Henderson and Cockburn (1993) found no evidence of increasing returns in a study of R&D in ten major pharmaceutical companies. But they did find that particular research programs benefit from being conducted within a large and more diversified research process, because of the opportunities of exploiting knowledge across heterogeneous program boundaries and the ability to capture spillovers within the firm and between firms (economies of scope). These results are supported by the Liebeskind, Oliver, Zucker and Brewer (1995) study of two highly successful new biotech firms that have extended their social networks to source scientific knowledge. Such networks promote organizational learning both through extending their scope and through contributing to the integration of knowledge within each firm.

Concentration and clustering are also typical of biotech and pharmaceutical industries. Feldman and Schreuder (1994) observe that ten of the world's twenty-five largest pharmaceutical companies are headquartered in a corridor between Philadelphia and New York City. They attribute this geographical concentration to the development of a unique network of inter-firm relationships of competence exchange, specialized inputs, labor pooling, and information exchange. There is evidence of cooperation rather than patent races (Cockburn and Henderson 1993). While a substantial number of university-based scientists participate in geographically bounded networks, more than two-thirds of the links between biotechnology companies and university-based scientists are non-local (Audretsch and Stephan 1995). Geographic proximity appears to matter for founders but less so for members of scientific advisory boards; it also appears to diminish in importance over time. Perhaps this has to do with the origin of discovery in the biotech field, often occurring in closely knit networks of trusted collaborators, particularly scientists working in the same organization being based on "particular hands-on experience with the discovering scientists" (Zucker, Darby, Brewer and Peng 1995). As Prevezer (1994) notes, local clusters of mutually supporting technologies often grow out of such initial clusters of 'seeds' into larger clusters with external and internal attraction. Such competence blocs take time to develop and diffuse geographically.

Large, centrally administered corporate organizations rarely excel at innovative behavior. They are good at improving existing products and processes and at establishing new 'subcontracting' firms in related technologies. But discovery of radically new devices or technologies does not come naturally to businesses engaged in repetitive production and people with administrative minds. Competence blocs share the internal synergies in competence exchange with large firms but their 'participants' are otherwise autonomously operated. They band together due to mutual horizontal attraction, not by hierarchical administrative imposition.

While new firms with fresh minds may be good at discovering and developing radically new things, they are notoriously bad at, and lack the resources for, carrying successful ideas to industrial-scale production. Most firms in biotech are still small by the standards of traditional industry. It is therefore interesting to ask how the biotech competence bloc will be capable of carrying successful volume products to industrial-scale delivery.[4]

One proposition would be that a lively, intermediary market between innovative firms and large-scale business would therefore form, where developed products are turned over to volume producers at a price. The development of such a market would also be very important for the national economy and a natural catalyst in a viable competence bloc (Eliasson 1986b). With many suppliers of new technology or products, the competitive situation of industries would improve. With a broad-based demand, compensation and other incentives for innovators and entrepreneurs would be high. Very little of such 'verticalization' of the competence bloc exists, however. Short-sighted large firms seem rather content with what is supplied, are happy to use the non-existing competition to cannibalize on the few naive scientist-innovators and are confident that their research laboratories will turn out whatever they need to guarantee future, long-term success. The biotech and pharmaceutical industries appear to exhibit the most vital innovation markets, perhaps because such innovative firms develop in university environments, are optimally sized and small, and because the innovations can be patented, a circumstance that should facilitate the transfer of innovations over the market and boost their price by making the property right well defined.

THE EXPERIMENTAL FOUNDATION OF ECONOMIC SUCCESS - POLICY IMPLICATIONS

Few of the studies referred to recognize the experimental nature of innovative activity and discovery, and the consequent high failure rate. This should be particularly relevant in biotech markets. The absence, or slow start, of the biotech industry in risk-averse Europe compared to the United States illustrates the point. Most development projects appear to fail, and not rarely because the source of financing dries up. Hence, one-

---

[4] It is also interesting to observe the recent race among already large pharmaceutical companies like Pharmacia-Upjohn and Glaxo-Wellcome to find partners, on the presumption that this will promote a steady stream of pioneering and profitable products at lower cost (*Business Week*, March 25, 1996, p. 24 ff.) . Cutting costs is the easy part, but the case for finding more new products is supported neither by theory nor by experience.

project start-up firms lead a risky life, and exit rates are high. The statistical correlations between star scientists and biotech successes reported above are also in part misleading in that they are observed between those few firms that have been selected (in the market) to survive and a selection of determinants for success. Causal conclusions for policy also require a study of failed attempts, in particular to check that the same determinants are not present also at business failure.[5] Until such careful research has been conducted, it should *not* be concluded that the breeding of star scientists also means the successful breeding of biotech firms.

It should also be recognized that failed business experiments are part of a natural discovery process (Eliasson 1995b, 1996a). In a business context they should be looked at as a standard cost for economic development. The number of business successes is positively related to the competence to set up clever business experiments, but also to the number of experiments performed. The more experiments performed, everything else the same, the larger the number of observed, failed experiments. Premature termination of a successful project is also a cost for economic development and a particularly large one when expressed in terms of lost income.

This allows us to draw a few conclusions on policy. *First*, the innovative discovery process is experimental, and the outcome of the experiments depends on the competence to set them up and discover outcomes, as well as to terminate bad experiments. This is not a game suited for large administrative organizations like big government or big firms, the mentality of which should be to do things right from the beginning (for instance writing law, *cf.* discussion in Eliasson 1990).

Second, the innovative process appears strongly dependent on a *competent* and patient *purchaser* in need of new solutions. Here government (defense, health care, or telecom authorities) and big firms have played important roles in supporting innovative activity in the past.

Third, and this takes us naturally into the next part of the paper, the market is never perfect. Government has an important innovative role in society at large in identifying where and when the market does not work, supporting institutions that make it work better and occasionally, but rarely, stepping in when no other solution appears to come forward. Antitrust policy is believed to be capable of serving that function by preventing dominant firms with large resources from developing into parasitic monopolies that predate on small, innovative competitors. This case is, however, not altogether clear in a world characterized by intense global competition. As long as government stays away from attempting to take over businesses looking for protection from market competition, it will do little harm by standing ready as a competent venture capitalist of last resort. It certainly did that in the case of Losec (see next section).

Innovations, by definition, cannot be planned. If you look for innovations, you always screen by known standards and are likely to miss radically new solutions.

---

5 Note in the case description below that the Losec substance was almost thrown out because of exactly this type of misinterpretation (selection bias) of a test.

Innovative activity is experimental, and new products and ideas are *discovered.* Great discoveries often start as a mistake or something thought of as an error that someone with a creative mind decides to take a closer look at. The art is to recognize new and superior solutions. Thus, the organization of innovative activity and the criteria for innovative choice are decisive for success.

Innovations often take a long time to reach the stage of industrial production. They are, therefore, often prematurely terminated when subjected to conventional economic criteria.

New ideas developed in large hierarchical firms often disrupt existing repetitive production and are therefore resisted. Hence firms, notably large firms, have a tendency to commit errors of the second type, i.e., to reject potential successes, because management does not understand new solutions, or – worse – does not discover them even when they are presented.

Great competence and creative imagination are required to discover new product ideas. Such competence has to be present among those who contribute resources during the discovery process, before income begins to flow. This capacity to understand and discover is part of competent venture capitalism.

Big firms usually apply systematic and standardized economic screening and, therefore, tend to cut long and unknown discovery processes prematurely. They, therefore, cut both bad and good projects early and are not prone to making type I errors.

Big companies often have large financial resources to commit long-term, but the (necessary) standardized management practices rarely embody the competence to select intelligently. This conflict is one of the most important management problems in economic development. The more complex and heterogeneous your activity, and the more resources you control, the less capable you are of intelligent and creative choice.

Small firms, new firms, and venture capitalists are more efficient than large firms in allocating competent judgement on innovative ideas, but they lack financial resources. The increased awareness of the risks associated with increased use of new medical substances and the more demanding tests for negative effects have created a business handicap for the small innovative firm. If a bridge between stodgy and cash-rich large firms and innovative, cash-starved small firms cannot be organized, we have a tricky long-term dilemma, often with a negative outcome.

The elaborate testing of new medical substances was instituted in the 1962 so-called Drug amendment in the U.S., after the Thalidomide incident. The purpose was to reduce wasteful use of inefficient drugs and the risks of serious negative side effects. Careful testing, of course, comes naturally as warranted procedure before new drugs are accepted, since they are going to be used on human beings (veterinary substances are not tested as strictly). Logically, however, we should think twice before accepting such an argument (Peltzman 1973). While you may save people from negative side effects, the long testing procedure means that sick people who have to wait much longer for the new drug, later declared free, suffer or even die. The zero risk (for side effects) procedure adopted by the FDA in the U.S. has come under increasing criticism recently for being unreasonable (*Business Week*, Jan. 30, 1995, p. 46 ff.). The argument is that a different risk balance is optimal. Peltzman (1973) in fact concludes that the welfare benefits of the 1962 Drug amendment are negative, and that the 'expert' judgement of the FDA process is not necessarily superior to that of market selection. On the matter of control of dangerous drugs from being released, furthermore, there is the argument that the U.S. court litigation system with very high penalties on firms that are negligent might be more efficient and socially less costly than the current regulation. Since firms undoubtedly understand the risks of their own substances better than centralized authorities and experts, stiff penalties might discipline the firms into more efficient and reliable testing and minimize the total costs of both failure and of taking new products to market. As the case now is, firms can pass the responsibility on to a government authority and the excessive costs on to the drug users. On the other hand, higher risks for the firms might raise the risk premium in their research budget and reduce their R&D spending. Whether this effect is larger than the costs mandated by FDA testing is an open question. Apparently, the design of regulation also becomes an element of the competence bloc we are studying.

THE STORY OF LOSEC

Astra's stomach ulcer drug Losec, perhaps soon to become the world's most sold pharmaceutical product, spent many years of uncertain development life in the laboratories. Between 1966 and 1984, Astra's top management tried to terminate the development project five times (Östholm 1995, p. 159; 1996). It survived thanks to a dedicated research team and two rescue grants from the Swedish National Board for Technical Development (then STU, now part of Nutek) in 1974 and 1976. Losec was the second of two major Swedish product successes during the 1980s; Ericsson's mobile telephone project, currently a business exceeding 40 billion SEK, barely survived a similar experience (Eliasson 1995a, chapter 11). This raises two questions:

- Is the experience of the Losec development program a standard experience in industry, notably in large firms?
- What would have happened to the two large firms, and to the macro economy, in the not unlikely case that both projects (Losec and mobile telephony) had been prematurely terminated?

To illustrate the uncertain paths embarked upon when attempting to develop entirely new solutions to a problem, we present the story of Losec in some detail. It is particularly important for the story to be told that the complexity and heterogeneity of the biological systems when developing new medical substances is understood, as well as the need to work across disciplinary borders.

This, in a few words, is the story of *Losec*.

Since the stomach content is strongly acidic, the stomach lining would be destroyed if it were not protected by a resistant mucous membrane. The reason for the acidic environment in the stomach is to protect us from ingested pathogenic microorganisms and to digest the food better. People have suffered from ulcers for hundreds of years. For a long time the only remedy was to avoid certain foods and drinks. It is still not understood why this protective membrane sometimes weakens such that the hydrochloric acid of the stomach can injure the stomach lining. Even though the existence of ulcers has been recognized since medieval times, it is still not understood why some people have ulcers and others do not, except that extremely high levels of hydrochloric acid tend to coincide with the development of ulcers, and low levels with healing of ulcers. The solution to stomach ulcers was, hence, to develop a drug that neutralized the acid such that the stomach lining could heal again. The medical need for such a product was initially not considered great at *Hässle* (since 1942 part of Astra) since such drugs (although not very efficient) existed already in the market. In 1957 Hässle, nevertheless, decided to develop a drug that was better than existing drugs that neutralized hydrochloric acid in the stomach, and in 1959 *Novalucol* was ready and became (together with related products) Hässle's bestseller during the 1960s. This success increased the resource flow Hässle could allocate to R&D. The ambition grew to develop a completely new biological product to treat peptic ulcers, namely a substance that inhibited the secretion of hydrochloric acid (Östholm 1995, p. 161). The plan was to eliminate the cause, not just the symptoms — then a new idea in the pharmaceutical business.

In the 1960s, pharmaceutical industry research was not highly regarded in academia. The 1960s was also a period of general hostility in the academic world towards capitalism and industry. To improve relationships between industry and academia, Hässle began to organize joint international conferences. The decision to initiate the search for a peptic ulcer drug along the new principles was taken in 1966 in the context of a conference organized by Hässle and Lars Olbe, a well regarded specialist at the Gothenburg University Hospital on the treatment of peptic ulcers in fistula dogs.[6] A method to inhibit or stop (not just to neutralize) the production of acid without the side effects of available drugs had to be found.

This was not easy, however. Several mechanisms and molecules cause acid secretion in the stomach, and one of the known molecules was gastrin. If one could create a molecule that looked like gastrin, it could compete for the receptor to which gastrin binds and thus, maybe, block acid secretion.

In 1964, the British biologist R.A. Gregory had identified the chemical structure of gastrin. ICI, a British chemical company, then tried to find substances similar to the gastrin molecule. Despite synthesizing almost 1,000 substances, ICI could report no positive results.

At Hässle, Lars Olbe had learned that local anaesthetics reduced the production of hydrochloric acid, maybe because they inhibited the release of gastrin. However, the available anaesthetics did not work in the very acid environment of the stomach. Olbe

---

6 Called fistula dogs for the surgically inserted tube in the dog's stomach to measure the amount of acid secreted.

therefore suggested that an anaesthetic that would work in the acid environment of the stomach should be developed. This required that an entirely new research area be opened up at Hässle.

Rats are the favorite test animals since they are cheap and easy to handle. Lars Olbe, nevertheless, advised against using rats as test animals, since they have different gastric juices from those of dogs and humans. Unfortunately, the researchers at Hässle persisted in working on rats for several years, citing publications to the contrary. Positive results could be reported in November, 1966. They, however, turned out in later experiments to harm the liver of dogs, and consequently could not be tried on humans (Östholm 1995, p. 168).

A new substance had to be tried. Some 150 substances belonging to the chemical group called carbamates, not expected to cause liver damage, were tested. In February, 1968, a particular substance from that group was chosen for human testing. When the doses were increased, however, a change in the blood of one test person, possibly indicating an effect on the liver, was observed, and the tests were stopped. This was in 1970. Corporate management was becoming skeptical, in fact firmly advising the termination of the project in the fall of 1970. But Ivan Östholm, heading the research team at Hässle, believed in the project, and insisted that it be continued. The researchers convinced top management that the project was worth continuing, even if they had to start from the beginning again. After four years of research, the team at Hässle had not produced any positive results. The arguments to continue the project despite negative results were (1) that the drug was still needed, (2) that the team had acquired valuable competence in the field, and (3) that they all believed strongly in Dr. Olbe's work. Continuation was therefore approved with a reduced budget. One also has to take into account that at the same time another drug - Seloken (see chapter 5 in this volume) - was being developed, and the research was going well. Astra wanted to allocate resources on something that appeared more promising. The research team now finally shifted to fistula dogs, although they had no idea of what to continue with, since nothing that was nontoxic seemed to work on humans. To be able to go on, Ivan Östholm and Hässle approached the large pharmaceutical company *Abbott* to obtain a larger budget through cooperation.

In 1972, the core of the research group on the acid secretion project was reorganized. Ulf Jungren, Sven-Erik Sjöstrand, and Lars Olbe found that they worked well together. In 1972, Sven-Erik Sjöstrand - the fourth project leader - was looking through old notes from conferences when he suddenly realized that two compounds, synthesized on different occasions, were very similar and had reported positive properties in inhibiting acid secretion. This discovery was a great step forward, and research now continued in a new direction. When the substances were tested on fistula dogs, they were found to be strong acid secretion inhibitors, and this became the starting point for an entirely new synthesis program which this time could use a new automated testing method. A computer-controlled stomach membrane from a pig submerged in a nutrient solution to generate acid speeded up the turnover of experiments significantly

(Östholm 1995, p. 176). Again, a new dimension to research, laboratory technology, was shown to be important.

Hässle was not alone. Other companies were also attempting to develop substances with which to treat stomach ulcers. James Black in the U.K. was trying a different approach. He had developed the beta blocker *Inderal* for heart diseases and was now trying the same drug on peptic ulcers. After synthesizing and testing some 700 hundred compounds, *Tagamet*, a histamine that blocked the secretion of acid, was discovered.

Black was awarded the Nobel prize for those two discoveries in 1988. With enormous resources, Black had developed a substance at *Smith Kline & French* (SK&F) that Hässle had almost developed with a fraction of the same resources, thanks to a unique collaboration with university researchers, using a method to prelimit the range of experiments by first trying to understand the biological systems mechanisms involved.

Black's discoveries, however, started a heated discussion in Astra as to its focus. Was there any point in competing with a large international company? Should Astra rather focus its efforts on an attempt to improve upon Black's idea and improve Tagamet instead of continuing with its unique concept to develop a substance that operated inside the cells that produce hydrochloric acid and minimized the side effects?

Fortunately, a reorientation to improve on Tagamet was prevented. *Glaxo*, having invested enormous resources, soon succeeded in developing the improved H2 blocker *Zantac*, which became a great success.

Hässle's ambition to discover a medicine that acted within the cells that produce hydrochloric acid continued, despite a negative corporate attitude. The compounds based on the lead compound discovered in 1972 that created the new ideas of how to proceed with the acid secretion project were effective inhibitors of hydrochloric acid secretion, but they were toxic on humans. The two compounds had similar structure. Based on this, almost 150 new substances were synthesized, varying the different regions of the compound, before a new substance was found in 1973 that appeared promising. This substance showed positive results in tests, and the partner Abbott continued to test. Trouble, however, did not end here. Again toxic effects on rats were found, and corporate Astra experts urged Hässle to terminate the project. A new derivative that appeared to be a strong secretion inhibitor had, however, been discovered during the safety tests, and corporate management was impressed by the Hässle team and allowed them to continue one more year.

More negative side effects were discovered in 1974. After much persuasion, Astra corporate management again reluctantly approved another year. The Hässle team, however, had now discovered what it was in the molecule that had the toxic effects and began to synthesize new variations on the compounds. In December, 1976, an effective acid secretion inhibitor was discovered, with no side effects. But company management was frustrated after four years of research but no useful substances. In this situation (1974), the Swedish National Board for Technical Development (STU) stepped in with a significant grant, provided Astra put up the same amount of resources. This helped the Hässle team for another year. However, the product was not yet ready. More resources were needed, and now Abbott wanted out. At the end of 1975 Hässle asked for more money from STU for 1976. The Hässle team did not get quite the amount of money they asked for, even though the evaluation of the project was very positive. One of STU's experts had the impression that the project was in a very early stage of research and he was not sure that it would lead to anything, even though he believed that the scientific knowledge created and the research was very interesting.

The reduced amount of money from STU arrived at a critical phase, and during 1976 the Hässle team found a new promising compound. Hässle now believed that they had the compound that would inhibit acid secretion efficiently without side effects. They were anxious to continue with animal safety trials and then go on with clinical testing as fast as possible.

Again, however, in June, 1978, dog tests showed negative indications. Therefore, tests on humans could not begin. The Astra research council, supported by two of its professors, urged that the project be terminated, until one member on the council (Björn Folkow; see Östholm 1995, p. 188) discovered that the test on dogs had been badly set up and that the negative results were wrong. Tests on human subjects could finally begin in 1979, and impressive results could soon be reported.

It turned out that the Hässle team was on to something much more promising than they understood at the time. In 1978, Schwartz (USA) had discovered the protein complex that acts as a pump secreting hydrochloric acid. The Hässle team now had a scientific foundation for their search for a substance that inhibited secretion of hydrochloric acid, and everything seemed to come together nicely. This 'hydrochloric acid pump' is situated in the lining of the stomach and regulates its acidity. The Hässle substance worked directly on the pump. The Hässle team was convinced that they had a winner, a substance that could now be tested on humans.

At the same time as the clinical tests were performed, the Hässle team continued to improve the substance. In 1979 they had around 10 different compounds to work with, all with very good acid secretion inhibition properties and no discovered toxic effects. They chose one - *Omeprazol.* Now that they finally had the compound, a way to administer it simply had to be found. It turned out that Omeprazol was an extremely difficult compound to work with. It was sensitive to light, water and heat! It could not be made into tablets, and upon contact with the hydrochloric acid of the stomach it was immediately destroyed! The Hässle team had to develop completely new methods for making and administering the drug. In 1981, large clinical tests were started in several countries and everything seemed good. Then came the disaster. In 1984 some cancer-like tumors had been found in rats that had been given Omeprazol. All clinical tests had to be stopped, and everybody was close to giving up. Fortunately, the team at Hässle tried to figure out where the tumors originated. They turned out to have been wrongly 'diagnosed' and had nothing to do with cancer and were only specific to rats, not humans. So the clinical tests could continue. In 1988, Omeprazol was finally approved for use in Sweden and several other countries under the name of Losec. Since then Losec has been a commercial success for Astra.

Recently a new bacterial species has been identified; *Helicobacter pylori*. The presence of *H. pylori* in the stomach is strongly correlated with ulcers in the patient. When this became known, it first looked as if the story of Losec would suddenly end. But *H. pylori* cannot be effectively targeted with antibiotics in the acid environment of the stomach. After Losec had removed the hydrochloric acid, antibiotics can easily kill the bacteria, and the ulcer patient gets well very fast. Hence, the bacterial involvement in peptic ulcers proved to be a boon rather than a threat to Losec. An entirely new development program around *H. pylori* has been opened up at Hässle and at an Astra research subsidiary in Boston, USA.[7]

CONCLUSIONS

The Losec story brings out four things that relate to innovative discovery processes in general. *First*, for new ideas to be recognized, significant and diverse (broad-based) competence has to be present among the resource providers (the venture capitalists). Large centralized organizations (government or private) engaged in repetitive production tend not to internalize such diversity and are thus bad discoverers.

Second, the discovery process is often very complex and the gestation period very long, not only wearing down expectant venture capital providers, but also turning profitability calculations negative, even at very late stages of development.

Third, the classical economic position on the policy role of government is to step in when the market has failed, except, of course, to support the efficient functions of markets. However, when the market is populated by large firms that hold the resources, the market has already failed in the classic economic sense; the public policy focus has to be reoriented towards also watching out for private non-market failure within the large business organizations. The Losec case illustrates this when Nutek (or STU) stepped in, and also shows that the solution does not lie in breaking up the large private firms (the conventional antitrust position). Nor does it lie in following the conventional policy advice of Arrow (1962) to socialize innovative activity. There may be other innovative organizational techniques to handle this important problem, such as the KaroBio business idea. So far, however, we have difficulties distinguishing such techniques from pure luck.

Finally (*fourth*), biotech and pharmaceutical discovery is extremely complex, stretching across several diverse academic and industrial fields and reaching beyond the capacity of individual teams and hierarchies to comprehend. They consequently have to be organized over the market to achieve a sufficient foundation of diverse competencies and alternative solutions.

---

7 Another illustration of competence bloc formation is the new biotech company star *Cortecs* that has developed a product to diagnose presence of *Helicobacter pylori*.

REFERENCES

Arrow, Kenneth, 1962. "Economic Welfare and the Allocation of Resources for Invention," in Richard R. Nelson (ed.), *op. cit.*, 609-626.

Audretsch, David B., 1995. *Innovation and Industry Evolution.* Cambridge: MIT Press.

Audretsch, David B. and Paula E. Stephan, 1995. "Company-Scientist Locational Links: The Case of Biotechnology," working paper, Wissenschaftszentrum Berlin für Sozialforschung and the Center for Economic Policy Research (CEPR).

Audretsch, David B. and Paula E. Stephan, 1996. *Biotechnology: Sources and Incentives,* Paper presented at the 1996 International J.A. Schumpeter Society conference, Stockholm, 3-5 June.

Business Week, 1994. "Biotech. An Industry Crowded with Players Faces an Ugly Reckoning," Sept. 26, 66 ff.

Carlsson, Bo (ed.), 1995. *Technological Systems and Economic Performance - The Case of Factory Automation.* Boston and Dordrecht: Kluwer Academic Publishers.

Cockburn, Iain and Rebecca Henderson, 1993." Racing to Invest? The Dynamics of Competition in Ethical Drug Discovery," mimeo., MIT Sloan School of Management, April.

Cohen, S.N., A.C.Y. Chang, H.W. Boyer, and R.B. Helling, 1973. "Construction of Biologically Functional Bacterial Plasmids *in Vitro,*" *Proceedings of the National Academy of Science USA,* 70, 3240-3244.

Cohen, S.N., and A.C.Y. Chang, 1977. "Revised Interpretation of the Origin of the pSC101 Plasmids," *Journal of Bacteriology*, 132, 734-737.

Dahmén, Erik, 1950. *Svensk industriell företagarverksamhet,* Band 1 and 2. Stockholm: IUI. Also published in English 1970 by the American Economic Association Translation Series under the title "Entrepreneurial Activity and the Development of Swedish Industry, 1919-1939". Translated by Axel Leijonhufvud. Homewood, Ill: Richard D. Irwin, Inc.

Eliasson, Gunnar, 1986a. "Kompetens, kommunikation och kunskapsuppbyggnad," in Gunnar Eliasson *et al.*, *Kunskap, information och tjänster*. Stockholm: IUI.

--------, 1986b." Innovative Change, Dynamic Market Allocation and Long-Term Stability of Economic Growth," working paper no. 156, IUI, Stockholm.

--------, 1987. *Technological Competition and Trade in the Experimentally Organized Economy*. Research Report No. 32. Stockholm: IUI.

--------, 1990. The Firm as a Competent Team, *Journal of Economic Behavior and Organization*, 13 (3), 273-298.

--------, 1993. *Den högre utbildningens ekonomiska villkor och betydelse* (The Significance and Economic Conditions of Higher Education), Rapport No. 2, Agenda 2000. Stockholm: Ministry of Education.

--------, 1994. *Högre utbildade i företag* (Highly Educated Personnel in Firms), Rapport No. 14, Agenda 2000. Stockholm: Ministry of Education.

--------, 1995a. *Teknologigenerator eller nationellt prestigeprojekt. Exemplet svensk flygindustri* (Technology Generator or National Prestige Project - The Case of the Swedish Aircraft Industry). Stockholm: City University Press.

--------, 1995b." The Economics of Technical Change - The Macroeconomic Consequences of Business Competence in an Experimentally Organized Economy," *Revue d'Economie Industrielle: Developments Recents*, Numéro Exceptionnel.

--------, 1996a. *Firm Objectives, Controls and Organization.* Boston/Dordecht/ London: Kluwer Academic Publishers.

--------, 1996b." Spillovers, Integrated Production and the Theory of the Firm," *Journal of Evolutionary Economics*, 6 (2), 125-140.

--------, 1996c. "The Competent Venture Capitalist," mimeo., Department of Industrial Economics and Management, Royal Institute of Technology (KTH), Stockholm.

Feldman, Maryann and Yda Schreuder, 1994. "Pharmaceutical Compound: The Origins of the Geographic Concentration of the Pharmaceutical Industry in the Mid-Atlantic Region," paper presented at the International Joseph A. Schumpeter Society conference, Münster, Aug. 17-20.

Giersch, Herbert (ed.), 1982. *Emerging Technologies: Consequences for Economic Growth, Structural Change, and Employment.* Kiel: Institut für Weltwirtschaft an der Universität Kiel.

Gross, Franz, 1983. *Decision Making in Drug Research.* New York: Raven Press.

Henderson, James, 1974. "A Note on the Economics of Public Intermediate Goods," *Economica*, 41, 322-327.

Henderson, Rebecca and Iain Cockburn, 1993. "Scale, Scope and Spillovers: The Determinants of Research Productivity in the Pharmaceutical Industry," NBER working paper No. 4466 (Sept.). Cambridge, MA.

Jaffe, Adam B., 1989. "Real Effects of Academic Research," *American Economic Review*, 79 (5), 957-970.

Laage-Hellman, Jens, 1987. "Process Innovation through Technical Cooperation," in Håkan Håkansson (ed.), *Industrial Technological Development. A Network Approach.* London: Croom Helm., 26-83.

--------, 1990. *Utveckling av ny biomedicinsk teknik. En nätverksstudie*, STU-info 768-1990, Stockholm.

--------, 1993. *Forskningsbaserat medicintekniskt företagande. En studie av kommersialiseringsformer, framgångsfaktorer, utlandsexploatering och Europasamarbete*, Nutekrapport B 1993:12. Stockholm: NUTEK.

Laage-Hellman, Jens and Björn Axelsson, 1986. *Bioteknisk FoU i Sverige. Forskningsvolym, forskningsinriktning, samarbetsmönster. En studie av det biotekniska FoU-nätverket 1970-85.* IM-gruppen, Uppsala. STU-information 536-1986, Stockholm.

Lichtenberg, Frank R., 1996. "The Effects of Pharmaceutical Utilization and Innovation on Hospitalization and Mortality," NBER working paper No. 5418 (Jan.). Cambridge, MA.

Liebeskind, Julia P., Amalya L. Oliver, Lynne G. Zucker, and Marilynn B. Brewer, 1995. "Social Networks, Learning, and Flexibility: Sourcing Scientific Knowledge in New Biotechnology Firms," NBER working paper No. 5320 (Oct.). Cambridge, MA.

Locke, Robert R., 1989. *Management and High Education since 1990 -the Influence of America and Japan on West Germany, Great Britain and France*. Cambridge: Cambridge University Press.

McKelvey, Maureen, 1994. *Evolutionary Innovation - Early Industrial Uses of Genetic Engineering*, Ph.D. dissertation, Linköping Studies in Art and Science No. 107.

Nadiri, M. Ishaq, 1978. "A Dynamic Model of Research and Development Expenditure;" in B. Carlsson, G. Eliasson, I. Nadiri (eds.), *The Importance of Technology and the Permanence of Structure in Industrial Growth*, Conference Reports 1978:2. Stockholm: IUI.

--------, 1993. "Innovations and Technological Spillovers," NBER working paper No. 4423. Cambridge, MA.

Nelson, Richard R. (ed.), 1962. *The Rate and Direction of Inventive Activity, Economic and Social Factors*. Princeton: Princeton University Press.

Nelson, Richard R., 1986. "Institutions Supporting Technical Advance in Industry," *American Economic Review*, 76 (2), 186-189.

Östholm, Ivan, 1995. *Drug Discovery*. Stockholm: Apoketarsocietén.

--------, 1996. *Nya skapelser – Losecentreprenörens recept*. Stockholm: Fisher & Co.

Peltzman, Sam, 1973. *Regulation of Pharmaceuticals Innovation: The 1962 Amendments*. Washington D.C.: American Enterprise Institute.

--------, 1975. "An Evaluation of Consumer Protecting Legislation: The 1962 Drug Amendments," *Journal of Political Economy*, 81 (5), 1049-1091.

Prevezer, Martha, 1994. *The Dynamics of Industrial Clustering in Biotechnology*, Paper presented at the International Joseph A. Schumpeter Society conference, Münster, Aug. 17-20.

Rehm, Hanz-Jürgen, 1982. "The Economic Potential of Biotechnology," in Herbert Giersch (ed.).

Rosenberg, Nathan, 1994. "Science - Technology - Economic Interaction," in Ove Granstrand (ed.), *Economics of Technology*. Amsterdam: Elsevier Science Publishers B.V.

Schumpeter, Joseph A., 1942. *Capitalism, Socialism and Democracy*. New York: Harper & Row.

SIND (National Board for Industrial Development), 1991. *Biotekniken i svensk industri - Bokslut och prognos inför 90-talet*. SIND 1991:5. Stockholm.

Stankiewicz, Rikard, 1986. *Academia and Entrepreneurs. Developing University-Industry Relations*. London: Pinter Publishers, Ltd.

Sölvell, Örjan, 1995. "Astra Hässle – Development of the Gastrointestinal Drug Losec," mimeo. 11D, Stockholm School of Economics.

Zucker, Lynne G., Michael R. Darby, and Jeff Armstrong, 1994. "Intellectual Capital and the Firm: The Technology of Geographically Localized Knowledge Spill-overs," NBER working paper No. 4946 (Dec.). Cambridge, MA.

Zucker, Lynne G., Michael R. Darby, and Marilynn B. Brewer, 1994. "Intellectual Capital and the Birth of U.S. Biotechnology Enterprises," NBER working paper No. 4653 (Feb.). Cambridge, MA.

Zucker, Lynne G. and Michael R. Darby, 1995a. "Present at the Revolution: Transformation of Technical Identity for a Large Incumbent Pharmaceutical Firm after the Biotechnological Breakthrough," NBER working paper No. 5243 (Aug.). Cambridge, MA.

-------, 1995b. "Virtuous Circles of Productivity: Star Bioscientists and the Institutional Transformation of Industry," NBER working paper No. 5342 (Nov.). Cambridge, MA.

Zucker, Lynne G., Michael R. Darby, Marilynn B. Brewer, and Yusheng Peng, 1995. "Collaboration Structure and Information Dilemmas in Biotechnology: Organizational Boundaries as Trust Production," NBER working paper No. 5199 (July). Cambridge, MA.

Zucker, Lynne G. and Michael R. Darby, 1996. "Costly Information in Firm Transformation, Exit, or Persistent Failure," NBER working paper No. 5577. Cambridge, MA.

APPENDIX 6: FIRM INTERVIEWS

(By either or both authors)

1. **Astra**, Södertälje.
2. **Karo Bio**, Huddinge (four interviews).
3. **Signal Pharmaceuticals Inc.**, San Diego (also IDEC, San Diego).
4. **Genentech Inc.**, San Francisco.
5. **DIACRIN**, Charlestown, MA.
6. **Center for Cancer Research** (Dr. Philip Sharp), MIT, Boston.
7. **Hoechst AG**, Frankfurt.
8. **Heraus Holding**, Hanau.
9. **Therexsys**, Staffordshire.
10. **Interactiva**, Ulm.
11. **Hässle, Astra**, Göteborg.
12. Ivan Östholm (two interviews).

7

MAPPING THE COGNITIVE AND INSTITUTIONAL STRUCTURES OF AN EVOLVING ADVANCED-MATERIALS FIELD: THE CASE OF POWDER TECHNOLOGY*

Anders Granberg

INTRODUCTION

This chapter is a report on a study focusing on the field of powder technology (PT). It is designed to achieve, simultaneously, two objectives: (1) to introduce a conceptual framework or analytical approach that may be used in charting the cognitive fields and institutional systems of different technologies, and (2) to present the main findings and conclusions arrived at by application of this framwork in the PT case.

The chapter is organized as follows: First, the starting point of the investigation - i.e. the field of powder technology - is defined and characterized. Next, a broad mapping of the PT field is undertaken in terms of the technical and scientific problems faced and the problem-solving competencies required. The mapping provides a basis for the identification of the Swedish PT system; the composition and structure of the system are described and the major strengths and weaknesses indicated. In the final parts of the paper, a recently launched national development program in the PT field is examined in the light of the relevant features of the technological system, and some concluding comments are offered on the nature of the investigation and the question of further research.

---

* In addition to funding within the STS Project, the author would like to acknowledge funding, also by NUTEK, of a monitor study (of which the present paper is a part) of the development and implementation of the Swedish Powder Technology Program (PT-90). The study is carried out by the present author in collaboration with Jens Laage-Hellman, the Industrial Market Research Group, Uppsala. Much of the industry data reported in this study has been collected by members of the Uppsala group. The findings of their industry-sector survey are presented in Laage-Hellman and Waluszewski (1992).

The study is partly based on information drawn from the available literature, broadly understood as including policy documents and program-related reports of various kinds as well as technical publications and relevant work in the field of 'Science and Technology Studies.' The second major source is that of knowledgeable and experienced people: some thirty interviews have been carried out with key individuals active in the industrial, academic, and government sectors of the Swedish PT system. In presenting here some of the material which has been collected and analyzed - and which is largely of a qualitative nature - extensive use will be made of graphic displays that provide overview and economy of expression. However, it will not be possible, within the limits of this paper, to comment on all details or aspects of the charts shown.

THE TECHNOLOGY

The present investigation differs from the other case studies within the STS Project in that, unfortunately, relatively little is available in the way of pre-compiled facts and figures on the conditions obtaining in the PT field, in Sweden and internationally. The dearth of survey data and statistical time series is particularly pronounced in the industrial domain. The situation is largely a reflection of the fact that powder technology has not - as an object of policy making and policy-related study - attained the prominence enjoyed by such broad fields as microelectronics and biotechnology.[1] It is also, to some extent, a reflection of the 'generic' or broadly 'enabling' character of powder technology, which makes the notion of a PT industry somewhat amorphous and elusive.

If, under these conditions, one wishes to identify and analyze the technological system, it is particularly important first to give careful consideration to the technology in question and to determine its scope and content. In approaching the system, one would have to proceed through the steps of (i) delimiting the field of technology, (ii) mapping the field in terms of problem areas or subfields, (iii) identifying the actors, i.e. the organizations or groups working in the various subfields, and (iv) finding out, exactly, what they do, how they do it, with whom they collaborate, etc. This may seem like a daunting task. Fortunately, the unravelling of the technological system is greatly facilitated by the fact that Sweden is a small country where the R&D community of any given field tends not only to be of limited size but also to be extensively covered by networks of formal and informal contacts ("everybody knows what everybody else is doing").

What is powder technology? Although extremely broad notions of the field may be found - encompassing the production and use of (nearly) all kinds of powders for

---

[1] It is worth noting, however, that the EC's "First Report on the State of Science and Technology in Europe", Brussels, undated (1988), includes powder technology in a small set of key "industrial materials and technologies" requiring renewed research efforts.

(nearly) all kinds of purposes[2] - it is more commonly seen, in contemporary policy contexts, as a materials technology oriented towards engineering-industry applications.[3] In conformity with that view, though with some specifications, *powder technology is formally defined here as the field of technical knowledge and competence underlying the development and production of powders and powder-based materials for engineering applications requiring advanced structural materials properties.*

It should be pointed out, first, that this definition - like our definition of a technological system (see chapter 1) - is based on a conceptualization of technology as a cognitive entity. Technologies, that is, are looked upon primarily as bodies of knowledge and knowledge-generating processes rather than as bodies of artifacts and physical fabrication processes. As regards the types of materials involved, it may be noted that so-called functional materials (which have useful optical, magnetic or electrical properties and which may, in some cases, be powder-based) are placed outside the PT field as delimited here. This is true also of those powder-based materials for structural applications, such as bricks and concrete, which are often referred to as 'traditional.'[4] The question of input materials, or the kinds of raw materials that might be used in the production of powders, is left open in the above definition. However, a survey of field activities - past, present, and planned - shows that we are in fact dealing with two major classes of materials, viz. metallic and ceramic materials, the powders of which are shaped and compacted into solid objects (components, billets, bars, etc.), either monolithically or in the combined forms of composites and compounds.

Powder technology thus comprises two major subfields - powder metallurgy (P/M) and engineering ceramics - with a common basis in broadly similar production processes, with specific linkages in some areas of shared problems and techniques, and with overlaps in terms of possible materials combinations. There are also, however, important differences and gaps between these subfields. Above all, what must be kept in mind when exploring the technological system and when considering questions of technology policy is the fact that they find themselves at different stages of development. Even though its potential is generally judged to be far from exhausted, powder metallurgy is now - through persistent development efforts and gradual advances over a period of some fifty years - to a considerable extent a mature, commercially well-established industrial technology. It is also, one should add, a field which has evolved as an outgrowth from the heavily dominant industrial tradition of metallic materials and metals-based design. Engineering ceramics, by contrast, is still, by and large, a field of research and experimental development, of prototype trials and minor

---

2 E. g. "Powder technology: a broad term encompassing the production and utilization of both metal and nonmetal powders", in *Metals Handbook* (1984).

3 This is the conception underlying the EC report (see note 1) and the Swedish Powder Technology Program (PT-90).

4 On the distinctions between 'advanced' materials, 'new' materials and 'traditional' materials, and between 'structural' and 'functional' materials/applications, see Cohendet, Ledoux and Zuscovitch (1988).

niche products. A long hoped-for, major breakthrough in the area of combustion engine applications has not materialized, and since the 1970s, when government programs were initiated in many countries, the field has passed through swings of optimism and disillusionment.[5]

Powder technology is part of the wider field of materials technology (or Advanced Materials Technology) and may thus serve as an illustration of some general features of that domain.[6] Most relevant in the context of this study are the following points: (i) Powder technology is, actually and potentially, an 'enabling' technology of great pervasiveness, entering into a very broad range of applications and products (cf. Figure 7-6). (ii) As in other materials fields - especially those centered around structural applications - development times tend to be long and drawn-out, typically involving slow-moving processes of substitution. (iii) Finally, a general rule much emphasized by observers of the PT field is that the effective development and introduction of new materials into practical use requires close links among the various actors involved, from materials scientists and engineers to component manufacturers and systems designers.

PROBLEM AREAS AND COMPETENCIES

So far, we have spoken of powder technology as "a technology". This is obviously somewhat misleading, considering the distinction just made between the subfields of powder metallurgy and engineering ceramics. Indeed, a closer inspection of the field makes plain that we are dealing here (as in the case of any technological field of some scope) with a large and complex cluster of technical and scientific specialties. In the following, an attempt will be made to chart the composition and structure of the cluster. The resulting 'PT map' is not, of course, the only possible overview chart of the field, though it is suggested that it is a well-grounded and useful one.[7]

The main conceptual premise of the mapping enterprise, as already mentioned, is the view of technology as knowledge and of technological development as a

---

5 The development of powder metallurgy and engineering ceramics is outlined in, repectively, Roll (1984) and Kingery (1986).

6 For a brief review of the salient features of advanced materials technology, see Granberg (1991).

7 The phrase 'technology cluster' has been chosen to underscore the point that we are dealing here not with a single, clearly delimited and well-integrated field of technology, but rather with a large and complex agglomeration of diverse technologies brought together by virtue of their ability to contribute, in various ways, to the solving of technical problems in the development and production of powders and powder-based materials for engineering applications. Of course, many of these technologies reappear in other clusters centered on other areas of application. The construction of pictures - static and dynamic, verbal and graphic - of the technology cluster is referred to as 'mapping.' The word is deliberately used to convey the point that (as in cartography, mutatis mutandis) there are questions of coverage, selectivity, scale, and projection that cannot be evaded, and, in particular, that there can be no such thing as wholly distortion-free map.

knowledge-generating process.[8] Within this cognitive perspective, particular emphasis will be placed on the problem-solving aspect of technological development. In other words, it is held that technology, or technical knowledge, like other forms of knowledge, grows by problem solving - by the identification, specification, and solving of technical problems.[9]

Figure 7-1 Powder Technology: A Schematic Representation of the Production Process

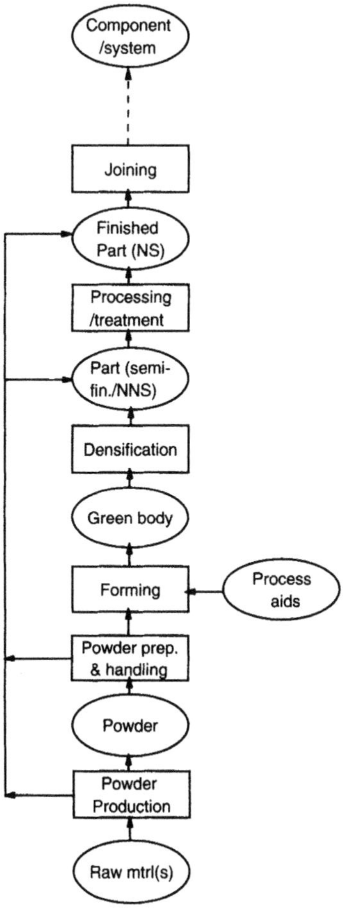

---

8 For a detailed exposition of the problem-solving approach to the exploration of technological fields, see Granberg (1987); for an application of that approach to a field of a very different nature from that of powder technology, see Granberg (1995).

9 Regarding the concept of a technical problem, it is suggested that the basic problem faced by the technologist is that of determining the means (or tools or devices or technical systems) by which particular functions can be effectively and efficiently performed. The twin concept of a technical solution can be explicated in terms of the articulation and corroboration of a design; see Granberg (1987).

What, then, are the major problems faced in the PT field? The black box of powder technology can be entered, most conveniently and in a very concrete fashion, by way of the production processes employed. A schematic picture of the overall production chain is presented in Figure 7-1. Starting with the input (at the bottom of the chart) of some raw materials, such as iron/steel or silicon, fabrication proceeds through a series of process steps (boxes) and intermediate results or products, including near net-shape (NNS) parts, up to the finished, or net-shape (NS) part, which may then enter into a more complex component or system. The lines to the left of the chain are to indicate the possibility of proceeding directly from powder to near net-shape or net-shape parts, e.g. through hot isostatic pressing (HIP) or spray forming. It is generally desirable - though especially so in the case of ceramic materials - to eliminate or minimize the processing of parts.

The production chain provides a basis for the identification of certain areas of technical and scientific problem solving. Taking a broad view of the chain as extending into the domains of the materials user, Figure 7-2 distinguishes five such areas, each relating to several links in the chain. Beginning at the user end, Area I is defined by the task of designing powder-based parts or components, usually within the constraints of systems design. This is, or ought to be, an overarching task which is stated in terms of certain end-user needs or specifications to be met, and which guides (and is partly guided by) the problem-solving efforts undertaken in other areas. The "ought" is most clearly expressed today in the attempts which are made - under such designations as 'integrated design,' 'concurrent engineering,' or 'simultaneous engineering' - to strengthen the ties between design and manufacturing, and which may also involve materials engineering. Area II is extensively connected with the various links of the production chain and comprises what can be regarded as the core problems of powder technology. Here, the focal tasks are those of the development, perfection, or modification of particular materials and fabrication processes and techniques. The control and coordination, by means of suitable hardware and software, of the various production steps are the defining problems of Area III, while the related problems of assuring materials and product quality throughout the chain make up Area IV. Finally, as a scientific and basic-technology domain broadly underpinning the production chain, Area V is centered on the tasks of specifying the microstructure of powders and powder-based materials, characterizing the materials in terms of relevant macro--properties (notably mechanical and thermal properties), and exploring the relationships between fabrication processes, microstructures, and macro-properties. It stands to reason that there will be large overlaps and ill-defined borderlands between this problem area and, above all, Area II.

What is the competence base required for a comprehensive and effective treatment of these problems? Or, more specifically, what are the technical and scientific disciplines or specialties that support - and feed upon - problem solving in the PT field? Set against the background of the preceding chart, Figure 7-3 gives an overview of the main knowledge-and-competence clusters involved. On the left-hand side are shown three smaller clusters: a design cluster (connected with Area I), a quality cluster (Area

IV), and a process-control cluster (Area III). On the right-hand side are found two major agglomerates: a materials technology cluster (Area II) comprising subsets focusing on particular materials, processes, and properties, respectively; and a materials science cluster grounded in various physics and inorganic chemistry specialties, with important subsets oriented towards structural explorations and 'microengineering' as well as instrumentation and techniques.[10]

Figure 7-2 Main Problem Areas

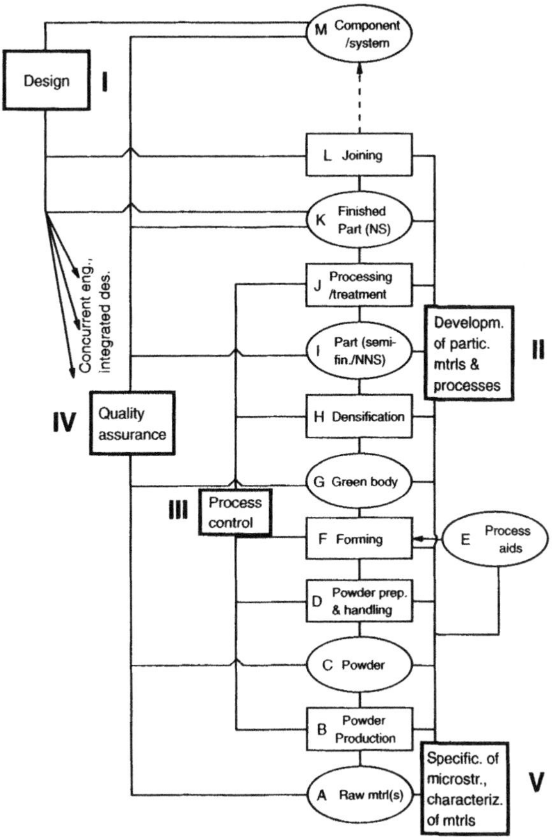

[10] The following acronyms and abbreviations appear in Figure 7-3: Atom. mod. = atomistic modelling; Autom. insp. = automated inspection; CAD/CAM = computer-aided design/computer-aided manufacturing; Contr. eng. = control engineering; Corros. = corrosion protection technology; El. microsc. = electron microscopy; Eng. des. = engineering design; Eng. mech. = engineering mechanics; FEM = finite-element method; HIP = hot isostatic pressing; Inorg. chem. = inorganic chemistry; Laser process. = laser processing; Metrol. = metrology; MIM = metal-injection moulding; Molecul. eng. = molecular engineering; NDT = nondestructive testing; P/M-Al = aluminum-based powder metallurgy: P/M-Fe = iron-based powder metallurgy; QA = quality assurance; Sensor tech. = sensor technology; SiC = silicon carbide-based ceramics; SiN = silicon nitride-based ceramics; Surf. sc. = surface science; Tribol. = tribology; Crystall. = crystallography; X-ray spectr. = x-ray spectroscopy.

Figure 7-3 **Major Clusters of Scientific and Technological Disciplines and Specialties Involved in the Development of Powder Technology**

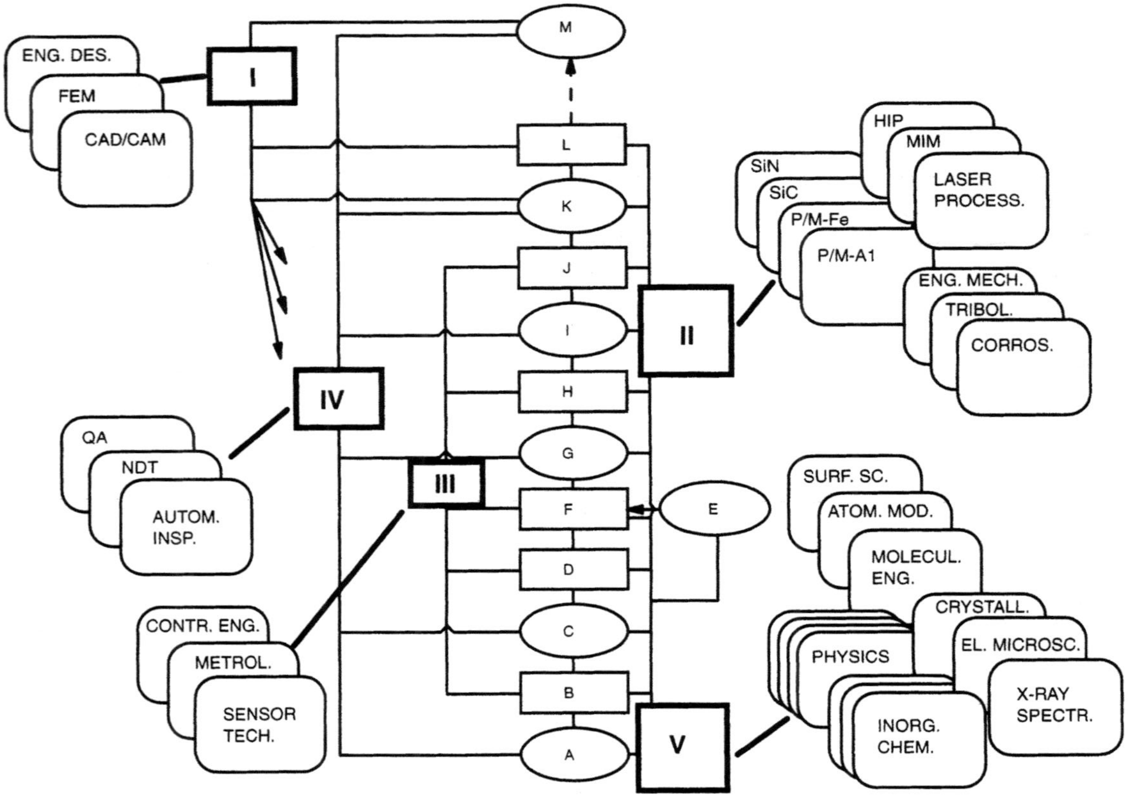

It would, of course, be presumptuous to claim that the PT map sketched here provides complete coverage of the relevant bodies of knowledge and competence. Although the ambition has been to identify the main problem areas and the principal components of the cognitive resource base, the clusters displayed could easily - in terms of specific content - be expanded or modified by using finer divisions or different sets of illustrations. Some arbitrariness seems unavoidable as there are no generally accepted rules for the dissection of technological fields. The absence from the chart of organizational and managerial competencies may be noted. The reason is not that they are considered irrelevant (on the contrary, they are recognized as important elements in the problem-solving processes of any field), but that no PT-oriented specialties of this kind have evolved.

In the development of the general field of materials technology and its relationships with other domains of industrial and academic problem solving, certain integration trends are discernible that bear also on the cluster configurations of Figure 7-3. First, there is the relatively recent tendency, already noted, to try to establish closer links between design, manufacturing, and materials engineering. In terms of the chart, this would entail a rapprochement between the design and materials technology clusters as well as a strengthening of internal cohesion in the case of the latter. A second line of development, which involves the two clusters on the right-hand side of the chart, is the gradual interpenetration and partial merger - over the past 20-25 years - of materials technology and materials science into a larger complex commonly referred to as 'Materials Science & Technology' or 'Materials Science & Engineering.' An important element in the integration process has been the emergence of atomistic modelling and molecular engineering as bridging specialties holding the promise of tailor-made materials, or 'designer materials', built from the ground up.[11]

At this point a brief aside on the evolutionary, or quasi-evolutionary, patterns of problem solving in the PT field seems appropriate. Technical problem solving, generally speaking, involves processes of trial-and-error learning where ideas are conceived, specified, and confronted with reality as laboratory models, prototypes or commercial products, and where the experience gained guides the development of alternative concepts or variant designs. In the production chain of powder technology (Figure 7-1), the generation of variety can be seen as taking place at all stages; there is thus an ever-growing diversity of powders and powder-based materials, forming and densification methods, processing methods, etc. In the search for optimal or satisfactory solutions, selection (or the reduction of variety) occurs both 'top-down', through the user needs or specifications that are to be met, and 'bottom-up', by virtue of the fact that the choice of raw materials and powders limits the options available higher up the chain. It would, of course, be desirable if the element of trial-and-error exploration could be minimized and the search performed in a more focused and cost-effective manner. The

---

[11] On the nature of the science cluster and the development of an integrated S&T complex, see Kranzberg and Smith (1988).

development of materials databases and improved characterization and screening techniques are contributing towards this end. Hopes for more fundamental advances, however - in powder technology and other advanced-materials fields - attach to the continuing efforts to improve the microscopic understanding and mastery of materials. The ultimate goal is the general ability to produce, by way of atomistic modelling and molecular engineering, materials with predictable (and desirable) macro-features. While some observers consider the progress achieved so far sufficient to warrant the designation of a "materials revolution", the goal still seems fairly remote and the actual development of new or improved materials appears still, to a considerable extent, to be driven by trial-and-error empiricism, including learning by producing and learning by using.[12]

Figure 7-4 The Powder Technology System: Main Actor Categories

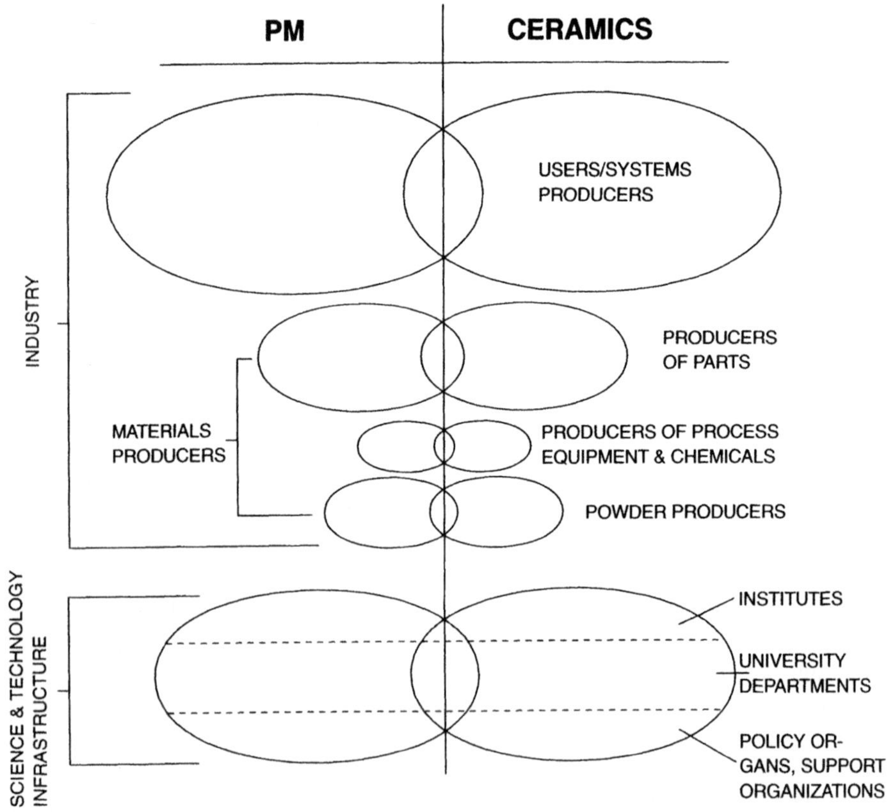

[12] For a critique of the 'revolutionary' views of, e.g., Forester (1988) and Cohendet et al. (1988), see Turner, Roy and Wield (1990). The possibilities and limitations of materials design by microscopic mastery are considered in Ashby (1987).

In this section, an attempt has been made to map - by means of a qualitative type of analysis - the composition and structure of powder technology as an evolving cognitive field. Though it is hoped that the analysis and the resulting chart may be of some interest in their own right, their role in the present study is subservient to the exploration of the institutional system of the technology. Thus, on a practical level, the chart has proved useful as a background grid and structuring aid in carrying out expert interviews and literature searches. Similarly, it has been instrumental in guiding the scans of the Swedish R&D landscape in the search for relevant actors. More fundamentally, it has provided a basis for the delimitation of the national PT system and for the characterization and appraisal of the system as a national competence resource.

THE TECHNOLOGICAL SYSTEM

Before embarking on the examination of the Swedish system, let us first consider the general composition of a hypothetical, complete PT system, be it a national or transnational entity. Given the preceding analysis of the PT field, it is reasonable to suggest that such a system will comprise certain major categories of actors. In Figure 7-4, these categories are displayed in a vertical arrangement which partly reflects the production chain, with a 'stack' of industrial actors extending from powder producers to user firms. The industry sector of the system is shown as resting on a scientific and technological infrastructure composed of noncommercial R&D organizations and policy agencies variously engaged in the cultivation of the PT field. A demarcation line separating the two subfields of powder metallurgy (P/M) and engineering ceramics has been drawn through the entire actor set.

While useful as a means of describing and structuring real-world systems, the scheme must not be interpreted in too rigid a fashion. For instance, overlaps may exist not only across the P/M-ceramics divide (as shown), but also vertically, between the different industry categories. Furthermore, it is clear that the relative size of the categories will vary greatly depending on the specific conditions obtaining in different countries and on the developmental stage or commercial status attained by particular technologies or products.

The Composition and Structure of the Swedish PT System

Now to the Swedish PT system - what does it look like, concretely? The institutional chart of Figure 7-5 provides a comprehensive overview of the actors involved (ca 1990) organized in accordance with the general scheme presented above. An important qualification regarding coverage must be made at the outset: the group of Swedish corporate users of PT materials/components is, in reality, far larger than is indicated by the chart, although it would be exceedingly difficult to ascertain the exact size and composition of the set. What has been included here are users in possession of an

Figure 7-5 The Swedish PT System: The Actors

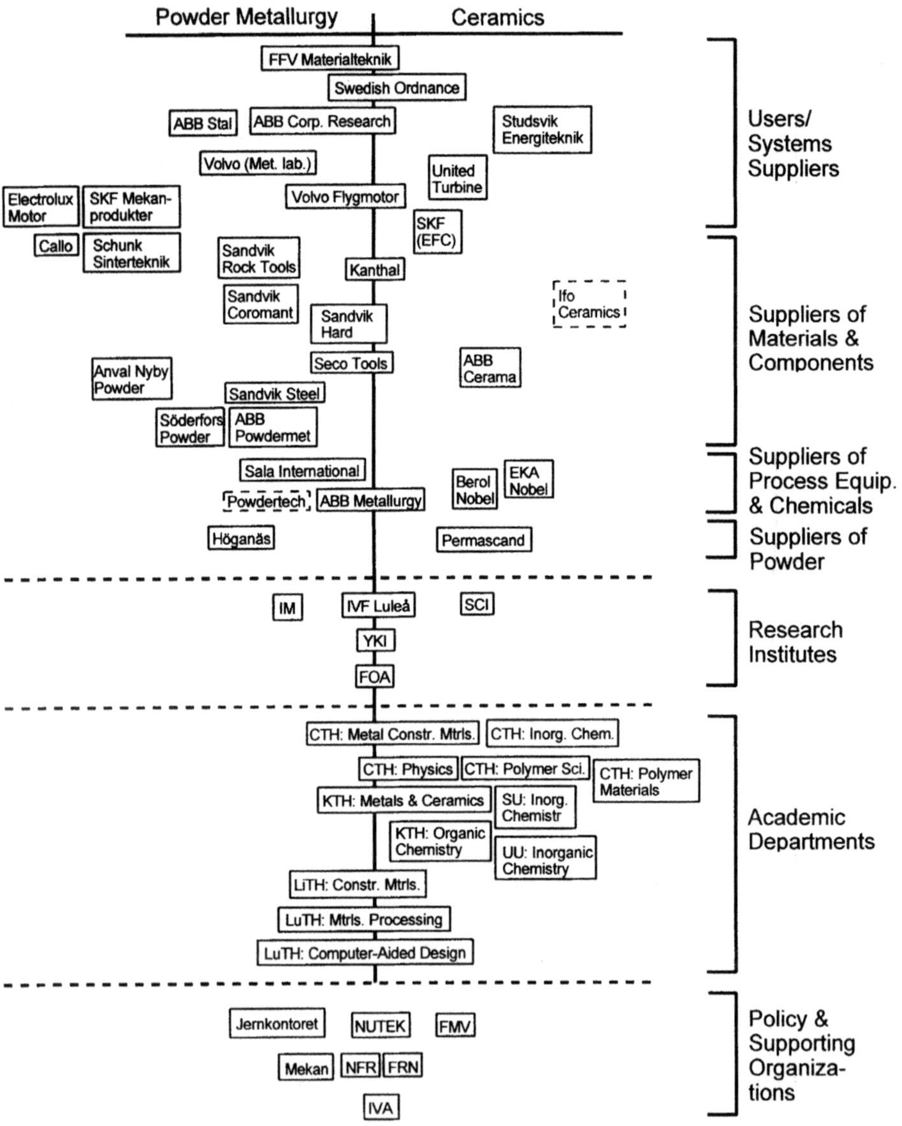

in-house R&D competence that enables them to undertake PT-related work of a relatively advanced nature (prominent examples are Volvo Aeroengines and ABB Stal). In addition to these first-echelon users there is, according to well-informed observers, "a large number" of engineering-industry firms, mostly small and medium-size, that use PT-based parts in their products though they do not actively engage in the development of powder technology.

In the general picture presented by the chart, certain features relating to the two subfields stand out. First, it can be seen that the P/M and ceramics fields are indeed, as one might expect, institutionally connected; a number of units are found straddling the center line, which means that they are active on both sides of the divide. Although they appear on all major levels of the chart, such institutional links are proportionately stronger in the scientific and technological infrastructure. (The question of inter-unit links or networks is still left open.) Secondly, a closer look at the distribution of actors would seem to indicate that we are dealing with a technological system that is, in a certain way, 'imbalanced.' The pattern can be described as one of diagonal concentration: a concentration of industrial actors on the P/M side and a concentration of infrastructure actors on the ceramics side. In other words, a relatively high level of industrial activity appears to be paired with a relatively low level of infrastructure activity in the case of powder metallurgy, while in the case of of ceramics the disparity is the reverse.

The picture may, on first sight, be somewhat puzzling. A detailed explanation would require a lengthy excursion into the history of powder technology, including both Swedish and international developments. For the purposes of this presentation, however, a brief interpretation of the general pattern on the following lines will suffice. Most important here is the fact, already emphasized, that the two subfields differ greatly in the extent to which they have matured into practical applications and commercial products. In Sweden, a country with old-established metals and engineering-industry traditions, the early-emergent possibilities of powder metallurgy were recognized and responded to by some materials firms, setting in motion a train of developments which has gradually - in the course of the post-war decades - increased the number of actors and the volume of activity. Drawing on existing competencies in iron and steel metallurgy, Swedish firms have thus been able to build strong niche positions in the P/M field, with a significant presence in powders, processes, and finished products. It should be noted that these development efforts - though accompanied and underpinned by a certain amount of academic and institute research - have been essentially industry-initiated, industry-driven, and industry-based. Some infrastructure support in the form of government programs has been provided, but was reduced during the 1980s, with reference to the relative maturity of the technology and the strong or adequate, independent capabilities of industry.

On the engineering-ceramics side, the situation portrayed in the chart is one which reflects a general, international pattern. As has been pointed out, this is a subfield of powder technology which still finds itself, by and large, in an R&D phase; numerous potential applications and products have been identified and are being explored, but few have been realized and successfully commercialized. In Sweden, as in many other countries, the 1970s saw the framing of government policies designed to broaden and reinforce the scientific and technological infrastructure of ceramics. Thus, a series of national R&D programs has been implemented and will be followed by further programs now in preparation. In parallel with the public-sector developments - and partly linked with them - Swedish companies have manifested an interest in the possibilities of engineering ceramics by engaging in monitoring, materials testing, or exploratory development. However, active industrial involvement has been limited mainly to some potential first-echelon users (e.g. United Turbine) and a few materials- and process-oriented firms (Permascand, ABB Cerama). The general conclusion, then, is that the 'imbalance' shown in Figure 7-5 should be seen as an expression both of the differential developmental characteristics of the major subfields and of certain distinguishing traits of the Swedish case. The pattern does not, one may add, reflect negatively on the national capabilities in engineering ceramics; indeed, by international standards, the Swedish competence base in that field is commonly judged to be relatively strong.[13]

The concentration of industrial activity on the P/M side is clearly indicated by the chart. This does not mean, though, that this subfield - which comprises a great variety of materials and processes - is fully or extensively covered by the actors in question. Rather, what emerges on closer inspection is a further concentration of efforts within certain materials areas. The focal areas can be identified as: (i) metal powders, notably iron-based powders (Höganäs, Anval); (ii) sintered materials (in Figure 7-5, the company cluster containing Schunk); (iii) fully dense materials (the cluster containing Söderfors), with a basis in the process of hot isostatic pressing, or HIP; and (iv) 'hard materials,' essentially cemented carbides (the cluster of Sandvik companies). In these areas, or in particular niches within them, considerable competence has been acquired in dealing with problems of the type previously (in Figures 7-2 and 7-3) grouped into categories "II-IV," with an emphasis on the improvement of known materials and processes. In terms of market share in their respective product domains, several of the Swedish P/M-materials suppliers have reached leading or high-ranking positions on a world level. As regards the development by user-industry firms of a P/M-(or PT-)related design competence (Area I), it has not been possible, given the nature of the actor category, to comprehensively and thoroughly assess the present situation. What may be safely asserted, however, is that advanced problem-solving capabilities of this kind - including the requisite knowledge base concerning relevant materials properties and processes - are to be found only among a few first-echelon users, such as ABB Stal and Volvo.

The Swedish PT system, as it exists today, encompasses a set of materials areas that differ widely in terms of economic or commercial importance. In table 7-1, industrial

[13] On the ranking list of 'advanced ceramics' or engineering ceramics, Sweden (including Swedish industry) is usually placed among the European leaders, after Germany and, perhaps, France; see e.g. Theulon (1989) or STU (1991a).

employment and sales figures (1988) are presented for the four P/M areas and ceramics; here, ceramics has been given a broader definition ("technical ceramics") which significantly increases the number of commercial applications.[14]

Table 7-1 Employment, Sales, and Exports in Five Powder Technology Product Areas, 1989

	EMPLOYEES	SALES (MSEK)	EXPORTS (%)
Metal powder	700	700	95
Sintered P/M materials	250	100	10
Fully dense P/M materials	300	400	90
Hard materials	4000	8000	95
Ceramics ("Technical ceramics")	150	270	30
Total	5400	9470	

What stands out is, first, the overwhelming (though expected) preponderance, by these measures, of the P/M field as a whole. The meager employment and sales shares of the ceramics field of less than 3% of the total would, of course, be further diminished by application of the more restrictive concept of engineering ceramics. Equally striking is the heavily dominant position of one particular area - that of 'hard materials' - not only within the P/M framework, but overall. This means that the Sandvik group alone accounts for some 85% of the total sales of the Swedish PT-materials/parts suppliers. A noteworthy feature emerging from the third column of the table (exports as percentage of sales) is the extremely strong export orientation of the supplier industry as a whole, with an overall export share of 92%. The exceptional cases of sintered P/M products and ceramics, which trail the other areas in terms of employment and sales, are distinctly deviant in that they are both heavily dependent on the home market.

Powder technology has been characterized in the preceding as an 'enabling' technology of great pervasiveness, de facto and potentially. Although the Swedish PT system is focused on certain selected areas and niches, it may still serve as an illustration of this aspect of the PT field. Departing from the focal areas listed in the above table (minus 'metal powder', which is excluded here on account of its being an input into the other P/M areas, and with 'ceramics' again restricted to 'engineering ceramics'), Figure 7-6 provides an overview of the kinds of products that flow from these areas and which enter into the applications and products of certain major receiving industries. Note that the transport equipment and machinery industries represent 'users'

[14] The data have been compiled as part of the background material for the Swedish Powder Technology Program ("PT-90") and appear in some of the early, unpublished program reports, e.g. "FoU-program för utveckling av metallisk och keramisk pulverteknik" (Slutversion 1989-02-09), 1989.

in the sense in which the word is generally employed in this paper, i.e. firms which incorporate PT-based materials and parts into their own products (or systems). The mining and construction industries, on the other hand, are essentially end-user groups with a practical need for PT-based tools (notably hard-materials tools) and equipment.

Figure 7-6 Focal Materials Areas with Illustrative Products and Industries

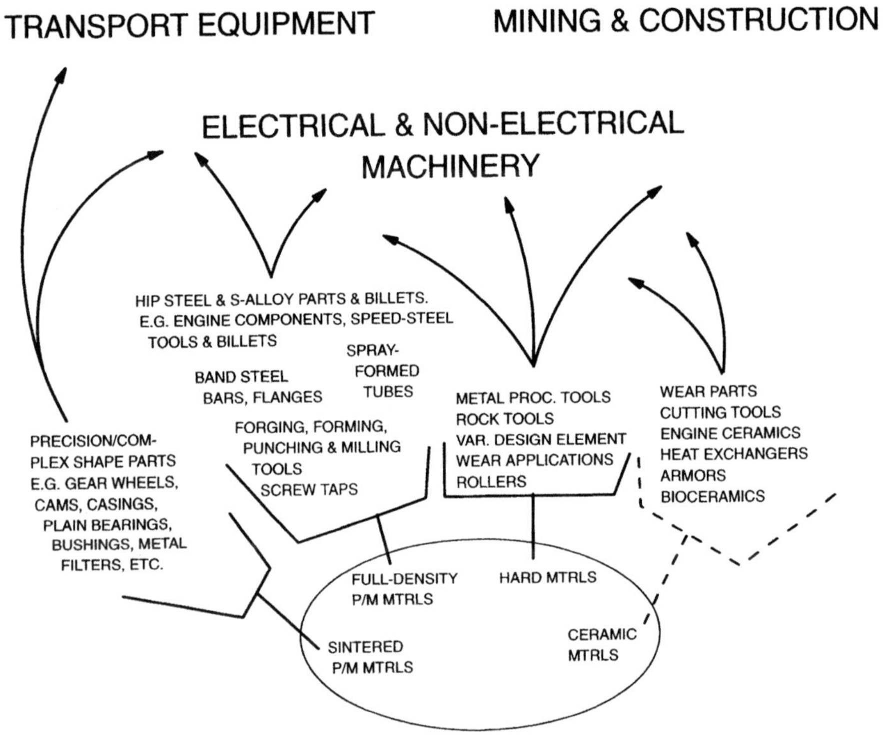

Let us turn again to the scientific and technological infrastructure of the Swedish PT system. The general distribution of actors and activities in favor of the ceramics field has been noted and commented upon. Still to be clarified are the nature and significance of the infrastructure as part of the national competence base. The relevant observations and conclusions can be summarized as follows. First, it is important to recognize the very limited scale of the overall PT effort undertaken by Swedish university- and institute-based groups. Thus, a man-year estimate for the P/M field gives a maximum of 20, while the total number of researchers working in the ceramics field is put at approximately 40 (Theulon 1989 and STU 1991a). In view of the severe resource constraints, the activities engaged in are remarkably diversified, with small-scale projects (single-investigator projects are common) addressing a variety of problems not only in the 'basic' domain of Area V, but also in Areas I, II, and IV. The research role of the universities includes, but is by no means confined to, the exploration of microstructures, the characterization of materials, and other Area-V tasks (such work is carried out mainly by some groups at the technical universities of Gothenburg and Stockholm, and at Stockholm University; they are complemented, in the field of surface science, by YKI, the Institute for Surface Chemistry). Academics are also, to a considerable extent, active in the applied areas of the field, in many instances in collaboration with a relevant institute (in ceramics, the Swedish Ceramic Institute, SCI; in powder metallurgy, the Swedish Institute for Metals Research, IM; and in production engineering, the Swedish Institute of Production Engineering, IVF). Accordingly, what emerges from a survey of the problem orientations and problem-solving competencies within the entire set of performing institutions - academic units and institutes - is an overall profile in which materials engineering and applications are more prominent than materials science and basic technology. These traits are likely to become more accentuated with the implementation of the recently launched, national PT program, "PT-90" (see Section 5).

A major portion of the infrastructure is now formally organized, within the framework of PT-90, as four university-based centers (Powder Technology Centers, or PTCs) involving not only academic departments but also institutes and interested firms. The centers are to operate primarily through collaborative R&D projects that provide linkage between the infrastructure and the materials-producing and materials-using industries. While there is no sharp and rigid division of labor, in their initial state the centers present themselves as competence clusters directed towards different portions or aspects of the PT production chain: the Stockholm center being chiefly concerned with the early fabrication stages, the Gothenburg center with the midsection, and the Luleå center with the later stages and the design connection; the Linköping center, finally, is more specialized, with a clear focus on nondestructive testing and quality assurance. As regards the applied profile of the centers and of the infrastructure as a whole, it should be pointed out that there is no close and simple correspondence between the particular traits of this profile and the specialization pattern of industry as outlined earlier. To be sure, there are some prominent common features, such as the shared concern with HIP techniques and the shared focus on silicon nitride ceramics. A close matching, however, is practically precluded by what has been described here as the pattern of diagonal concentration and by the relative diversity of infrastructure activities. Thus, discrepancies appear, on the ceramics side, due to the low level of industrial activity and interest in the 'immature' areas explored by university and institute researchers, and, on the P/M side, due to the limited research appeal of the commercially well-established areas of this subfield.

One very important way in which the infrastructure may support and shape the development efforts of industry is, of course, through education. The innovativeness of supplier and user firms, and their ability to correctly assess and effectively exploit the possibilities of new or improved types of materials, are crucially dependent on the appropriate education and training of their technical personnel. In a broader innovation perspective, the provision of basic and continuing education to the design and production engineers of the user industry stands out as a key task. It is evident, furthermore, that the educational needs associated with powder technology or other advanced materials fields cannot be satisfied by the universities alone and that contributions from different parts and levels of the educational system are required. In the case of the Swedish infrastructure and the PT activities so far engaged in, the emphasis has been clearly on the performance - by universities and institutes - of the R&D function. While this has yielded educational benefits in the form of research training (involving doctoral-thesis work as well as informal learning-by-doing), no concerted effort has been made prior to the PT-90 program to achieve a broader dissemination of knowledge through the channels of graduate-level curricula, extramural courses, industry workshops, etc.

The Networks of the Swedish PT System

A technological system has been defined here as a network of actors operating in a particular field of technology; more specifically, it was stipulated that such a system be defined in terms of knowledge or competence flows rather than flows of ordinary goods and services. The elucidation of the network aspect of the Swedish PT system is thus an important objective of the present study. In the brief account that follows, only the basic features and general patterns of the systemic links will be considered. In terms of network 'nodes', the focus will be on the actors presented in Figure 7-5, i.e. the first-echelon users, the materials/parts suppliers, and the infrastructure actors. For obvious reasons, it has not been possible to include in the network analysis proper the large and loosely connected fringe of 'passive' or potential users. A basic point which emerges from our survey of the central categories of PT actors is that linkage and cohesion within the national collective is provided not by a single, homogeneous network, but by a variety of overlapping/interrelated networks; three main types can be distinguished.

There is first the 'economic network', defined as a set of commercial actors connected through buyer-supplier relations (our definition of a technological system does not imply that flows of ordinary goods and services are irrelevant). This is a formal, transaction-based type of network, the size and stability of which may vary considerably. Although its principal nodes are located in the industry sector, it may connect with the infrastructure through the purchasing of research, educational, and consulting services. The Swedish PT system comprises a number of economic networks involving a broad spectrum of actors, from powder makers and equipment suppliers, through producers of powder-based materials and parts, to user firms. It is hardly surprising, given the pattern of 'diagonal concentration' noted earlier, that actor links of this kind are to be found mainly on the P/M side. What is more significant here is the fact that the Swedish firms operate in buyer-supplier networks that are, to a very large extent, international. The materials industry, we have seen (Table 7-1), is heavily dependent on foreign users: overall, the export share of the materials and parts supplied amount to more than 90% of the sales value. At the same time, the Swedish engineering-industry users appear to be similarly dependent on foreign suppliers of finished

PT-based parts and components. Again, the lack of statistical data makes the appraisal of the general user situation hazardous. However, industry observers with an inside view of the field are agreed that only a minor portion of the finished PT parts used in mechanical and electrical engineering applications are in fact purchased at home and that the key link in this context - i.e. the production of sintered parts - is a weak one; apart from a very small producer, Callos, there is only one domestic supplier of such parts, viz. Schunk, a subsidiary (with no R&D activities of its own) of a German firm. It stands to reason that these conditions, which are summarized in Figure 7-7, significantly limit the scope for the development of strong user-producer links within the Swedish PT system.

Figure 7-7 The Import/Export Dependence and the Narrow Sintered-Products 'Waist' of the Swedish PT System

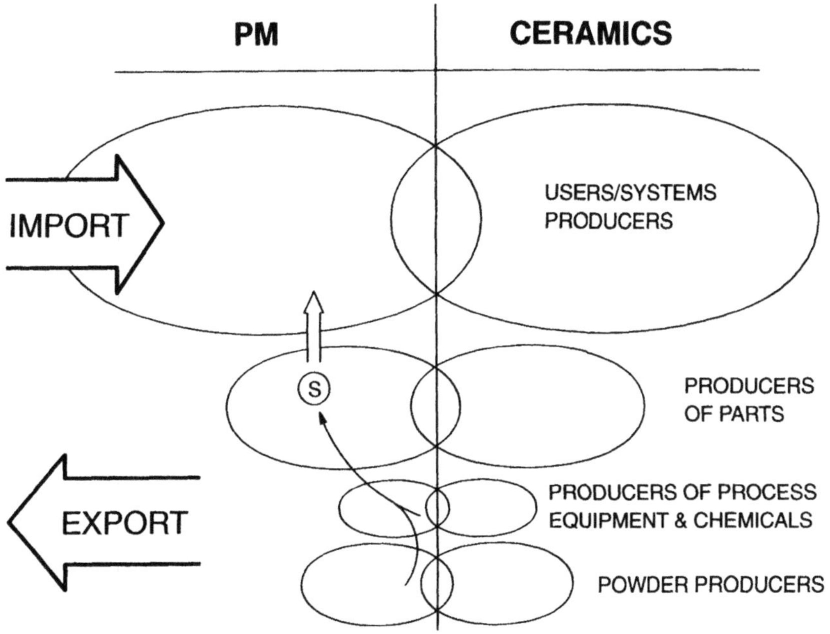

The second type of network, which frequently overlaps the first, is the 'network of technical problem solving,' where different actors addressing a shared R&D task are interlinked by relations of collaboration and exchange. It is the problem-solving network that provides the core links of the technological system as an entity "defined in terms of knowledge or competence flows rather than flows of ordinary goods and services". Networks of this type are usually developed within the formal framework of a project or program, but are often expanded through the establishment of informal links with actors outside the group of 'official' participants. Also, it is not uncommon for such networks, once formed, to persist beyond the original task; they may thus continue to operate, or be reactivated, in subsequent R&D ventures, perhaps with some changes in size, composition or structure. It should be noted, moreover, that these are networks which are not tied to any particular part of the system; they may fall within, or cut across, the categories of users, producers and infrastructure actors.[15]

The principal actors of the Swedish PT system (Figure 7-5) are connected, directly or indirectly, by links established within a large number of problem-solving networks. As far as formal participation in organized collaborative R&D efforts is concerned, the dominant pattern is that of joint problem solving by pairs or triplets of institutional actors, though there are also examples of larger cooperatives, including cross-sectoral ones. The PT system can thus - from this point of view - be seen as operating as a loosely connected aggregate of 'mini-networks.' In a broader perspective, however, where both formal and informal relationships are considered and where the recurring or persistent patterns of collaboration and exchange are more clearly discernible, the system presents itself as a more cohesive or comprehensively integrated entity due to the workings of larger, overlapping networks. On the background of the economic networks outlined above, the following features of the problem-solving networks stand out. First, the vertical span of the linkages which they create within the national system is a wide one, extending from policy agencies (as promoters and organizers of cooperative R&D) and academic science departments to first-echelon users. Second, as regards the P/M-ceramics divide, the distribution of networks over the two subfields is, in this case, more even, the industrial strength of the P/M field being counterbalanced by the R&D-intensity of the ceramics field. In the former field, Jernkontoret, the cooperative organ of the Swedish iron and steel industry, has played an important role as a supporter and coordinator of industrial R&D projects, notably through its special P/M program "Teknikområde 80" (TO-80). Third, although it is by no means uncommon for companies and noncommercial actors to engage in collaboration and exchange with foreign partners, the aggregate international linkage thus established is still comparatively weak. In other words, the problem-solving networks are, to a far greater extent than the economic networks, nationally based and circumscribed. Finally, it is evident that these networks are also, generally speaking, more strongly influenced by government action. In many cases, they are indeed the direct results of programs or major projects initiated and organized by national policy agencies.

The third type of network distinguished here may be called 'the national community network.' This is an informal network that links, by way of personal contacts, individuals active in the various subfields and specialties of a given technology and a given country. Based on a common interest in the development of the field, it is an extensive and loosely knit, though relatively stable, type of network which cuts across the different sectors and levels of the technological system. The Swedish PT system is supported by a broadly inclusive ("everybody knows everybody") community network, the nodes of which are formed by the leading practioners - such as chief engi-

---

[15] A detailed picture of the problem-solving networks in the industry sector is provided by Laage-Hellman and Waluszewski (1992).

neers, senior researchers and R&D managers - of the various organizations involved. The P/M-ceramics divide again manifests itself in a relative paucity of cross-links, especially on the industry side. Within the two subfields, however, extensive webs of interpersonal relations have evolved by which knowledge of who-is-who, who-is-where, and who-does-what is diffused through the system; in this context, representatives of the institute sector have often performed a bridging or mediating role vis-a-vis industry and academia. The community network is also, of course, internationally connected through numerous informal links between the community members and their foreign colleagues. The establishment and maintenance of such links takes place partly within the framework of international associations, conferences, symposia, etc.

Figure 7-8 The HIP Network

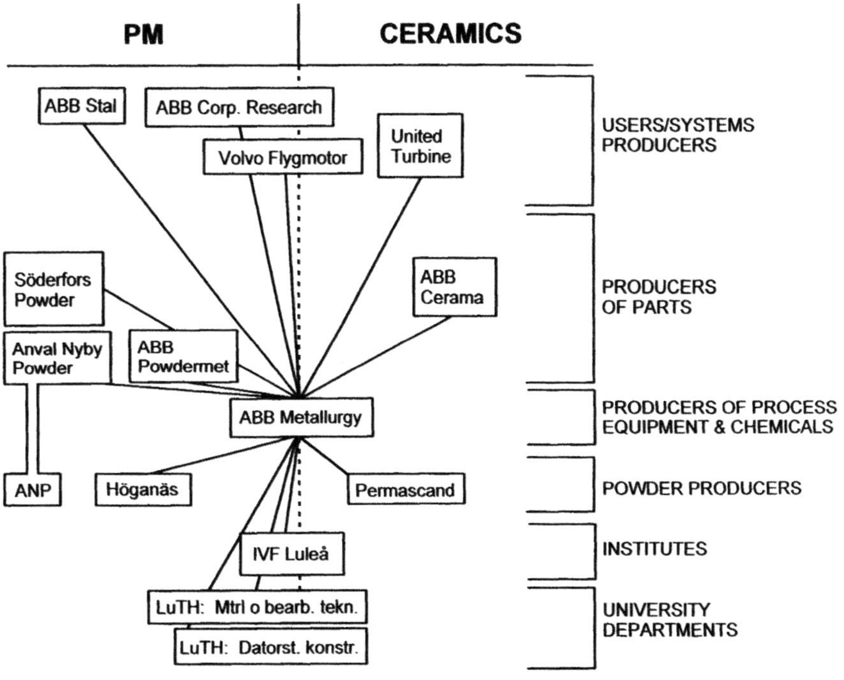

The three types of network are closely interrelated. The community network, for instance, is both a basis for and a result of economic linkups and joint problem solving. Relationships of the two latter types will, in turn, often appear in mutually reinforcing combinations, notably in industrial networks that tie together users and producers of components and equipment. An important example - within the Swedish PT system - of a network of this kind, involving problem-solving as well as economic links, is the 'HIP development block' shown in Figure 7-8. This is a relatively large and durable network which is grounded in a common concern with the tools and processes of hot isostatic pressing (HIP). Straddling the P/M-ceramics divide, it has evolved over an extended period of time, with a starting point in the early-1950s development at ASEA (now ABB) of powerful isostatic presses. Although the initial efforts were directed towards the fabrication of synthetic diamonds, other possibilities were perceived and the field of application was subsequently expanded into powder metallurgy and engineering ceramics. Today the tools themselves (the so-called Quintus presses) and the particular technology which they embody have matured and stabilized; the HIP processes, however, are still undergoing change as old production methods are improved and new variants are developed. The central role of the HIP tools in the emergence and growth of the network is emphasized in Figure 7-8, where the equipment maker (now ABB Metallurgy) is displayed as the hub in a circle of users and R&D actors, incuding some infrastructure actors. It should be pointed out that the economic and problem-solving links that exist, or have existed, within the circle are not shown in the chart.

The Strengths and Weaknesses of the Swedish PT System

To conclude the characterization of the Swedish PT system, the major strengths and weaknesses of the system will be briefly summarized. First, on the positive side, should be mentioned the fact that the Swedish materials industry has developed a significant presence in the field - that a number of firms have acquired advanced competencies and built strong commercial positions in certain areas or product niches; that the R&D and production activities engaged in are broadly extended over the PT chain; and that the actors are partially interlinked and supported by economic and problem-solving networks, with ties also to the national infrastructure and the international environment, in particular to foreign users. As regards domestic users, actual and potential, the existence of a relatively large and diversified engineering industry sector, including engine and transport equipment manufacturers, must be considered a major asset. Also to be mentioned are the enhanced (though, in an international perspective, still comparatively modest) R&D and educational capabilities of the infrastructure - notably on the ceramics side, which has been a favored area in terms of technology policy support - and the valuable bridging functions performed by the institutes in relation to industry and academia. Finally, as a positive feature that partly reflects the small-country status of Sweden, one may point to the fact that an extensive technological community network has evolved which facilitates the development of collaborative relations of a more formal nature, in R&D projects, national technology programs, and commercial undertakings.

The weaknesses and limitations of the system are also, in part, attributable to the constraints typically faced by small industrialized countries. Thus, it is hardly surprising that the Swedish PT suppliers are able to satisfy only a minor portion of the product and technology needs of the national engineering industry, nor is it surprising that the home market merely accounts for a minor portion of the sales of most suppliers. Accordingly, there is limited scope for the growth of industrial networks or development blocks within the system. Such growth is further constrained by a particular feature of the Swedish PT system, viz. its narrow 'components waist' in the key area of sintered products (cf. Figure 7-7). A significant strengthening of domestic user-supplier links would thus require a broader and more R&D-intensive materials industry base in that area. Efforts to expand the national supplier base beyond its present areas of (largely P/M) specialization and to develop the PT competencies of the user industry would, however, be hampered by another structural characteristic of the system. This is the 'diagonal pattern', noted earlier, where a concentration of industrial activity on the P/M side is combined with a concentration of infrastructure support on the ceramics side. Attempts to branch out from existing technological strongholds and product niches into other areas of the P/M field would clearly benefit from a more evenly balanced infrastructure. Finally, a general weakness of the infrastructure - not only in powder technology, but in other fields as well[16] - lies in the small size and limited internal research funds of most academic units. Lacking adequate resources and 'critical mass', it is difficult for these units to maintain a standby capacity in areas which are not targeted in current programs, to effectively engage in broad-spectrum monitoring, and to respond in a flexible and timely manner to new possibilities and demands.

SHAPING THE TECHNOLOGICAL SYSTEM

In the conceptual framework employed in this study, a national program supporting the development of a particular technology can be seen as an intervention in the corresponding technological system. The assumption here is, of course, that supportive policies are not devised at zero point and that some system, however embryonic, will be in place at the time of implementation. Note also that a policy agency which enters into a given field will ipso facto be an actor in the system of that field (cf. Figure 7-4). The aim of the intervention, then, is to strengthen the system in some respect, e.g. by adding missing parts or reinforcing weak parts, or by promoting the development of new or tighter links among the different parts of the system. The systemic scope of the policy measures taken may vary greatly, from highly selective programs focusing on particular actors (such as a strategically placed institute with possible bridging functions), through mid-range programs designed to strengthen certain sectors (such as the academic sector or the infrastructure at large, in the case of a typical technology-push program, or the user industry, in the case of a program emphasizing technology diffusion and information support), to 'holistic' programs where a combination of measures is applied with the aim of improving the performance of the technological system as a whole. Obviously, the collection of policy tools that might be used in a program of the latter type is very large. It is reasonable to suggest that any

---

16 See Granberg (1995); for an appraisal of the conditions obtaining in one of the PT subfields, see STU (1991).

program, whatever its scope, should be based on a careful appraisal of (i) the targeted technology: its structure and dynamics as an evolving transnational body of knowledge, and its economic and social significance as an expanding set of practical applications and products; and (ii) the technological system: its anatomy as an institutional complex, held together by a variety of actor networks, and its strengths and weaknesses as a national entity engaging in the development, diffusion, and exploitation of the technology. As already mentioned, the field of powder technology is now, in Sweden, the target of a major national program entitled "Powder Technology 90" (or "PT-90"). In this section, the main premises and objectives of the program will be outlined along with some of the problems encountered in the course of its development and early implementation.[17] The examination of the program will be guided by the basic assumption that we are dealing here with an attempt to shape, in certain ways, the Swedish PT system.

The discussions that were eventually to produce the first drafts of a national program in the PT field began at STU, the Swedish National Board for Technical Development, in 1987. Although the talks were initially confined to the question of a successor to the ongoing Engineering Ceramics Program, the concept of a broader support effort that would encompass powder metallurgy was introduced and accepted at an early stage. The expansion was grounded in the expectations, or hopes, that the affinities and overlaps between the technologies of engineering ceramics and powder metallurgy would provide a basis for mutual support and cross-fertilization, and that the infrastructure of the P/M field would benefit from a closer association with the more dynamic and academically better established research activities in the ceramics field. There was also the practical consideration that a single, integrated program would be particularly appropriate in the case of a small national PT community where separate, small-scale programs may be difficult to justify and where the organizational and managerial tasks of a joint effort would not be overly complex.

After an extended period of preparatory study and discussion, followed by a shorter planning phase, PT-90 was officially launched in January 1991. The final version of the program was, of course, in some respects a result of practical compromises and adaptations; it can also, however, be looked upon as a conclusion drawn from appraisals of both the technological field and the national technological system. Apart from the P/M-ceramics connection, the features of powder technology that were emphasized were certain basic traits shared by many materials fields, viz. the 'enabling power' or potential pervasiveness of the technology, with implications for industrial competitiveness and economic growth; its interdisciplinary character and the attendant need to promote the interactive development of a large variety of technical and scientific specialties; the importance - in particular projects - of fostering close links

---

[17] This section is largely based on program documentation and interviews with the planners and R&D practiticioners involved in the development and implementation of PT-90.

between materials R&D, production engineering, and product design; and the long development times involved and the need for patient money and risk sharing.

As regards the Swedish situation and the evaluation of the national PT system, the main conclusions arrived at can be summarized as follows. First, concerning industry, it was noted that Swedish materials producers had built strong positions in parts of the P/M field, especially in iron-based powders and hard materials, and that an advanced HIP competence relevant to both powder metallurgy and ceramics had been acquired. It was also maintained that the domestic user industry was potentially large, but that the exploitation of the technology was seriously impeded by insufficient knowledge about the properties and possibilities of PT-based materials. As to the infrastructure, the view was expressed that the breadth and volume of education and noncommercial research in powder metallurgy were not adequate to the task of maintaining and expanding the industrial strongholds created in that field. Although the conditions obtaining in the ceramics field were considered less problematic, the judgement passed on the PT infrastructure as a whole was that a broader and more durable knowledge base, underpinning the activities of industry, was needed.

The goals and guidelines of the national program subsequently agreed upon are largely derivable from these premises. A detailed examination of PT-90 cannot be undertaken here; only a brief outline will be given of its core concepts. Thus, as regards the principal aims of the program, what stands out in the final documents and public announcements are the points (i) that the integration of materials development, product design, and production engineering will be promoted; (ii) that particular emphasis will be placed on the needs of the user industry; (iii) that R&D linkups and information exchange between materials users, on the one hand, and materials producers and materials researchers, on the other, will be supported (this is referred to as the "bridging of the technology gap"); (iv) that the infrastructure will be reinforced, especially on the P/M side, and closer links developed between the infrastructure and industry; and (v) that, in dealing with these tasks, an effort will be made to capitalize on existing technological strengths and exploit the possibilities offered by an integrated P/M-ceramics program.

The means by which the program objectives are to be accomplished are multiform. However, three main types of policy measures can be distinguished. The first and major component of the PT-90 strategy is the support of collaborative R&D projects involving firms, institutes, and academic units. The projects, which should be product-oriented and relatively large-scale, will be jointly funded by government, i.e. STU/NUTEK (in 1991, STU merged with two other government agencies and became NUTEK, the National Board for Industrial and Technical Development), and industry. A second component, closely connected with the first, is the creation of a set of university-based centers - called Powder Technology Centers (PTCs) - with membership open not only to academic departments, but also to institutes and firms. The centers are to play a key role in the initiation and performance of collaborative R&D as well as in the provision of research training. The third component, complementing the R&D-related initiatives of the first two, is the development and implementation of sub-programs in the areas of education, technology diffusion, and information support. The proposed measures are directed primarily towards the present and future design engineers of the user industry. As oftentimes happens, the financial design put forward by the planning group of PT-90 had to be reworked. Thus, the funding request that entered the internal review and budgetary process of STU/NUTEK was for agency allocations totalling 150 million Skr (approx. $25 million) over a five-year period, with the assumption that an additional 120 million Skr would be provided by industry. What emerged from the process was a reduced and modified scheme, with agency contributions amounting to 119 million Skr distributed over seven years. It also became increasingly clear, as the program moved through the planning phase into early implementation, that the principle of joint government-industry project funding on a fifty-fifty basis would have to be handled in a flexible, pragmatic manner. Concerning the time frame of PT-90, one may ask why a longer-term support effort (say, over a 10-year period) was not proposed. One answer is that the PT-90 planners realized that such a proposal would have represented a radical, 'overambitious' departure from the established program format of the agency and that approval would have been unlikely. In addition, it was assumed that it would be possible - following a common policy practice - to incorporate important extensions and follow-ups in later programs.

What are the chief characteristics of PT-90, viewed as an intervention into the national technological system? It is evident, first of all, that we are not dealing here with a typical technology-push program, or an attempt to stimulate innovation by way of enhancing the infrastructural supply of technologies and competencies, even though measures towards that end are included among the policy tools of PT-90. Nor can it properly be described as a demand-pull program designed to generate user demand for the relevant technologies and competencies, although user support and knowledge transfer are indeed important elements of the program. In other words, PT-90 is not an example of a selective policy intervention, focusing on a particular part or sector of the technological system. Instead, the basic aim of the program is the more comprehensive ('systemic' or 'holistic') one of strengthening the system as a whole by attending to all of its major components and the way in which they interact. The breadth of the enterprise manifests itself in different dimensions: (i) horizontally, in the coverage and hoped-for cross-fertilization of powder metallurgy and engineering ceramics; (ii) vertically, in the ambition to achieve closer integration across the production chain and to engage the entire range of actors, from infrastructure units to user-industry firms; and (iii) with regard to the spectrum of activities involved which, as we have seen, encompasses the basic categories of R&D, education, and technology diffusion. Given the breadth and systemic orientation of the program, it is hardly surprising that it is also pervaded by a strong concern with the bridging of internal gaps and with the development and effective use of actor networks, notably networks of the problem-solving type.

The framing and implementation of a technology support program of this nature poses particular problems and challenges. For instance, in dealing with a relatively large and heterogeneous collectivity of actors, a variety of - sometimes conflicting - interests and demands must be harmonized and coordinated. The active cooperation of industry, including the commitment of corporate resources, must be ensured. The quandaries and constraints faced in seeking to advance a national system that is highly dependent, economically as well as technologically, on the international environment must somehow be coped with. The participants in the program must be provided with a clear and coherent picture of what is, in fact, a complex, multi-purpose and multi-level undertaking - and so on. To this should be added the consideration that a program such as PT-90 is likely to be more sensitive to unanticipated changes in its design premises (e.g. in the form of budget reductions or the withdrawal of certain actors) than is true of the more traditional, push-oriented materials technology programs run by STU/-NUTEK. In the latter case, where the programs are largely composed of independent projects carried out by academic and institute researchers, such changes can usually be absorbed without major disruptions; in the former, however, the result may be that vital parts or linkages in a promoted system/network will fail to materialize.

In the course of preparatory study and planning, the relative emphasis placed on different program areas and program concepts has shifted along certain lines. Thus, the weight attached to the engineering-design perspective and the role of the user industry has tended to grow at the expense of producer-industry concerns. At the same time, the theme of the so-called technology gap (separating users from producers-cum--infrastructure) has become increasingly prominent while that of the P/M-ceramics nexus has become more subdued. As to the balance struck between the two subfields of powder technology, a shift in the overall support effort favoring powder metallurgy is discernible; this shift is, of course, closely connected with the aforementioned changes. We may also note, without going further into the matter, that both the organizational structure and the specific policy measures of PT-90 (e.g. the different types of R&D projects envisaged) have been redesigned or modified at various points in the development process.

The development of any technology program can be considered from many different angles. It may, for instance, be viewed as a creative process where ideas are generated, elaborated and evaluated in a continuing discussion among program planners, consultants, potential participants and others, though one or a few individuals with imagination and relevant experience are likely to play a key role; it is not difficult to identify such individuals in the case of PT-90. It can also be seen as a 'political' process involving conflicts and compromises between representatives of different parts of the technological system, pursuing their special interests as they seek to influence the definition of the program; in the short history of PT-90, front lines of this kind have become apparent, e.g., between powder metallurgy and ceramics, and between academia and industry. Yet another way of looking at the development of a program is to focus on the elements of learning and adaptation - of learning about the technological system as well as the political and economic limiting conditions of a support effort, and of adapting the program design(s) to the perceived realities. The evolution of PT-90 readily lends itself to an interpretation in these terms. Thus, it is clear that the basic program concept and its specification through successive design variants are to a large extent grounded in the insights and experience accumulated by the planners concerning the technological system and the various constraints faced by the policy initiative. The 'diagnosis' of the strengths and weaknesses of the system, arrived at by way of preparatory studies and consultations, is a case in point. The lowering of expectations regarding the financial role of industry and the reconsideration of funding principles can be seen as adjustments to a constraint (the reluctance of industry to firmly commit itself) that became increasingly pronounced in the discussions held with potential corporate participants. Reinforcing that constraint was a major change in the general economic framework of the policy initiative: preparation of the program took place during a boom period - implementation, however, began in a deepening recession.

While some of the difficulties encountered are attributable mainly to changes in the external conditions (a further example of this is the budget reduction suffered by PT-90), others are more clearly related to certain features of the program and/or the technological system. The latter category encompasses a great variety of problems, large and small, experienced - or, at the present stage, anticipated - by planners or participants or both. However, a rough division can be made into two major problem complexes. One of them is centered around the question of the articulation and communication of the program content. It stands to reason, considering the systemic and network-building ambitions pursued, that it is of vital importance that all parties involved be provided with a coherent and sufficiently detailed picture of the planned effort, including its specific ends and means as well as its overall orientation, structure, and rationale; a picture that will clearly indicate the roles to be played within the larger context by different actors and types of activities, from user-firm design engineering to academic materials research. It is also important that information about changes in some part/aspect of the proposed program - such as the project guidelines - is effectively conveyed to the participants. Given the scope and complexity of PT-90, the task is not a trivial one. In the planning and early implementation phases of the program, shortcomings in the performance of these functions have been a source of considerable friction, impeding the mobilization of actors and the organization and startup of specific activities. Thus, in all sectors of the national PT system, a majority of the participants interviewed report having experienced significant problems in obtaining up-to-date information about the general design of the program and/or the particular means of implementation.

The second problem complex arises from the confrontation of certain major program aims with certain structural characteristics of the technological system. PT-90 has been characterized here as a broad, systemic policy intervention that seeks to promote integration within the materials-design-manufacturing triangle, along the production chain, and across the "technology gap". These aims, we should note, are to be accomplished by steps and measures that fall within the proper domain of action of a national (rather than international) technology policy agency. What makes the enterprise appear problematic - and in certain respects questionable - are mainly the following features of the Swedish PT system.

First, the ambition to integrate production chains and to bridge technology gaps by promoting cooperation and exchange among relevant actors, and to do so by drawing, as far as possible, on existing industrial strengths, must be set against the fact that activities in the area of hard materials (the overwhelmingly dominant area in terms of industrial output and world-market shares) are essentially in-house; the steps from materials preparation to final product, that is, are controlled by a single firm, the typical products being end-user tools rather than components for client-firm systems. This means that the need for integration and 'gap filling', should it arise, is most likely to be dealt with as an internal company issue and not as a problem of external linkage or networking.

Second, these program objectives must also be confronted with the fact that the technology gaps which (presumably) hamper the user industry are bound, to a large extent, to appear at/outside the limits of the national PT system, due to the heavy dependence of that industry on foreign suppliers. Such international technology gaps are not, however, directly targeted by PT-90, though the expectation is expressed that the monitoring of the international research front and the transfer of basic technology will be enhanced by way of infrastructural support. A significant reduction of the external dependence of the user industry would require a broadening - underpinned by relevant infrastructure efforts - of the domestic materials-industry base into such materials areas as superalloys and aluminum and titanium alloys, or such process areas as powder forging and metal-injection moulding. The ambition of PT-90 is indeed to encourage certain expansions of this kind, at least as far as the acquisition of basic competencies is concerned. There can be no assurance, however, that program measures towards that end would in fact, in a longer-term perspective, lead to industrial production or other commercial activities. The question also remains open whether the program might, in some way, alleviate the problem of the narrow 'waist' of the Swedish PT system in regard to sintered steel components; as noted, the absence of a strong producer of such components, with a significant in-house R&D capability, is a major impediment to the development of vertical, economic and problem-solving links/networks within the system.

Third, the systemic or integrative aims of PT-90 can be seen as encountering more favorable conditions of realization in the field of powder metallurgy than in engineering ceramics. The special problems that arise in the latter case are mainly due to the relative immaturity of the field, or the fact that the user industry is still in essence a collectivity of potential users and that the (small) materials industry still has little to show in the way of marketable products. The technology gap is therefore wider and deeper and more difficult to bridge. Not only is there a general lack of relevant knowledge and experience on the part of the users - and a correspondingly weak articulation of needs - but there is also an inadequate supply of information from the side of the materials researchers and engineers; the materials experts are still, in many instances, unable to provide reliable data of the kind requested by the design engineers. Regardless of how the specific issue of the ceramics-technology gap is dealt with, an adaptation of PT-90 to the differential conditions of the two subfields seems unavoidable, such that a stronger element of technology push be allowed for in the ceramics-related parts of the program.

In a national policy perspective where the maintenance or enhancement of industrial competitiveness and economic growth are focal concerns, the general orientation of the program and the objectives discussed above may seem well-chosen: the strengthening of complementary competencies in the national PT system and the promotion of networks or development blocs exploiting the complementarities is to ensure that the employment- and income-generating potential of the field is realized, as far as possible, within the country. One should bear in mind, however, that Sweden is a small nation and that there are limitations to this policy concept which must be accepted. Thus, while a broadening of the national competence base in PT materials and PT process appears both desirable and feasible, it would be unrealistic to expect a materials industry to evolve that could cater to all PT needs of the user industry. Also, the possibilities for materials firms to grow on the strength of the home market alone are generally small. This means that the import/export dependence of the user/supplier industry is likely to remain high - even if the program is successfully implemented - and, consequently, that a significant proportion of the problem-solving partners of the firms will be located outside the national PT system. Although collaboration and exchange within the system is promoted by the program and is facilitated by geographical and cultural proximity, the actual partnership choice must, of course, be made by the individual actor.

CONCLUDING REMARKS

One aim of this paper has been to put forward, albeit in an indirect manner, a conceptual approach to the exploration of certain cognitive and institutional aspects of the development of modern, science-based technologies. The basic elements of the approach - to recapitulate - include the general view of technology as knowledge and the notions that technology advances by problem solving; that technologies can be decomposed into clusters of problems and related competencies; that problem solving takes place in an institutional entity - the technological system - comprised of partially interconnected organizations or actors; and that national support programs targeting a given technology can be interpreted as interventions in the corresponding national system. These concepts and ideas have been applied in a study of powder technology in the setting of Sweden, focusing first on the cognitive field, then on the technological system, and finally on a relevant policy intervention, now in an early stage of implementation.

It is hoped that the case study will have demonstrated the usefulness of the conceptual scheme in mapping the composition and structure of the technological field and the technological system, and in elucidating the possibilities and limitations of the national support program - but that is for the reader to decide. A concluding comment seems appropriate, however, on the purely qualitative nature of the analysis carried out in the preceding. Two features, both of which could be considered problematic, may be noted. First, it is clear that this is an approach which calls for a determined and time-consuming effort on the part of the analyst to open the technological 'black box' and examine its contents. Depending on the size of the technological system, the task of identifying the actors and collecting the requisite information may also be a laborious one. Secondly, it is evident that an element of subjectivity must be accepted in an analysis of this kind, due to the inherent difficulty of rigorously assessing the validity of conclusions arrived at by qualitative modes of reasoning. Although one cannot hope to eliminate these problems, it may be possible to reduce their significance. In principle, this could be achieved by the collection and analysis of quantitative data that would complement and/or partially replace the qualitative material that would otherwise be the exclusive empirical base of the study. The mapping of the field and system of a given technology could be facilitated and enhanced by, or checked against, bibliometric data and patent statistics; R&D and industrial-economic statistics, when available, could be drawn upon in similar ways. Whether and to what extent this is feasible depends, of course, on the particular case at hand. It is the intention of this author to explore these possibilities - especially with regard to bibliometrics - in a study of another advanced-materials field, viz. that of high-temperature superconductors.

REFERENCES

Ashby, M.F., 1987. "Technology of the 1990s: Advanced Materials and Predictive Design," *Phil. Trans. R. Soc. Lond.*, A 322, 393-407.

Carlsson Bo (ed.), 1995. *Technological Systems and Economic Performance: The Case of Factory Automation.* Boston and Dordrecht: Kluwer Academic Publishers.

Cohendet, Patrick, Marc J. Ledoux and Ehud Zuscovitch, 1988. *Advanced Materials: Economic Dynamics and European Strategy.* Berlin: Springer-Verlag.

Forester, Thomas (ed.), 1988. *The Materials Revolution.* Oxford: Basil Blackwell.

Granberg, Anders, 1987. "A Conceptual Framework for the Study of Generic Fields of Technology." Research Policy Institute: Lund.

Granberg, Anders, 1991. "Aspects of Monitoring Developments in Structural Materials R&D", *Research Evaluation,* 1 (2), 107-114.

Granberg, Anders, 1995. "Mapping an Evolving Technology Cluster: The Composition and Structure of Factory Automation," chapter 4 in Bo Carlsson (ed.), *op. cit.*, 89-117.

Granberg, Anders, 1995. "The Academic Infrastructure of Factory Automation," chapter 5 in Bo Carlsson (ed.), *op. cit.*, 119-164.

Kingery, W.D. (ed.), 1986. "Ceramics and Civilization: High-Technology Ceramics, Past, Present, and Future", vol. 3. Westerville, OH: American Ceramic Society, Inc.

Laage-Hellman, Jens and A. Waluszewski, 1992. "A Pilot Study of the Powder Technology R&D Network - the Industrial Perspective," IM-Group, Uppsala.

Kranzberg, Melvin and Cyril Stanley Smith, 1988. "Materials in History and Society", in Thomas Forester (ed.), *op. cit.*, 85-118.

Metals Handbook, 9th Ed., Vol. 7: Powder Metallurgy, 1984. Metals Park, Ohio: American Society for Metals.

Roll, Kempton, 1984. "History of Powder Metallurgy", in *Metals Handbook*, op. cit.

STU (Swedish National Board for Technical Development), 1991. *International Evaluation of Swedish Structural Ceramics Research*, STU-1991.04.20. Stockholm: STU.

STU (Swedish National Board for Technical Development), 1991a. "Resultatredovisning: Vad STU gjort för utvecklingen av materialtekniken och dess användning" (Reporting Results: What STU Has Done for the Development of Materials Technology and Its Use), *STU-info* 839-1991.

Theulon, H., 1989. "High Technology Materials, Recent Materials," *STI Review*, 6 (December), 7-72.

Turner, Calum, Robin Roy, and David Wield, 1990. "Materials: A New Revolutionary Generic Technology? Conditions and Policies for Innovation," *Technology Analysis & Strategic Management*, 2 (3), 221-233.

8

GENERAL PURPOSE TECHNOLOGIES, INDUSTRIAL COMPETENCE AND ECONOMIC GROWTH
– with Emphasis on the Diffusion of Advanced Methods of Integrated Production*

Gunnar Eliasson

SUMMARY

This chapter surveys the recent flow of literature on spillover effects to all industry from R&D spending in specific industries and firms. Particular attention is paid to spillovers from so-called high-tech industries to other industries. Spillover effects are significant and appear solidly established by accepted econometric methods. The interpretations can, however, be discussed, based as they are on the naive R&D notions of 'modern' macro growth theory. This paper takes the interpretation down to the micro level, within firms, *to study the channels through which technology is diffused.* Only at that level can economic growth be understood in terms of the nature of R&D spending, the nature of human-embodied competence in technological change, and market competition in the local innovation and diffusion process.

R&D spending in manufacturing consists largely of the cost of acquiring (*imitating* and *introducing*) new technology developed in leading-edge (high-tech) firms. Such

---

* Earlier versions of this paper have been discussed at seminars at IUI and the Department of Industrial Economics at the Royal Institute of Technology (KTH), Stockholm, and at Saab Aircraft. This is an updated version of the paper published under the same title in Dec. 1994 as TRITA-IEOR 1994:2, KTH, Stockholm. A similar but shorter version has been published as "Spillovers, Integrated Production and the Theory of the Firm," *Journal of Evolutionary Economics*, 6 (2), 1996, 125-140.

I am particularly grateful for discussions with and comments by Bo Carlsson and Ishaq Nadiri who have read the whole paper. Comments by Pontus Braunerhjelm, Lennart Elg, Billy Fredriksson, Ulf Edgren, Anders Granberg, Kai Hammerich, Staffan Jacobsson, Bo Karlsson, Staffan Laestadius, Rikard Stankiewicz, Per Storm, Pentti Vartia and Synnöve Vuori have helped me to improve this final version.

firms, developing and using frontier technology in their production, therefore often serve other firms as 'higher institutes of technical education and research' and 'infrastructure suppliers' of complex applications-oriented production and product development know-how, which regular technical universities are incapable of providing. This paper, hence, is chiefly concerned with the diffusion of advanced production and product development know-how in large-scale production, employing the bulk of the manufacturing labor force, and less with innovation-driven R&D and growth through entrepreneurship and the establishment of new firms. The new econometric results on spillover effects give credence to *advanced private R&D spending* as a generic infrastructure capital.

Diffusion of new technology occurs through the application of best-practice techniques by advanced firms in the production of other firms along four avenues: (1) the *mobility of competent people* throughout industry, (2) the growth of technological applications-oriented *competence blocs* around particularly advanced firms, (3) the introduction of new technologies in other firms, and their effects on existing production and (4) learning *in subcontracting* firms through competent purchasing or vice versa, *from* advanced specialist subcontracting firms or engineering consultants. From a national policy point of view, it is, of course, important that new technology is diffused to domestic firms rather than to firms in competitor nations. This is, however, a narrow focus of policy concern since all advanced economies (including the United States) draw significantly on foreign sources of technology. Advanced technology is very much an *international pool of opportunities available only to those able to access it*. To be successful, firms must develop the *receiver competence* needed to transform externally available technology to internal industrial production competence.

Most econometric studies reported on are based on industry data, not firm data. The results are reviewed here as reported in the original source, but they are also reinterpreted from the micro-to-macro perspective. In doing so, I take as examples *integrated production* and the systems competence to organize complex networks of engineering consultants and subcontractors. Information technology and the aircraft industry are used for illustration. The aircraft industry in particular represents frontier technology in mature engineering industries, the industrial backbone of large firms in the mature industrial economies. *The aircraft industry already today employs the tools, materials and product technologies that engineering firms generally will have to know in the future*. The analysis shows the importance of other forms of knowledge capital than R&D investment, and of dynamic market mechanisms to support the diffusion of new technology. The most important such mechanism is the *market for competence* embodied in people, the labor market.

One implication of the studies reported on is that the educational performance and the capacity of labor markets to allocate competence are decisive for the growth and economic welfare of nations. The educational attainment is intimately tied to the technological performance, the most advanced firms providing valuable educational services to all other firms. *Spillovers are technical and commercial as well as organizational*. Only the internationally most competitive firms diffuse internationally competitive technology to domestic firms, and *efficient learning by firms can only occur in competitive markets. Firms improve their performance in response to challenges from the best competitors*, and they do it predominantly *through organizational change.* Leading-edge firms are created only in direct confrontations with the best producers in international markets. One important conclusion of this analysis is that *to become and stay excellent in one industry, such as mechanical engineering, the country must have some of the internationally most advanced producers of that industry represented domestically, as educational and research institutions.*

INDUSTRIAL GROWTH BASED ON SELECTION, COMPETITION AND ORGANIZATIONAL CHANGE - THEORETICAL BACKGROUND

Standard economic analysis of economic growth assumes given institutions and a given organization of the economy and then looks for a technological explanation. This approach reflects a misunderstanding of how new technology[1] is introduced in firms. Productivity growth at any level of aggregation depends on how competition in markets forces firms to improve their resource allocation. Since firms use organizational change as an instrument to improve performance, changes in institutions and the organization of markets become critical elements in macroeconomic growth. Technology sets the upper limits to what a firm or an economy can achieve. Where underneath these upper limits the firms or the economy operate is determined by the efficiency of markets to allocate resources (Eliasson 1978, 1994d). *Selection* mechanisms play a critical role in that process. Hence, there is *no way to keep technological and economic performance apart.* Let me explain in terms of Figures 8-1 and 8-2.

Figures 8-1 and 8-2 show Swedish production units (divisions or firms) in manufacturing in 1991 and 1993 ranked by labor productivity.[2] Each firm is represented by a vertical bar (four of which are shown in the figures). Apparently, the 'diversity' of this population of firms is great. Some of this diversity can be explained by differences in capital intensity. The difference between the value added per employee (i.e., labor productivity) and the wage cost level (the almost horizontal jagged curve in Figure 8-2)

---

1 Technology, knowledge, and competence are terms frequently used in this paper. *Technology* will always be used in its original meaning as knowledge about technical matters. *Competence* emphasizes the ability to apply such knowledge. It can be embodied in human beings, teams, and organizations. *Information* is used to mean communicable technology and knowledge as distinct from *tacit* knowledge that cannot easily be communicated. But communication requires a *receiver competence* in the human being or organization. In order to reduce the terminology I have chosen to broaden the concept of technology to include also *organizational technology*. It is probably here that confusion may arise, but this usage is perfectly logical. Reorganization is a method used by firms to raise performance in response to competitive challenges. An important part of the story of this paper is *that successful reorganization requires significant management competence*. In a factory context, the technical content of organizational techniques becomes obvious. For a discussion of these concepts, see Eliasson (1990b, Chapter I).

2 Curves such as those in Figures 8-1, 2 and 3 are referred to as Salter curves, following Salter (1960).

is the gross operating profit per employee. Some of this profit is necessary to cover capital costs. But after having corrected for that and having ranked the firms by rates of return (Figure 8-3), the performance spread is still enormous. A reshuffling of resources (for instance of labor) from the worst to the best existing performers would produce massive improvements in macroeconomic performance. Hence, competitive mechanisms that generate such reallocation of resources cause economic growth (Eliasson 1991b). This is exactly what occurs in a macro economy with functioning markets that push and induce resources to move. The fact that this reallocation does not occur immediately, weeding out all low performers and flattening the Salter curve such as those in Figures 8-1, 8-2, and 8-3, is explained by lack of competence in firms to implement new technology (receiver competence - see Eliasson 1986a, 1990a).[3]

Figure 8-1 Ranking of Swedish Manufacturing Firms in Terms of Labor Productivity, 1991 and 1993 (output in 1980 prices)

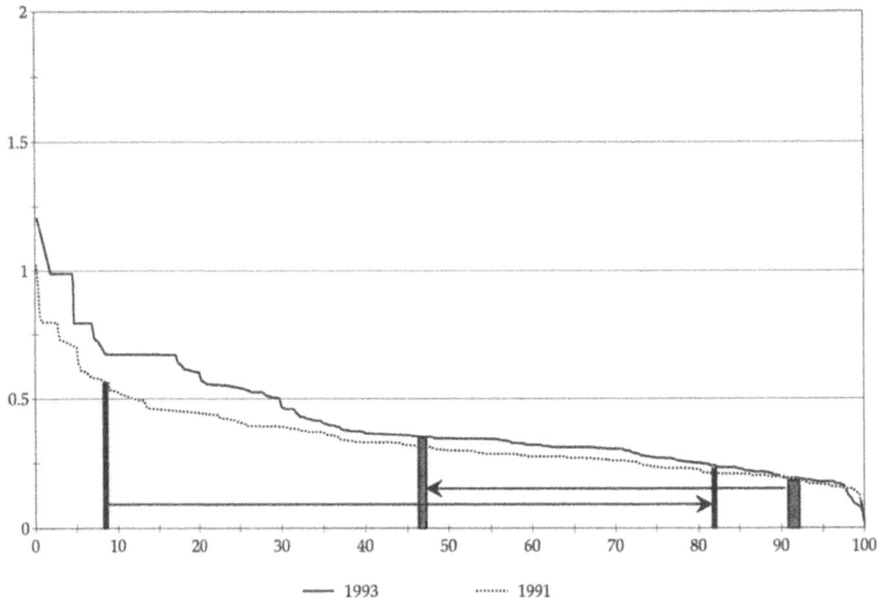

Note: The four columns represent two firms that have increased (shaded) and lowered their relative performance, respectively. The same firms also appear in the following diagrams.

Source: Albrecht *et al.* (1992).

---

3 There is also the more subtle additional explanation that a dynamic system has to exhibit significant structural diversity as in Figures 8-1, 8-2, and 8-3 to remain stable (Eliasson 1984, 1991b). It will collapse if the Salter curves somehow become too flat. In the real economy there is no such thing as a static equilibrium.

Measured ex-post performance, as in Figures 8-1 – 8-3, in fact, strongly underestimates the true ex-ante performance spread, because it does not show new, not yet introduced technology that will upgrade performance through new *entry*. Competition through new entry and reorganization of existing firms also force low performers to exit, freeing resources for more efficient use elsewhere.

Figure 8-2 Ranking of Swedish Manufacturing Firms in Terms of Labor Productivity and Wage Costs, 1991 and 1993

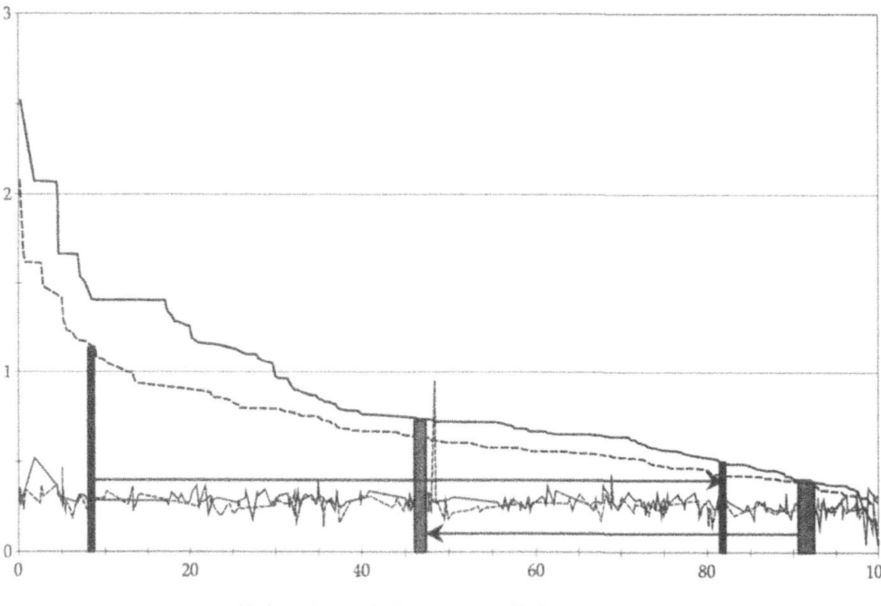

Source: Planning Survey (see Albrecht *et al.* 1992).

Thus, the upgrading of economic performance occurs through four different investment mechanisms that can only be understood if explained in terms of *market competition* at the micro level (Eliasson 1994b, 1995b). See Table 8-1.

Table 8-1 The Four Investment Growth Categories

1. Entry
2. Reorganization of existing firms
3. Rationalization of existing firms
4. Exit

Source: Eliasson (1992, 1993a).

Each firm (identified by a vertical bar in the figures; the rankings of two firms are indicated) is threatened from all directions. The better firms (high up to the left) can pay more for resources through bidding up prices. The not-as-good performers can improve and become better, challenging firms that currently happen to be superior. This occurs through *reorganization* and the introduction of new technology and through *rationalizing* existing production lines. Reorganization often extends beyond the boundaries of the firm, involving acquisitions and spinoffs of parts of firms or whole firms intermediated in the market for mergers and acquisitions. But also *new firms* can *enter* from the upper left and establish themselves as autonomous competitors, and firms that cannot cope fail and *exit* (a disinvestment) at the low, right-hand end. Hence, *competition causes economic growth* (Eliasson 1993a,c, 1995b,c) *through the four investment mechanisms* in Table 8-1. The intensity of competition in each market then depends on the shape of the distribution of firms in terms of performance, the upper left being important for setting high standards and the long tail determining how smooth to the macro economy the 'creative destruction' or the exit process will be.

Figure 8-3 Rankings of Swedish Manufacturing Firms in Terms of Rate of Return, 1991 and 1993

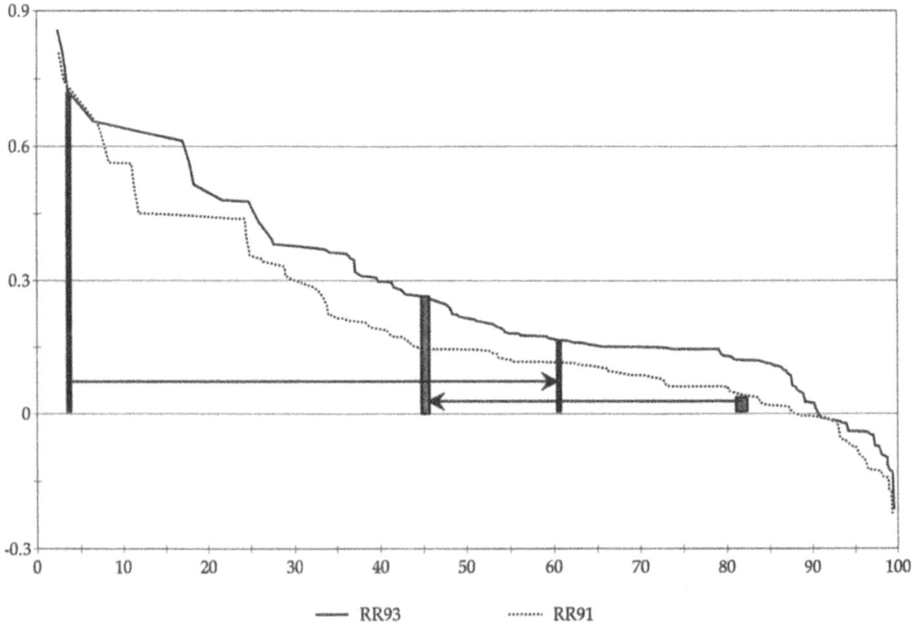

The capacity to innovate, improve, and compete hinges critically on the abilities of firms to *learn* and to *combine* techniques innovatively. This in turn depends very much on the *diversity* of the technological and commercial environment of each firm. In a viable market, the top commercial performers set the standards. When the whole population of firms is homogeneous, the competitive pressure is low and there is less to learn than from a rich, diverse and competitive environment. If the market lacks top performers, the standards are set too low. Also, the ability of the whole downstream range of firms to learn from the best performers requires significant *receiver competence* which is typically accumulated through R&D investment (Eliasson 1986a, 1989, 1990a,b). Pavitt (1979) observed that in the mid-1970s Swedish industry exhibited a range of technologies far greater than in most industrial nations, and only slightly smaller than in the three leaders, the United States, Japan, and Germany. Pavitt noted a heavy concentration of technological know-how in mechanical engineering industries, quality steel, forest products and telecommunications. Judging from Jacobsson and Philipson (chapter 2 in this volume), using a somewhat different methodology, Swedish manufacturing firms reorganized between 1981 and 1989 and narrowed their focus to metals and mechanical manufacturing (the technology upon which the industrial revolution was once founded), pharmaceuticals (a new technology), and telecommunications. The consequence has been a reduction in the diversity of the domestic learning environment for all Swedish firms. At the same time, this development also reflects the success of giant international firms in the 1980s, based on global economies of scale, a development contrary to that in other nations (Carlsson 1992).

This reorganization reflects the balancing of specialization and utilization of national comparative advantage on the one hand and various forms of economies of scale on the other (see Table 8-2). Firms can combine factor inputs to gain internal economies of scope, and systems of firms can support each other through technological spillovers (infrastructure effects). Such systems can be configured to make innovation and organizational learning more efficient than if each agent had to operate on its own. Endogenization of (3) in Table 8-2 within a financially defined cluster can be seen as the definition of a firm as a competent team (Eliasson 1990a). If a corresponding scale economy can be organized at the national level, there is a case for industrial policy through the creation of a *technological system*.

Table 8-2 Factors Determining the Organization of Production

(1) (National) comparative advantages or disadvantages
(2) Specialization along (1)
(3) Clustering of scale economies in firms (competitive advantages) or in technological systems

The competence of a firm to learn and innovate depends critically on its staffing and, hence, on the educational level of the working population and the possibility of firms to staff themselves with talented and well educated people. On this score the last decade shows a dismal development in Sweden; a lowering of the quality of Swedish education, and according to OECD statistics, one of the lowest shares among the industrialized countries both of university graduates by age group and share of engineers and natural scientists in the labor force. Recent statistics also show a continued disinterest among the young to take on the more difficult study programs in engineering and natural sciences (Eliasson 1993b, 1994a,c,e), a situation that Bishop (1989) suggests implies low *labor quality* in an international comparison and slow productivity growth.

The mechanisms of Table 8-1 are driven by dynamic *competition* in markets. To be effective, dynamic competition requires *free access* to markets through new entry and competitive reorganization and improvement (Eliasson 1991b,c). A large firm can launch a new product or start a new activity as a new division. An individual or a team of individuals can organize a firm to exploit a new idea or technology. These are two forms of *entrepreneurship* without which *competition* will be lacking, as pointed out already by Adam Smith (1776), and without which there will be little or no economic growth (Eliasson 1991c, 1993a, 1994b).

PROBLEMS, HYPOTHESES AND CONCLUSIONS

Joseph Schumpeter argued that new firms enter the market and old firms attempt to improve their performance, thus imposing harm on those firms that can not improve, often forcing them to exit from the market. Schumpeter regarded this 'creative destruction' as an important positive mechanism in economic development.

Caballero and Jaffe (1993), taking note of Schumpeter's idea of 'creative destruction,' add as a second feature a 'public-good' aspect of Schumpeter's innovators: they create economy-wide increasing returns (see also Eliasson 1989; Romer 1986). Caballero and Jaffe (1993) then, however, go on to straight-jacket their insights into a neoclassical general equilibrium setting. They estimate their model and come up with some startling results. The annual rate of creative destruction of capital was in the range of 2 to 7 percent for the 1970s. Technological obsolescence has increased over the century from about 3 percent per year to about 12 percent per year in 1990.[4] They also find that the rate of diffusion of knowledge is rapid with an average lag between one and two years.

Finally, they note that the potential of spillovers from new ideas to new knowledge generally has been declining over the century, meaning that a decline in the effective potential stock of knowledge has taken place, leaving less to new investors to build on and resulting in an observed decline in the average private productivity of industrial

---

4 A similar result on physical capital for Sweden 1870-1975 was obtained by Bentzel (1980).

research. To maintain the same rate of economic growth as before, a growing R&D share in output is needed.

Caballero and Jaffe (1993) tell a story of increasing costs of supporting future growth in high value-added production, of the importance of competition, not least to force resources out of low-productivity activities, and of the increasing difficulties of supporting the relative wealth of an already rich nation (see Eliasson 1989). What does this mean for the industrial innovation base of an industry dominated by large firms in old technologies? Are the rich industrial nations facing (1) a traditional problem of diminishing returns to R&D investment, (2) a substitution of R&D for hardware machinery , or (3) a relative fall in the returns to R&D investments due to learning of the same technologies of competing firms in other nations?

By contrast, Nadiri (1993) finds (1) that here is no credible evidence of diminishing returns to R&D investment and inventive activity in general, (2) that there is a strongly positive relationship between R&D investment and total factor productivity growth in firms, and (3) that there appear to be strong R&D spillover effects at both the firm and industry levels. The implication is that the potential for future rapid total factor productivity growth among already highly industrialized countries should be great, provided they spend enough on R&D. This leaves the substitution hypothesis (2) above, which implies *industrial restructuring*.

The question then is (asked by Nadiri 1993, p. 31 ff.): with such high rates of return on R&D, why don't firms spend much more on R&D? *There is a clear case for underinvestment in R&D*, he concludes. Among the possible reasons he mentions risks and difficulties of appropriating the rents from new technology. To this we can add insufficient entrepreneurship and point to the differences in this respect between Europe and the United States: Europe is systematically incapable of capitalizing on the enormous international pool of new technology to build new industries.

The conclusion offered in this paper is that even if Nadiri (1993) is correct, the main reason for not seeing a full-scale realization of the indicated productivity potential of industrial R&D is the *lack of appropriate applications competence (receiver competence)* throughout industry and (perhaps) also a lack of incentives and competition making sure that the potential is realized. It appears that in new technology areas even the leading performers are not taking full advantage of the business potential opening up and the rest of related industries barely at all, due to lacking receiver competence (Eliasson 1994c). There could also be a case of social and political resistance (Eliasson 1994b) to the rapid structural change associated with rapid technological transformation of industries.

In the rest of this paper I will review research on the externalities associated with industrial knowledge accumulation from the perspective of *micro-based selection theory of economic growth* (Eliasson 1993c, 1995b, 1995c) in Table 8-1.

LOCAL LEARNING FROM THE INTERNATIONAL TECHNOLOGY POOL - THE ORIGIN OF TECHNICAL CHANGE

The post-war period has created a pool of new technology of enormous dimensions and diversity that has still only been marginally exploited commercially. Learning about the content of this pool of technology and its commercial applications takes place locally and for each agent (individual, firm) from a platform of more or less professional and always specialized competence (Eliasson 1990a, 1994b). Even firms in the large and most advanced industrial economies, including the United States, draw a significant part of their technology from abroad (Eaton and Kortum 1995a,b), and the ability to capture the opportunities in this international technology pool depends on their receiver competence. Many multinational firms have become specialists in that intelligence business (Eliasson 1991a).

A number of researchers have attempted to formulate more or less well-founded theories for how such technologies are turned into economically viable industrial expansion, most of them paying due respect to Joseph Schumpeter's ideas of innovation, entrepreneurship and creative destruction, and some of them using the idea of a synergistic development bloc.[5]

It is also well established that R&D spending in manufacturing is substantial compared to machinery investments and contributes significantly to total factor productivity growth (Mansfield 1968; Eliasson 1971; Bernstein and Nadiri 1989a). Furthermore, it is complementary to and exhibits higher returns on the margin than hardware capital. It is, however, also observed that social returns to R&D investments are far above private returns, reflecting a market imperfection in Arrow's (1962b) terminology, such that generated knowledge is not perfectly exploited industrially (Bernstein and Nadiri 1991). A large part of the literature is concerned with the theoretical and empirical aspects of this inefficiency. Nadiri (1993) observes the very large social returns to R&D investment because of spillover effects and suggests that it reflects a 'substantial underinvestment in R&D'. The Japanese post-war miracle, for instance, has been explained in many ways. For a long time the popular political explanation was based on the classical central planning argument: central coordination by MITI (the Ministry of Trade and Industry). MITI argued for the early systematic building of 'feeder' industries like steel and shipyards to supply cheap inputs for industrial expansion (The Boston Consulting Group suggested that for Sweden in 1979).

---

5 See Dahmén (1950), and Carlsson and Henriksson (1991). Researchers have a tendency to invent their own terms. Besides 'development blocs' we have 'industrial clusters', 'technological systems', 'technological networks' or 'spillovers', 'general purpose or generic technologies' etc. Arguments and analyses are rarely precise enough to warrant a new term for each author, and none of the new features of the concepts invented that I have seen are used to carry analytical results beyond the original idea of a 'development bloc' of Erik Dahmén or a 'technological system' (see Carlsson 1995b, and Carlsson and Stankiewicz 1991) or a 'technological competence bloc' (Eliasson 1995a), the latter two emphasizing synergistic competence building, while Dahmén's development bloc had a pronounced physical investment definition. Thus, whenever I use another term, it is only for verbal variation.

It has, however, been pointed out (Carlsson, Dahmén *et al.* 1979) that such central planning devices only work if industrialization imitates an earlier industrialization phase of an already advanced economy like the United States. When a country is operating on the technological frontier, on the other hand, new promising industries cannot be identified centrally, and attempts in that direction involve picking winners, for which the central public bureaucracy is quite incompetent (Eliasson and Ysander 1983).

Even if central guidance or planning have completely fallen out of favor (even with politicians) after the final collapse of the Soviet empire, subsidizing industry of all kinds still remains a popular show of industrial political prowess. Firms and industries are supported financially to (1) sooth interest groups such as farmers and unions, (2) hide unemployment or (3) support industries that are classified as advanced, hoping thereby to create advanced industries. The third category falls under the heading of this thesis. The British and French taxpayers lost billions on the commercially hopeless supersonic jetliner Concorde. The European Airbus project is receiving billions of ECU from Brussels to beat Boeing, the dominant producer of large commercial aircraft, said in turn to benefit tremendously from the building of military aircraft. National firms in high-tech industries increasingly become acquired by the state as they gradually fail, and so on.

The important question is whether such subsidies contribute at all to industrial high-tech performance (Fölster 1991). A related question is whether subsidies that include incentives to cooperate in R&D will increase R&D activity, notably high-spillover investments (the commitment hypothesis) and/or make it more efficient. Katz (1986) established theoretically that when there are spillover effects from R&D, when profit sharing is feasible, and when competition is low, then cooperation may be socially beneficial. But how can anybody know beforehand? Fölster (1993b) finds that subsidies that require cooperation in the form of result-sharing agreements increase cooperation but decrease incentives to conduct R&D. On the other hand, subsidy provisions that require cooperation (such as EUREKA) but not result-sharing do not increase cooperation but increase R&D investments. Fölster suggests that the result-sharing agreements might increase the efficiency of R&D but could also mean that cooperation reduces competition and, hence, also investment. (The non-result-sharing arrangements, on the other hand, might mean that firms simply apply for subsidies for programs where they intended to cooperate anyway).

The Sematech case provides an interesting illustration. 14 leading U.S. semiconductor producers, with a U.S. government subsidy of $100 million (not very much) founded a consortium in 1987 to boost U.S. competitiveness vis-à-vis similarly subsidized Japanese consortia. Irwin and Klenow (1994b) test the commitment and the sharing hypotheses. They found that whereas the commitment hypothesis provides a rationale for government subsidies, the sharing hypothesis does not. They also found that Sematech induced the member firms to cut their overall spending by some $300 million per year, providing support for the sharing hypothesis. Theoretical analysis of these matters allowing for spillovers, input and result sharing, other forms of cooperation, and various incentive schemes not only depends critically on the particular choices of assumptions but tends to escalate into impossible complexity whenever modest realism is introduced. One clear conclusion that can be derived from these theoretical studies is that incentives to organize R&D efficiently and to spend more rather than less on industrial R&D are different in private and in government-supported R&D, depending upon how the support content has been defined.

More important than technological competition stimulated by government are the circumstances under which technology is developed (Eliasson 1995a). Competitive technology is both economic and technological (call it industrial), and industrial competence never develops in a subsidized or protected and (for the same reason) academic environment. The conclusion to be drawn later in this text is that the technological spillovers which have been documented econometrically and which will be surveyed below all refer to spillovers in a competitive market environment. The same resource input in a non-market (protected or subsidized) environment would not have produced effects of the same magnitude.

The possibilities of capturing the results from R&D investments have been frequently referred to in literature. There are two sides to this problem: (1) to protect the inventing firm from cheap imitation and patent infringement, and (2) to charge for the otherwise free spillovers. There is a troublesome dilemma when easy imitation and patent infringement make industrial activity rich in spillovers privately unprofitable; this reduces investment in the same activity, thereby costing the nation in lost output growth.

Patent protection varies in kind and quality between countries. The protection offered depends on the kind of products and production. Eaton and Kortum (1995a) show that patents by the same firm and for the same product vary between countries in the quality of protection they offer. For example, Portugal and Ireland offer notoriously poor protection.

The geographically localized diffusion even of patented knowledge noted by Jaffe, Trajtenberg and Henderson (1992) is only cold comfort when it comes to protection from systematic imitation and patent infringement of firms that are highly competent in the business. Aoki and Prusa (1995) make an interesting theoretical analysis of the differences between the U.S. and Japanese intellectual property protection systems. Under the Japanese system, the patent application, with all its information, is publicly disclosed 18 months after the filing date. The patent office's examination of the substance of the application usually does not begin until after the public disclosure. Information contained in the U.S. application, on the other hand, is published only *after* the patent has been granted. Aoki and Prusa (1995) show theoretically that the Japanese system will make firms inclined to make smaller improvements in quality than they would under the U.S. system. Their study also shows that industry profits will be higher with the Japanese rules but that consumers will prefer the U.S. rules.

These differences in protection also have consequences for the spillover and coordination argument. The Japanese system should stimulate information sharing via licensing and joint R&D ventures, a greater research intensity, an earlier expected date of success, and more profits. On the other hand, there should be fewer major inventions with great spillovers to outsiders. No wonder then that firms appear to prefer information dissemination (Aoki and Prusa 1995, p. 18).)

This whole discussion raises the wider problem of the trade-off between the long term and the short term. Do the benefits of innovation stay within firms as resources to innovate further and grow, or are they competed away and passed on to consumers? Apparently, they have done both in the computer industry (Greenstein 1994b).[6] But what is the situation in other, more 'normal' industries? This trade-off is largely intermediated by the institutions determining access to intellectual property rights. The contract technology has a very different legal foundation in different countries. Apparently, the development of an economically *efficient legal system and contract technology regulating and enforcing the rent-sharing rules between firms and between the firms and the government* is a high-priority issue, in particular for a country that has the intellectual foundation and the ambition to be a future front runner in economic development. Economic theory so far has little to offer on this point but still much more than has been understood by policy makers.

This is not enough, however. While it has been observed (see, for instance, Bernstein and Mohnen 1994) that foreign (e.g., Japanese) firms may be more efficient than U.S. firms in tapping the international technology pool, largely created in the U.S., few have made a point of the fact (Eliasson 1990a, 1991a; Carlsson 1995b) that industrial traditions and competence systems might differ significantly in their organizational ability (*receiver competence*) to appropriate externally developed technology. This *leakage* of technology *may explain why it is becoming increasingly costly in terms of new research to generate new knowledge* from the existing pool of knowledge (Caballero and Jaffe 1993) while at the same time no apparent diminishing returns to industrial R&D can be observed (Nadiri 1993). If this is true, there is a *strong policy argument for making sure that industries with high receiver competence and strong downstream technical applications competence exist in a country.* This is a strong argument for nurturing firms as institutions for higher technical education and research.

Zucker, Brewer, Darby and Peng (1995) offer an interesting explanation to the (sometimes) high returns to intellectual capital, namely the ability to exclude others from accessing the rents through learning or imitation, by forming a hierarchy or otherwise creating organizational boundaries. Within this 'firm' a common interest in achieving the maximum rent, a mutual trust, develops that keeps critical, economically valuable information about 'how to repeat the experiment' inside the organization. The authors find this typical of biotech firms, particularly those formed around a 'star' scientist or a pair of star scientists. The economic rents that can be captured so to speak overrule the academic urge to publish; at least not all the information needed to exactly

---

6 Greenstein's (1994b) study attempts to measure the increases in user/buyer's value of computers, notably in the form of large product quality improvements that are not captured in 'constant quality' computing prices.

repeat the experiment will be published. The empirical testing of this hypothesis is also interesting. (See further chapter 6 on the biotechnology competence bloc.)

One of the standard explanations for the slow diffusion of new technology has been the vintage model, featuring firms with sunk costs in old technologies making the old technologies competitive on the margin with new superior technologies. This cannot explain all, Jovanovic and MacDonald (1993) argue, pointing to the slow spread of the diesel locomotive. Firms differ above all in their use of technologies "because it is costly for them to overcome information barriers," or rather, to use their own terminology, because they do not know enough to understand how to do it efficiently. If *receiver competence* were not a problem, the diffusion of spillovers would be more intense, and faster.

GENERIC TECHNOLOGIES LEVERAGE UP INDUSTRIAL EXPANSION

The argument for policy to support learning in the previous section becomes more forceful if we consider the possibilities of scale economies from new technology appropriations. The feeder industry argument has a modern reincarnation in the form of *technological systems* (Carlsson and Stankiewicz 1991; Carlsson 1995a), based on *general purpose technologies* (see Bresnahan and Trajtenberg 1992; Helpman and Trajtenberg 1994) that may also feed into the technological systems of a multitude of other industries. This idea goes all the way back to economic history studies of the waves of generic innovations that created long waves of economic development, such as the steam engine, railroads, electronics, etc. The central idea is the creation of positive externalities.

Generic or general purpose technologies in recent times are computer and information technology, biotechnology and new materials (see previous chapters). They all represent particular core technologies in potential industrial development blocs (Dahmén 1950). These technological systems extend the idea of an industrial development bloc to encompass also the supporting institutions and the general quality level of the labor force and of the scientific base — the competence bloc — of a country.[7] As a consequence, education and basic research have become items on the industrial policy agenda.

One policy question is whether the positive influence of a new generic technology on the economy requires the local presence of those industries, or whether the beneficial effects could come through learning or imports of technology rather than developing and using the technology or the new product in indigenous production. Was it necessary to have a steam engine and a locomotive industry to run railroads efficiently? Is an advanced chips (semiconductor) industry necessary for success in computers or information industry? For the same reason, do you need an advanced information

---

7 The concept of a competence bloc is defined and discussed in chapter 6.

technology industry to become proficient in the use of computers? Is it possible to establish the same positive generic information effects at some lower level, intensive in the use of the same technology, like software or aircraft industry?

There is no simple answer. It all comes back to the existence of sufficient *receiver competence*. This competence differs substantially among individuals, firms and countries, as do the requirements between industries. The long delay between productivity effects and the advance of computing technology did not depend on a slow diffusion of advanced new hardware in U.S. industry (Greenstein 1994a)[8] but on the slow development of user or receiver competence, a phenomenon observed already in Eliasson (1980) for robot installations in Swedish manufacturing firms. Berndt, Morrison and Rosenblum (1992) have observed the corresponding change in the competence composition of the U.S. manufacturing labor force. They find that increases in the high-tech composition of capital are positively related (by sub-industry) to rapid increases in the employment of white-collar, educated non-production workers, more competent than the earlier staff in putting the more advanced equipment to efficient use.[9]

More to the point of this paper: Is it possible to build industry on the basis of new results from scientific research without having the same science fully represented in the country? Can the labor force be trained in the new engineering tasks if the corresponding research is not represented? Can a competent staff of engineers in such an industry be obtained and upgraded without having the industry represented locally?

I will choose *information technology* (industry) and the *aircraft industry* as cases, partly because they are highly advanced and have been subjected to several studies, partly because the aircraft industry represents a technological merger between information industry, new materials technology, and sophisticated mechanical engineering production; it is well versed in the range of technologies in which the future engineering industry in advanced countries will have to excel. The purpose is to identify particular *private investments in technology* as infrastructure or public goods type of capital with considerable external effects or positive spillovers on other production.

On this score, it is interesting to observe that Sweden — and all of Europe for that matter — has more or less failed to develop a viable computer industry, while there are several successful telecommunications companies.

The above argument raises several questions. For instance, what kind of competencies are involved? How do they differ in *innovation*-driven compared to *imitation*-driven growth? Can the firms of a country choose technological style and stay

---

8 'Average practice' in computing advanced in the United States as rapidly as 'best practice,' lagging it by a maximum of 6 or 7 years (Greenstein 1994b).

9 As a result, a parallel statistical decline in labor productivity has been recorded. This depended, first of all, on the absence of productivity effects for reasons given by Greenstein (1994a). Then, as productivity effects appear, they will be econometrically (and wrongly) established as linked to physical installations of capital.

with it for ever, or do they need a well-rounded mix of different competencies to be able to radically change technology and to reorganize in order to stay on a fast growth track?

With respect to imitation there is a whole range of possibilities, reflecting the fact that the most sophisticated form of imitation is quite similar to innovation. There is a world of difference between the learning of a firm to produce a simple component of a sophisticated product, and the long Japanese learning history in the automotive industry or the design and building of complex civilian or military aircraft. Each new aircraft design involves a multitude of innovations, some of which are created locally, while others are created through the combination of various traditional technologies acquired externally. Most observers of economic development are not aware of the fact that a considerable innovative element with substantial value-creating capacity is involved in the organization of complicated production and the design of a complex product, even though one can find *no* new technologies in any of the components. The commercial value lies in the combination, creating synergistic cost reductions in production and/or new, commercially valuable properties in the product. This systems competence is the hallmark of a few companies in some of the advanced industrial economies. Hence, the transistor or micro chip industries are very different from the aircraft industry.

One would expect each stage of industrial development to build on a particular spectrum of competencies. At which level of complexity of applications is the global technology base optimally exploited locally and national diffusion of new technology at its maximum?

INDUSTRIAL RESEARCH IS A MATTER OF ORGANIZATIONAL LEARNING IN DIVERSE, DYNAMICALLY COMPETITIVE MARKETS

The development of new industrial technology embodies many ideas and involves many people. It is a complex interactive process that cannot be distinguished from the diffusion of new technology (Jovanovic and Rob 1989). Above all, the ability to combine ideas innovatively and successfully carry them through the various phases of development to industrial production not only requires the ability to conceptualize the whole process (the business idea, *experiment*) but also a prior and parallel building of competence to learn and assimilate the various technologies (*receiver competence*; Eliasson 1986a, 1989, 1990a, 1990b, p.17). This innovation process will not occur if not forced by competitors, setting high standards and demonstrating that superior technologies have already been realized commercially elsewhere (Eliasson 1994b).

As observed by Bernstein and Nadiri (1989b), intra-industry R&D spillovers generate a sequence of effects. (1) The production costs of the receiving firms decline; (2) the structures of receiving firms adjust; and (3) the R&D spending (often) declines, since it is more cost efficient to imitate (through R&D spending) than to develop new, own technology. There are many situations in which industries that do not trade or engage in joint production nevertheless benefit through borrowing each other's knowledge capital. The optimal arrangement for efficient knowledge diffusion is a diverse technology environment (*heterogeneity*; see Eliasson 1984; cf. Henderson 1994) in which all firms can learn from one another. The more common situation is that of one or a few leading technology-generating firms that (1) train people, (2) establish complex supplier networks, and (3) introduce particular technologies. The large firms are usually the first to introduce new machines. They can afford it and often start on a small experimental scale.[10] In general, however, the *technologically advanced large firms often service the rest of the economy with early experimental introductions of new industrial technology*, and diffuse the know-how, mostly through turnover of people, to downstream and perhaps less advanced firms, especially their subcontractors. There is a growing literature which demonstrates that R&D spending in particularly advanced firms contributes, through spillovers, to productivity development in other firms (cf. the advanced user argument in Chapter 3 above).

In an innovative industry where technologies and structures change rapidly, however, *the imitating firm will always introduce old technology*. To be commercially viable, the imitating firm has to introduce something new, for instance large industrial-scale manufacturing of products based on the new, imitated technology. Leading-edge, high-technology firms — to use a fashion term — then become a significant part of the technological system of a country, provided, of course, that the country has firms that are competent to exploit the domestically developed technology. In the extreme case, if firms in foreign countries possess the most competent imitators, domestic industry might even suffer from rapid technological advance in their own research institutes and technologically advanced firms. Japanese firms have been particularly efficient in basing large-scale production for global markets on technology developed in the United States (Eliasson 1986b; Bernstein and Mohnen 1994). During the last few decades, Swedish multinationals have been particularly efficient in industrial intelligence when it comes to accessing the global industrial technology pool and building industrial scale production on their learning experience (Eliasson 1991a).

WHICH GENERAL PURPOSE TECHNOLOGIES ARE CREATED THROUGH INDUSTRIAL RESEARCH?

The conventional theory of economic growth pictures differences in per capita income or aggregate productivity levels between countries as the result of differences in physical capital per worker and the savings ratio.

It is, however, also widely believed that there is an 'empirical' tendency for per-capita income or aggregate productivity levels to converge among countries because laggard economies 'can learn from' technologically advanced economies. There are

---

10 This was observed already in the late 1970s by the *Swedish Government Computer and Electronics Committee* (DEK) to be the case for new robotics techniques.

different versions of this catch-up hypothesis. The (*first*) original catch-up hypothesis was first suggested by Gerschenkron (1962) and formulated by Abramovitz (1986). The (*second*) 'new growth' theory of Romer (1986) and Lucas (1987) plays with aggregate models of knowledge-based factor accumulation that allow for the possibility of persisting production or growth rate differences because of limited capacity of learning or transferability of stocks of knowledge between countries. A *third* approach (e.g., Olsen 1982) emphasizes institutional sclerosis factors of special interest groups that slow down growth in already rich countries (negative learning). While the Abramovitz hypothesis explains convergence as followers catching up with the leader, the Olsen hypothesis leads to the same result via the rich nations slowing down. The new growth theory, then, allows for something in between. Nadiri and Prucha (1992) investigate these propositions on a cross-country (although fairly homogeneous sample) statistical material. They find that disembodied technical change is the prime explanatory variable for output and production growth. This result is compatible with the catch-up hypothesis through learning or through spillovers between countries.

Having observed (econometrically) that all countries, even technologically very advanced countries like the U.S., draw significant parts of their technology from abroad, the catch-up hypothesis has been reformulated in terms of an international pool of advanced technology (Eaton and Kortum 1995a; Nadiri and Prucha 1992). Eaton and Kortum (1995b) pitch the conventional hypothesis of differences in physical capital per worker against the hypothesis of differences in available technology. They conclude that the rate of closing of 'technology gaps' between five leading economies since 1950 can be explained by the ability to adopt more productive technologies regardless of whether these are developed at home or abroad. This, however, still leaves the problem of explaining how the international pool of technology develops. More micro-oriented studies on technology diffusion by above all Nabseth and Ray (1974) and Mansfield (1985) addressed the same learning problem at a more technical level.[11]

The empirical literature on technological spillovers started to grow with the various studies on technology diffusion from creative university environments (e.g. Nelson 1986) and got 'theoretical' inspiration from the new growth theory (Romer 1986; Lucas 1988). Earlier studies by Nadiri (1978, 1980a,b, 1993 and 1988, 1989a,b, 1991 with Bernstein) were, however, instrumental in moving the whole field onto a constructive micro- and sector-econometric basis. In particular, they introduced the idea of *private infrastructure capital* and not only public infrastructure capital as the technology generator. The most recent results, in fact, show private infrastructure being closer to

---

[11] The theoretical spillover literature is quite another matter. De Bondt (1995) distinguishes between three generations, beginning with decision-theoretic analyses of rivalry between and the timing of innovations (see Kamien and Schwartz 1976). The current and third generation method in the analysis of innovations uses game theoretic analysis of spillovers in oligopoly. This theoretical third extension derives naturally from the previous (second) generation concern with asymmetric innovation races or games. Typical of all these theoretical analyses is a strict and faithful foothold in static equilibrium.

the market than public infrastructure capital, giving rise to larger positive spillovers and stronger productivity effects than public infrastructure capital (see below). The hypothesis formulated and successfully tested is that general purpose technologies can be exploited locally for specific commercial applications. Such local applications require inputs of resources for learning and appropriation. R&D expenditures are typically devoted to learning and imitation. One should, therefore, expect R&D expenditures to increase around a general purpose technology.

The early investigations by Bernstein and Nadiri (1988, 1989) of the spillover effects to all industry from particular high-tech industries[12] shared considerable cost-reducing and rate-of-return-increasing diffusion effects, emanating from R&D spending in a particular group of five high-tech industries studied.[13] Increased capital spending was a complementary effect among the receiving industries and the linkages were quite specific in the sense that spillovers from each receiving industry emanated from a narrow range of industries. Thus, for instance, R&D spending in the scientific instrument industry increased performance in the chemical industry. The so-called new growth theory has generated a sequence of later private infrastructure studies along these lines (reported on below).

This means that the optimal technological system designed to generate spillovers should be quite broad-based. Technology, even basic technology, goes far beyond the university laboratory. There is also the free-rider problem. Not only do some firms (at no cost) receive technological spillovers from other firms that have to cover their R&D costs (infrastructure effects). Some of the receiving firms even appear to reduce their own R&D activity. This effect has, in fact, been econometrically established for Japanese firms in response to R&D investment in U.S. firms (see Bernstein and Mohnen 1994). A similar result was econometrically established (see above) for the Sematech project. A typical finding (see Henderson 1994, using country data) is that diffusion effects from traditional industries are short-lived and die out within a few years, while effects from high-tech industries persist longer.

Geographical proximity has often been cited as important for the speed of transmission of new technology. It is viewed as one reason for the concentration of both R&D and economic activity in general (Eaton and Kortum 1995a; Vuori 1995). Jaffe, Trajtenberg and Henderson (1992) have even found that patent citation tends to be geographically localized to countries or even states in the U.S.; geographical localization, however, fades over time. Even high technology among the major research countries diffuses much more slowly between countries than within them (Eaton and Kortum 1994). Foreign countries are important sources of technology; except for the United States, OECD countries derive almost all of their productivity growth from abroad (Eaton and Kortum 1995a). The 'ability' of a national economy to benefit from

---

12 As distinct from university research to industry as in Nelson (1986) and Jaffe (1989),

13 They are scientific instruments (SIC38), chemical products (SIC28), non-electrical machinery (SIC35), electrical products (SIC36) and transportation equipment (SIC37).

the international pool of technology, however, depends critically on the *receiver competence*, not least the competence of firms to coordinate a large number of specialist consultants and subcontractors (Eliasson 1986a, p. 57, 1989, 1990a, 1995a) or the absorptive capacity (Cohen and Levinthal 1989) of the firms of the economy. The bulk of R&D spending in firms is oriented towards increasing the receiver competence (Eliasson 1986a, pp. 81 f). Eaton and Kortum (1995a, 1989) observe that the country's level of education is significant in explaining this ability.

Besides geographical proximity, *technological proximity* has been investigated. An interesting and important observation is that knowledge spillovers are not restricted to a closely related regime of technology space (Jaffe 1986; Jaffe, Trajtenberg and Henderson 1992). Some industries are more self-sufficient (e.g., pharmaceuticals and biotechnology) than others when it comes to the in-house supply of knowledge (Vuori 1995), but the general results are that few or no firms are self-sufficient even when it comes to critical *complementary* knowledge.[14] The obvious conclusion is that this competence in acquiring, coordinating and combining externally available, complementary technology will become of critical importance for many firms in the advanced industrial economies (Eliasson 1995a).

An additional conclusion appears to be that not only do the western firms underinvest in R&D; geographical and learning barriers (not only transactions costs in the traditional sense) also limit access to a vastly underutilized pool of internationally available technology. Since R&D spending in firms is strongly oriented towards building receiver competence, the two conclusions coincide. Firms may not underinvest in new technology creation, but they underinvest in enhancing their receiver competence or capacity to learn (i.e. to identify, acquire, integrate and carry to industrial scale production) internationally available technology. It is important to keep this distinction clear since each type of investment has different policy implications, and R&D normally measures investment in both.

Combining the two proximity dimensions, Feldman and Audretsch (1995) conclude that technological diversification within the same science base, not specialization, leads to increased innovation.[15] They find that geographical proximity appears to matter, at least in biotechnology. However, its influence is not overwhelming; more than two-thirds of the links between biotechnology companies and university scientists are non-local.

Closeness to application and to corporate research also implies closeness to *markets* and *competition*. Since no study really models competition and market rivalry as a source of learning and technological spillovers, such observations and conclusions are indirect. Multinational firms are often quoted as vehicles for technology diffusion (Ethier and Markusen 1991; Eliasson 1991a; Fors 1993). Hines (1994) observes, interestingly, that technology imports and local (foreign subsidiary) R&D appear to be

---

14 Cf. IBM's problem in acquiring advanced telecommunications know-how (Eliasson 1995a, 1996).

15 A similar result was obtained by Granstrand and Sjölander (1993).

substitutes. He finds that the higher the taxes levied by the foreign government on royalties paid to the American parents, the greater the R&D intensity in the affiliates. This may look as great policy advice to the developing countries to tax foreign subsidiaries more heavily. But such advice is probably completely wrong. The quality of local R&D may be quite inferior, especially if combined with import protection, and hence, generate little or no growth in international competitiveness. Several studies (e.g. Coe, Helpman and Hoffmaister 1995) indicate strongly that the import of a large variety of intermediary products and capital-embodying foreign technology boosts local productivity. For developing countries "trade is a vehicle for technological spillovers." The authors conclude that total factor productivity growth in these countries is "positively and significantly related to R&D in their industrial trade partners and to their openness to trade with industrial countries." In addition, restrictions of trade between similar regions such as Europe and North America unambiguously reduce worldwide growth. They even quote the Grossman and Helpman (1989a,b) suggestion that making it easier for a less developed country to imitate technologies developed in an advanced country will speed up growth worldwide, by locating production to the less developed country and freeing high-quality resources in the advanced country for the development of new technology.

Quite naturally, academics have been investigating spillovers from their own research institutions, to the extent that they have neglected the importance of spillovers from corporate research laboratories (Eliasson 1995a). The most important spillover mechanism from university research appears to be related to the mobility of educated and talented people (Eliasson 1994a). Several studies (Nelson 1986; Vuori 1995; Zucker, Darby and Armstrong 1994; and others) observe that the academic links vary between industries and that they are particularly strong in biotechnology. An interesting question is whether this is because biotechnology is a young, science-based industry which has not yet been technologically structured, or whether it is a different industrial technology than the others (Eliasson 1989). Zucker, Darby and Brewer (1994) and Zucker, Darby, Brewer and Peng (1995) suggest (1) that biotechnology tends to be commercialized around innovative academic 'star' scientists, and (2) that the intellectual capital created in that technology is particularly suited to be contained within an organization, a firm, through a mutual interest (a 'trust') among the scientists that makes it possible to maximize the rent to be shared. Jaffe (1986) has cautiously suggested that university research *causes* industry R&D and, hence, operates as a catalyst. This again, if correct, has significant industrial policy (or rather educational policy) implications. Above all, it strengthens the earlier observation of the importance of established links between university and corporate research.

The spillover and diffusion effects between firms and industries studied so far have been restricted almost exclusively to 'infrastructure-type capital' that can be measured, or has been measured; most studies refer to R&D investments in U.S. industry.

The important task now is to

(1) find out whether R&D investment is a good measure of general purpose technologies producing general, positive, macroeconomic infrastructure effects;

(2) investigate the nature of such investments in general purpose technologies, such that they can be used for policy design; and to
(3) identify the channels through which they are diffused, received and implemented.

THE CHANNELS OF TECHNOLOGY DIFFUSION

The diffusion of new technology from private research and production establishments follows five different patterns:
(1) Competent people move in the labor market (*labor market turnover*).
(2) New firms are established by people leaving research labs in large firms or universities (*innovative entry* and *entrepreneurship*).
(3) Independent firms appropriate new technology through licensing or learning (*imitation*).
(4) Subcontractors learn advanced technology from leading user firms (*competent purchasing*,) and vice versa: systems-coordinating firms learn from specialist engineering firms.
(5) Technology diffusion is stimulated within large firms through the above four avenues (*internal technology transfer*).

All five technology transmission mechanisms require attention to the organization of markets, the dynamics of competition, and the internal economy of firms. Understanding the growth effects requires explicit micro-to-macro analysis of the dynamic interaction of firms in markets, of the internal dynamics of firms, and of how this affects firm behavior and the macro economy.

The internal behavior of firms is particularly important, since it may both facilitate and hinder the diffusion process. Three important factors to consider are:
- receiver competence
- customer (product) competence.
- internal resistance

Receiver competence, as we have repeatedly emphasized, is critical for the adoption of new technology in product development and in manufacturing processes. It is internal to the firm. To build receiver competence requires significant investment. *Customer (product) competence* refers to the contribution of the customer in developing new products (see e.g. Tryggestad 1995). Competent users of new technology have often been important 'teachers' and 'risk sharers' of firms developing new products and technologies. A classic example is a national telecommunications agency teaming up with a local telecommunications equipment producer to develop new technology. The Telia-Ericsson relationship in Sweden is one example. In studying the diffusion process, the *internal resistance* at all levels has to be factored in. Radically new technology always shakes up internal management fiefdoms and will be resisted. It may conflict with other, strategic ambitions of top management, as between business information systems and mobile telephones in Ericsson in the second half of the 1980s (Eliasson 1995a), and therefore also be discouraged. All these circumstances complicate the relationships between the development, diffusion, and adoption of new technology. Above all, awareness of these complications tells us that we are concerned with much more than R&D capital in a production function.

The Nature of R&D Spending in Large Firms

Pure innovation is a rare thing; it is based on a unique idea and normally not associated with much measured resource use. Most research, in academia as well as in business, involves adding knowledge to an existing knowledge base, recombining existing knowledge bases or (more common in industry) integrating different research fields or testing out marginally changed designs.

The bulk of measured R&D expenditures in a firm is devoted to the gathering and integrating of such knowledge, and more and more frequently through external engineering consultants. This, to a large extent, means devoting resources to intelligence on what other, often superior, competitors are doing in the market (Eliasson *et al.* 1985; Eliasson 1986a, p. 81 ff.; 1989). A small country in particular, but also the very large countries, cannot maintain all technologies in-house. Large, multinational corporations in advanced technologies, therefore, are important national receivers and repositories of globally available technology that is put to use within the company and then diffused to the rest of the economy (Eliasson 1991a). New industrial technology development normally means that the stock of firm knowledge is combined (merged) with new technological developments elsewhere to upgrade an existing product or process or to create a new integrated product technology. Computing has replaced mechanical switching in telecommunications, opening the gate for digital communications networks. Mechanical technology, instrumentation and electronics have merged successfully in mechanical engineering, forming increasingly sophisticated products (electronic stoves, cars, aircraft). Digitalization of information systems in mechanical products like automobiles and aircraft has made modularization of products possible which, in turn, makes maintenance and further production development easier. Aircraft production has typically been an advanced technology generator. It has not only been an early user of advanced mechanical engineering technology, digital information technology, and new materials. It has also been the forerunner in integrating these technologies in product design, engineering and production, thus employing currently the technologies that future engineering firms in advanced economies will have to employ (Eliasson 1995a). Hence, the existence of advanced production methods in advanced business firms in a country is essential for its capacity to absorb, exploit and diffuse new technology (von Hippel 1980; Carlsson and Jacobsson 1991) within the country. In the later stages of bringing new technology to industrial scale, the *application side* becomes more important than the innovation side. Large international firms represent a particularly important instrument in that respect when it comes to exploiting the technology potential associated with traditional production for competitive global markets (Eliasson *et al.* 1985). Large advanced firms, furthermore, are often forerunners when it comes to experimenting with new technology and new equipment and the hiring and training of new personnel to support such introductions. They, therefore, represent a source of new technology and industrial know-how for other, less advanced firms.

Applications Know-How Economically More Important than Basic Research

While research is sometimes referred to as the combination of ideas into new ideas (Jovanovic and Rob 1989), the practice of industrial research that generates faster total factor productivity growth is all *applications* and combinations of known technology with, perhaps, some new idea. Only a few firms in each industry are carriers of the very best-practice production methods and the most advanced products. Often the supreme performers reside in foreign countries. This means that the bulk of production uses methods that are inferior to the best and produce less sophisticated products than the best. This entire range of behind-the-frontier firms, hence, is poised for learning about better practice already implemented elsewhere. If they do not, they will soon be overtaken technically and commercially by other firms. This learning is concerned with acquiring applications know-how and frequently not even with development work. In these firms, basic research has no meaning and does not occur.

Spillovers through On-the-Job Learning

On-the-job learning (Lundberg 1961; Arrow 1962; Alchian 1963) is a particular form of knowledge diffusion that has attracted surprisingly little attention as a transmission mechanism in the recent spillover literature. Irwin and Klenow (1994a) serve up some startling results on spillovers in the semiconductor industry. Firms learn three times more from an additional unit of their own cumulative productivity than from an additional unit of another firm's cumulative production (spillover learning). There is no observable difference between Japanese and other firms in the capacity to learn, nor is there a difference in spillovers between firms in one country and in different countries. On the other hand, "intergenerational spillovers" between consecutive new families of products [dynamic random access memory (Dram) semiconductors] are nonexistent. According to the Irwin and Klenow hypothesis, firms are essentially unable to transfer the know-how from one generation of Drams to another. They have to 'relearn' from scratch. In an industry with a rapid turnover of products, this is a scary situation. The currently dominant firm may easily be thrown off the market with the next generation of products, which has in fact happened several times in the U.S. The positive thing is that a come-back is also possible, although not easy.

Education Builds Competence to Learn

The welfare of a nation is based on the relative competence (to that of other nations) of its people and firms. This obvious fact has long been understood (see, for instance, Westerman 1768), even though this general understanding has not been obvious from the reading of the large volume of literature on economic growth produced in the post-war period. The 'bulk' of competence put to use in production (Eliasson 1994a) has been passed down to the current generation from earlier generations, *making education* generally defined, or *knowledge transfer, the most important production activity*. If that competence or knowledge, its accumulation and its allocation could be measured and explained, we would have the necessary starting point for a general theory of economic growth (Eliasson 1989). What we soon learn, however, is that most of that critical transfer of knowledge occurs as a natural part of production. On-the-job learning is as much a 'spillover' process as are R&D investments. R&D spending is, in fact, to a large extent an investment in learning.

International Technology Diffusion

Bernstein and Mohnen (1994) observe that growth in productivity at the aggregate level depends significantly on the stock of accumulated R&D of the nation. With increased international trade, this stock of accumulated knowledge becomes international rather than national. They fail, however, to recognize the enhanced ability of the multinational corporation *to appropriate internationally available technology locally* (Eliasson 1991a). Bernstein and Mohnen investigate how U.S. R&D capital accumulation affects the production structures, physical and R&D capital formation, and productivity growth in the Japanese economy, and vice versa. They find, interestingly, that the Japanese firms (or rather industries) have been more efficient in absorbing and exploiting U.S. technology than are U.S. industries in absorbing and exploiting Japanese technology. U.S. R&D capital accounts for some 60 percent of Japanese total factor productivity growth, while Japanese R&D capital has only contributed some 20 percent of U.S. productivity gains. In large measure (my conjecture), this depends on the different orientation of U.S. and Japanese industries. Japanese industries are closer to the industrial application stages and efficient in understanding and absorbing new technology created elsewhere, much as the large Swedish multinationals have been, at least during earlier decades of the post-war period (Eliasson 1991a). In the 1950s and 1960s, U.S. firms were said to be adept in picking up new technology developed in Great Britain, where firms were slow or unable to commercialize their own inventions. Lately, U.S. firms have been more concerned with basic research but have often been unable to commercialize and appropriate the rents from their own internally developed technologies (Eliasson 1994a).

International technology diffusion both substitutes for domestic R&D and requires domestic R&D (investments in receiver competence). You need your own R&D to be able to appropriate internationally available technology. But technology diffusion *from* the U.S. appears to have been so generous that the Japanese industries have in fact decreased their own R&D intensity, according to Bernstein and Mohnen (1994). Two 'spillover regimes' are needed to explain this. Either Japanese firms have increased their effectiveness in international technological intelligence and/or the sheer volume of technological flows from the U.S. and elsewhere has made it possible to enjoy significant economies of scale. The complementary explanation is that access to U.S. technology through intelligence has been so cost efficient, compared to developing the same technology on their own, that Japanese firms have decided to cut their own development programs. However, Bernstein and Mohnen only investigate available data and give no interpretation of this kind.

They also find that private returns to R&D capital are around 17 percent in both Japan and the U.S. and that the existence of international spillovers increased social returns to R&D by a factor of three and a half to four times. This appears to be a lot, if one can believe the numbers.[16] In one sense, this suggests that considerably more should be invested in R&D to appropriate (for local use) more of the pool of available technology (see Nadiri 1993 on 'underinvestment'). The same results, however, also indicate the significant underutilization of the existing stock of technology and the enormous potential for competitive change when some firms, or the firms of a particular country, manage to improve their competence to access that pool. At the same time, however, a realistic analysis should also consider what this more efficient exploitation will mean for incentives to invest in new knowledge creation, especially if the commercial appropriation takes place in other firms or economies than where the new technologies are developed.

These results are supported by Coe and Helpman (1993) who estimate the effects of international R&D spillovers. They find large effects of such domestic and R&D capital stock on total factor productivity. Foreign R&D capital stocks have particularly large effects on the smaller countries. About one-quarter of the worldwide total factor productivity effects of R&D in the seven largest countries "are appropriated by their trade partners." Again, the neoclassical macro method used does not allow any explanation of how these effects accrue. Several explanations can be proposed; the one that best fits the reported evidence appears to be a model where *local R&D is more devoted to learning than it is to creating new technology.*

---

[16] They also find that a 1% increase in the international spillovers from the U.S. decreases Japanese labor intensity by 3.5 percent, while the corresponding effect on the U.S. firms is insignificant. Why this should be the case is a mystery to me.

This result should also be compared with Vuori's (1992) finding that returns to own R&D appear to be lower than generally thought. On the other hand, she finds that the returns to technology diffusion are high and that without substantial own research efforts, firms will not benefit from research in other firms. Also, technological diffusion is more important for small than for large industrial economies.

THE FORMATION OF NEW INDUSTRY THROUGH COMPETENCE ACCUMULATION - THE CASE OF SWEDEN

Following the new growth theory tradition, most studies on R&D have adopted the notion of generic innovation-driven new technology, originating in R&D and diffusing through industry by way of multiple applications. Silicon-based technologies and biotechnology are often quoted as examples of an already realized such technology and of a new technology to come, respectively.

General Historic Background

New technology that involves radical change usually diffuses through the establishment of new firms and takes a long time to mature in the form of large firms, industrial scale production, and significant employment effects. Above all, the road to success is always littered with mistaken business ventures. The bulk of new technology, however, occurs through piecemeal upgrading of the existing technology base. Volume effects occur much faster if new technology can be diffused through applications in existing industry rather than through the roundabout way of new entry. Successful diffusion serves as an upgrading of the competitiveness of the existing industrial base and prevents failure. There are, however, two important aspects on the innovation process to keep in mind. In the long run, radically new technology is needed for industry to stay competitive in the sense of supporting sustained growth (Eliasson 1972), because existing technologies are being imitated by firms in other countries, thus competing down earlier technology rents. Radically new technology, as a rule, has to be introduced through new entry (Eliasson 1991c, 1995c). Large companies, on the other hand, base their efficiency on large-scale routine production and are not very good environments either for innovative research or for the rapid acceptance of radically new technologies and the consequent organizational adjustments. Swedish big business, nevertheless, exhibits vintages of success stories in that respect over the past century, but also spells of slow (subsidized) failure, notably in shipbuilding (Carlsson 1983a,b).

The turn of the century witnessed a remarkable burst of entry of new firms in Sweden in mechanical and electromechanical engineering as new inventions were turned into innovative businesses by entrepreneurs: the ball bearing, Dahlén's lighthouse, electromechanical telephone switches, heavy electrical machinery etc.[17] The growth of these and other firms, as well as later spin-offs from existing firms, created a varied and for that time quite advanced technological base that slowly fostered the growth of Sweden's currently successful large companies as well as many firms that were established but later failed. Progress was slow, however. Glete (1993) talks about the failing experiments with new industrial technology in the early industrialization

---

17 Also see Jörberg (1961, 1983) on the learning of new technology from foreign firms by Swedish firms.

phase, ending in 1890, and the great industrial break-through during 1890-1920. Most of Sweden's later success stories in industries were founded before 1890 and were part of a much larger population of not well recorded, subsequently bankrupted firms. This technological competence base was sufficient to support a rapid transition of Swedish manufacturing in the post-Second World War period from a dominant footing in basic industry to a dominant mechanical engineering base. This, however, also means that the Swedish industrial base still is dominated by the technology on which the industrial revolution was founded (Glete 1994). The large automotive (including trucks) technical competence bloc was part of this post-war transition, as was the shipbuilding industry and a budding aircraft industry.

Large Firms as Technical Universities

The innovative phase of new entry around the turn of the century might not have been possible without the competence base built earlier and the availability of a skilled labor force in Sweden, trained in existing handicraft industries during the preceding century or two (Eliasson 1980). The shipbuilding industry and the automotive industry would not have been possible without a constant increase in the supply of skilled laborers and engineers and the creation of a varied components and parts supplier system. The war period created a new industrial situation where technology and skills not previously existing had to be rapidly developed. An aircraft industry was created that now consists of one firm and a network of suppliers in Sweden and abroad. Each step in the development built on earlier steps, gradually expanding the competence base of industry. Some of this upgrading of competence took place in technical institutes, but most of it occurred in *firms*, notably firms operating in highly competitive international markets. In many respects, these firms *served as higher institutions*, not of learning but of *competence development*, representing a form of *infrastructure capital*, and also supplying related industries with competent labor.

While the development in the early part of the 20th century was innovation-driven via the establishment of new firms, the post-war Swedish development has rather taken place through increasing the level of sophistication and organizational technology in the already large firms, based on the by then old mechanical engineering technology of the industrial revolution. Resources generated in basic industries were gradually transferred to the fast-growing and at that time sophisticated engineering firms. Step by step these firms expanded into larger scale industrial production, the firms reinforcing each other's development within the same technology and creating large supplier networks related to the same large (and with time international) firms. Bo Carlsson (1992, 1996) has observed that this expansion continued among the large international firms through the 1980s. Swedish industry, hence, followed an opposite development pattern to that in other advanced economies, where scale of production and size of firms followed a downward trend through the 1980s. For a while, at least, it looked as if Schumpeter's

(1942) notion of the increasingly successful firm based on systematic industrial research would eventually come to dominate each market in Sweden.

Since these firms have become increasingly dominant, they have also become increasingly important for the entire economy, making it more exposed than before to sudden competitive change in international markets. Hence, technologies related to the continued improvement of these industries, such as factory automation, have been a current cause for industrial policy concern.

Swedish industry requires extremely high competence in order to manage large international firms engaging in advanced product development, complex production, and global marketing. The critical industrial policy issue is that this development might suddenly come to an end *in Sweden* if the Swedish economy becomes incapable of supporting a sufficient number of leading-edge, large firms that operate successfully in the international markets and that not only create high value-added production on their own but also serve as providers of technology (spillovers) and skilled labor to related industries. Sweden currently has such industries, the aircraft industry being one of the best examples. But other problems loom over the horizon. The most important is perhaps the negative development of the level of education and skills of the labor force, but also (and perhaps because of that) the difficulties of maintaining advanced R&D activity and an advanced support system for subcontractors. A particular problem, more specific to the aircraft industry, is the arbitrary political conditions in global product markets where governments are dominant customers and where subsidies to local (private) producers to reduce costs artificially are the rule rather than the exception (Eliasson 1995a). Such circumstances create a rather special incentive environment for private producers and a difficult policy problem for governments that believe in the benefits of open competitive markets. A general conclusion of this study is that *positive spillovers are not generated by economically inefficient firms, even though they are classified as high-technology firms.* Technological spillovers embody a significant, probably dominant economic element. The following three sections illustrate that.

LEARNING COMPLEX, INTEGRATED PRODUCTION[18]

Any successful firm excels in some particular aspect of product development, production, or marketing. Most of what the firm does may be simple and trivial, but its locus of excellence is what matters, as for example in quality control of chain

[18] In this and the following sections it becomes important to be clear about some definitions. First of all, throughout the paper *production* has been broadly defined to include all value added-creating activity, including styling, design, product development, marketing, and distribution. The entire design, construction, and product development phase, prototyping and testing, as well as the post-production phase (including later consulting, learning to use, maintenance and continued design improvements), are together referred to as *engineering* which therefore becomes the important element of complexity and competence use, especially in its interaction with the making of the product (physical production). Some companies are excellent at engineering but badly organized when it comes to physical production, and vice versa.

production to ensure that the probability of failure of any link is negligible (Laestadius 1994). As long as the learning of that particular quality control technology is difficult, the firm that has acquired the competence may have at least a temporary quality monopoly. This particular competence niche will be safe until others catch up or an entirely new technology develops (for instance, synthetic fiber chains) — but then the company may not be able to change technology, reorganize and stay competitive. More complex technological advantages may be more difficult to imitate — for instance the ability to integrate product design, product engineering, and production in such a way that costly adjustments due to misunderstanding and bad communication between different phases of production are minimized. Similarly, complex *integrated production* requires an organization of assembly that minimizes failures along the line and makes sure that quality controls are enforced. The most costly organization is to fix quality defects after delivery of the final product to final inspection, or after delivery to the customer. Some complex systems such as public telephone switches or aircraft simply are not allowed to fail.

Even the design of a detail in an automobile or an aircraft often exhibits a high degree of complexity. It is, then, not difficult to understand that the design of the entire product soon becomes overwhelmingly complex and requires advanced engineering technology. When the design, construction, prototyping and production problems are seen as a whole, there are thousands of different ways of organizing this solution, and only some of them are economical. The ability to conceptualize the whole at an early stage becomes the important, dominant competence in this business. Technology and economics merge at all levels in the 'realization' of a complex product. Above all, complex *integrated production*, as a complicated production organization is sometimes called, can be organized in many different ways ranging from the making of the entire product, including all its details and components in the same factory, to a sophisticated organization of subcontractors. Integrated production, therefore, can also be seen as the *'systems competence' to work together with many engineering consultants and subcontractors*. The more decentralized the production (through markets), the more intellectually demanding the total systems responsibility. The boundaries of the firm organization and the market become dynamically dependent on management competence, technology, and competitive conditions in markets (see Eliasson 1996b).

For instance, when (1) everything is done within one firm and one factory, designers can monitor prototyping and production and correct designs as they are introduced and tried. Even this interaction between designers, engineering departments and production facilities in large firms is often difficult to organize efficiently, and ex post adjustments are always costly.

An additional organizational difficulty arises when some of the coordination is organized through the market and (2) the coordinating firm designs products or components for production by a subcontractor. The difficulty escalates again (3) when the systems contractor also specifies the product for the subcontractor, who is responsible for both design, technical solutions and production.

When (4) product designs and production methods are subject to steady and rapid innovative change, a further dimension of complexity is added, and the optimal balancing of subcontracting arrangements against internal production becomes a critical competence element of the entire production organization. Even for simple products there is no way to achieve analytical control of the whole, and the basic competence is organizational; it is embodied in the team structure of the organization. It 'sits in the walls' as it is often expressed among engineers. The design team of the aircraft industry is an illustrative example,[19] but this problem is typical of a growing number of firms engaged in manufacturing complex systems products. The organizational technology itself, in fact, is about to become the dominant technology base of large engineering firms among the mature industrial nations. It is very difficult to learn for firms in industrially less developed economies. It, therefore, constitutes a temporary knowledge monopoly capable of generating high value added per unit of labor input. This, in turn, illustrates the nature of *organizational learning* of competence that does not diffuse easily beyond the local technological system within which it is developed, and where the appropriate receiver competence exists (diffusion and spillover through turnover). Firms whose competence is based on complex organizational learning find that if they have lost competence or lagged behind, it takes very long to rebuild or catch up. It may in fact be impossible. The competence to define and specify the complex product, including methods of production, obviously is more important for established, large engineering firms than for small, newly started innovative firms. With integrated production, thus defined, the following advantages can be achieved:

(1) When the capacity exists to take a *holistic* approach, costly mistakes can be avoided (design flaws, bulky engineering designs, badly organized production processes).
(2) Styling, product and engineering designs, and testing (i.e., engineering) and production can be integrated synergistically.
(3) Difficulties of maintenance and further development of the product can be anticipated already at the development stage.
(4) Among many possible ways of organizing the whole, the best can be chosen.
(5) Simulation techniques (*computational prototyping*) make it possible to find the best specification at an early engineering stage. Large cost reductions are possible.
(6) *Quality control* becomes more effective. Costly *ex post* adjustments are avoided.

Holistic integration of many technologies all the way from early product conceptualization, engineering design, and testing to industrial scale production is decisive for cost-efficient production of complex products.

The technique to avoid costly *ex post* quality corrections is to make sure that the critical managers have the intellectual capacity to look at *production as a whole*. This

---

19 For concrete illustrations see Fredriksson (1994).

ability can be fostered theoretically in systems analysis, but the applications side is far more difficult and requires more than principles. Such *holistic organizational competence is more difficult to imitate than simple sequential production.* Firms that are capable of developing such organizational competencies or technologies will therefore possess a technological robustness in competitive markets against the imitative challenges of rapidly learning firms in newly industrialized countries. This organizational competence is applicable to a significant part of traditional industry in the mature industrialized countries. To maintain future growth capacity, an already advanced economy needs many such firms. It may continue to grow rapidly if it has the technical and economic competence to exploit the opportunities and the social and political competence to cope with the often disruptive structural change that follows (Eliasson 1994b).

As we have already observed, positive spillovers only occur in frontier firms in each technology and market. Frontier and commercially successful firms only develop under dynamically competitive market conditions. The aircraft industry belongs to the most advanced producers when it comes to the development and use of the organizational technology. A significant part of the technology concerns the art of working synergistically with many technical consultants and subcontractors.

THE MARKET FOR COMPETENT PURCHASING AMONG MANY SUBCONTRACTORS

Complex integrated production is not only a matter of internal production organization. The combined organization and technological demands soon become overwhelming and the firm has to decentralize its total production through markets. The efficient use of a combined internal hierarchy and a decentralized market organization of production requires the development of a particular systems competence not only to cooperate with specialized producers in external markets but also to coordinate the entire operation for a particular purpose. Economic and technical circumstances determine which mix of market and internal administration organization that is optimal. Thus *the organizational structure is subjected to a steady change as market circumstances change, requiring flexibility in management and organization.*

A developed market for subcontractor production also offers a rich mixture of firms that are specialists in particular technologies, that prefer long-term contract arrangements, that engage in simple, low-tech subcontracting production and are specialists in flexible adjustments and frequent changes of product designs and customers. *An economy without an advanced and differentiated subcontracting industry will not easily become proficient in the new type of industrial production organization that is developing in advanced economies.*

In fact, a significant part of *technology diffusion* occurs through the competent purchasing from new, small and medium-size firms by large high-technology firms that specialize in product development, production coordination via markets, and distribution. Technology diffusion, however, also goes the other way: from high-technology specialist engineering consultants and subcontractors to the purchasing firm. The integration effects occur over the entire system. The coordinating firm purchases functions, not products or components. The more advanced the entire product and its production, the more demanding the services of subcontractors. The capacity to take a total systems responsibility therefore becomes a high-technology organization competence in itself.

The argument of this paper is that this systems competence is (1) not easily imitable by firms in less advanced nations; (2) it is part of the generic technology and diffusion mechanisms of the entire economy. Furthermore, (3) it defines the next development phase of the mature engineering industry of the advanced industrial countries. If those firms are not able to cultivate commercially successful frontier firms in this technology (e.g. aircraft industry) and a rich supply of advanced engineering consultants and subcontractors,[20] the system will lack the critical technology generators for the next generation of engineering industry (Eliasson 1995a, 1995b). The firm responsible for total systems coordination will be closest to the market, notably to the competitive international markets. This firm, therefore, will be ahead of its subcontractors or fail if it cannot match its competitors. It will be confronted with competition first, learn how good the best competitors are, and be forced to learn to survive. Purchasing by the most advanced systems firms, therefore, is perhaps the most advanced 'technical education' a country can offer its firms, and a career in such a firm is often the best 'higher engineering education' a talented student can obtain (see further Eliasson 1990b, 1991d).[21]

Such customer competence contributions can, of course, also go the other way, i.e., from an advanced specialist subcontractor to the systems coordinator. 'Education' occurs predominantly through specialists, the setting of standards, and selection. Subcontractors that are not up to standards are simply not asked to participate. A new subcontractor who joins the system must be constantly upgraded to systems standards through learning and support. The art of *complex integrated production*, hence, *is the art of working efficiently together with many engineering consultants and subcontractors in designing, developing and producing* a sophisticated product. The selection of engineering expertise and subcontractors becomes decisive for the success of the entire system.

---

[20] The entire subcontracting industry does not have to be located in the country of the system's main firm(s). In fact, the advanced international engineering firms have been rapidly learning to access the 'global technological system' (Eliasson 1991a). The problem is rather practical. The more an already international firm has to rely on international expertise and subcontractors, the more footloose it becomes, and the less interesting it is for it to keep its intellectual locus of operation where it was originally located.

[21] This customer (product) competence contribution has often been misrepresented in literature as emanating only from public purchasing, notably defense procurement of high technology products. But as shown in previous chapters, esp. ch. 3, this has nothing to do with whether procurement is public or private but with the existence of competent buyers.

The development so far has been that the systems coordinator attempts to compose the product from as few complex part systems as possible (modularization). The technological demands on subcontractors then increase radically and a small country, even if advanced, becomes too small to supply all the specialist competence required. Coordinating firms have to access critical competence globally (Eliasson 1991a). This picture is general. In well-established systems, such as automotive products, subcontractors have managed fairly well to convert and raise their standard to become international suppliers of automotive subsystems. In the aircraft industry, the standards are much higher and the consequence for a small country like Sweden has been that its aircraft industry often has had to go abroad to find advanced sub-component subcontractors.

THE ADVANCED FIRM AS A HIGHER INSTITUTE OF TECHNOLOGY

The more complex the product or the production organization, the more difficult it is to convey the required holistic competence by way of traditional teaching techniques. This is especially so if teaching and research are fragmented into separate departments with little incentive to cooperate, which is typical of the academic world. Under such circumstances, little teaching and little research addressing the whole are offered, which is typically the case when it comes to *systems engineering*, as distinct from systems analysis. Systems engineering is a field that falls between the chairs at technical universities and the know-how typically resides in firms applying various forms of complex integrated production. Such production technology represents the peak of excellence in large-scale engineering production. Sweden has only got a few advanced firms in that field.

The main reasons for bringing this up in this context are that such complex production and products are gradually becoming the main area where the production competence of mature industrial economies remains a fairly well preserved competence monopoly, and that the establishment and growth of additional complex, high value added production in an economy depends critically on the *existence in that country of the most advanced forms of such production as institutions of higher technical education.*

Diffusion of Technology through Turnover of Engineers

In an early study, Gustafsson and Lindvall (1978) investigated the transfer of human-embodied know-how from the Saab-Scania aircraft division to other industries through turnover of technical personnel. They observed that aircraft production may operate as a technical school and that the wage development in the aircraft industry is slower than in other industries (everything else the same), but also that the wage increases significantly when the trained individual changes jobs. This is compatible with the hypothesis (my interpretation) that employees in the aircraft industry recognize the personal investment in learning that they receive on the job and are content with a somewhat lower wage than the market wage. After they have been 'educated,' they move on to other firms that pay better. This loss of trained people is still a fact for the Saab-Scania aircraft division in Sweden, especially to industries needing technicians with an ability to think holistically about the entire production system. Gustafsson and Lindvall (1978) also observed that such turnover of personnel was not restricted to technicians. Workers in the aircraft industry learned many advanced skills, notably work sequences requiring extreme precision, skills that could not be acquired elsewhere. This notion is compatible with the general hypothesis of this study, that *to become and stay excellent in industries such as mechanical engineering, the country has to have some of the most advanced producers of that industry represented domestically, as educational institutions.*

The Subcontracting Industry Also Requires Competence

The aircraft industry (Saab Aircraft and subcontractors) is probably the most advanced subcontracting constellation of firms within the Swedish mechanical engineering industry. This industry is dominated by the very large Swedish multinationals, but it also includes a large number of not so advanced firms, notably the subcontracting industry (Braunerhjelm 1991). The mechanical engineering industry is the bulk employer in Swedish manufacturing. If it fails, employment and real wages will suffer significantly.

The more technologically advanced the final producer, the higher the technological demands on its subcontractors. While the subcontracting industry around Swedish automobile production has survived surprisingly well in Sweden, subcontractors to the aircraft industry have often failed. Apparently the aircraft subcontracting industry is based on more advanced technology than automobile subcontractors. While the rent-generating capacity should be larger among aircraft industry subcontractors, if their competence is high, the possibilities of achieving volume production are smaller and require that subcontractors establish themselves in the international market early on. Since customers in that market are usually government bodies or government-operated firms, this is not easy. In the context of Swedish military aircraft production, the Swedish government has also demanded that Saab look for the lowest cost source of components. Hence, most subcontractors around Saab aircraft components have been shut down, and the remaining ones are typically the largest firms. This is unfortunate for three reasons. The structure of high-tech subcontractors is thinner than it could have been, and the spillover effects from advanced aircraft assembly in terms of both technology and employment increasingly diffuse out of the country. Third, the tough demands on cost savings by the Swedish government have increased the dependence of Swedish industry on the largest firms even more.

The Social Returns to the Aircraft Industry

The spillover or infrastructure effects of an aircraft industry, however, still exist and can be estimated. If the industry had been one hundred percent civilian and operating in the private market, there would have been no methodological problem. The effects would have been all positive. This is, however, rarely the case in the aircraft industry. If it has no military part (as Fokker in Holland), it is often partly or wholly owned by the government and heavily subsidized. Hence, the social returns to the aircraft industry is an important policy issue.

In the Swedish case, the military aircraft side complicates the computation of social returns. There is a civilian part that builds on spillovers from the military side and that also itself creates spillovers back to the military side and to other firms and industries. I will therefore use as a benchmark the alternative that the Swedish military could have been satisfied with using a military aircraft developed elsewhere, but that it would then have demanded that production take place on license in Sweden. Production and licensing costs would then define the military value of the plane to Sweden. In the actual case, Saab both developed and will produce the new fighter plane, but we can safely assume, for the comparison, that production in Sweden of the Swedish designed aircraft would be cheaper than production of as many foreign designed aircraft in Sweden plus licensing costs.[22]

Two forms of positive or negative social value accrue to Sweden. Assuming that the two aircraft solutions (fighter planes) are of identical military value, we have a positive social value if

a) domestic production costs are lower than licensed production costs plus licensing costs;
b) net technological spillover effects are larger than the development costs.

The social value of development work consists of value added created in other firms outside the development project that would not have been created in the absence of the project (spillovers) minus the negative costs to the economy incurred if resources are pulled away from other activities.

As observed by Fölster (1993a), most studies of spillover effects observes, do not recognize the negative effects that occur because R&D programs draw resources from other areas. Fölster then designs a method to account for these indirect effects and applies it to the Swedish military fighter project JAS. This calculation assumes that the positive effects under item (a) above are zero. Fölster interviewed related firms about

---

22 This setting for comparison may be too simple. There is no product comparable to the JAS plane, capable of landing on regular but reinforced roads, etc. Modifying any foreign product to achieve such performance would have been almost as costly as designing a new plane. Modifying the Swedish defense strategy to fit available foreign plane designs would also have been prohibitively costly, as would the adoption of a standard European defense strategy, using available planes, since then defense costs in Sweden would increase to counter the now very large 'standard' invasion potential that under Swedish special design strategy, with JAS type planes, requires the use of different and costly designs of an invader. Therefore, the assumptions made for comparison should be acceptable.

the consequences of JAS development on them and showed that the JAS project has generated positive spillover effects on other industries, net of the military development costs and indirect costs in the form of lost alternative uses of technical personnel, etc., of 15 billion SEK, to compare with direct project development costs of 12 billion SEK. If the military value of the produced fighter planes is equal to or higher than the production costs, the development of the planes has been costless for Sweden, since the spillover effects are significantly larger than the development costs. These spillovers, he concludes, are far above what one can expect from subsidies.

The spillover effects would probably have been significantly smaller if the same development costs had instead been removed from the commercial environment of a corporate development laboratory and placed in a public or academic research lab. The large spillover effects have to do with JAS being a leading-edge technology project. *No such effects would be obtained from employment-promoting subsidies in simple production such as the enormous subsidy program of Swedish shipbuilding in the 1970s* (Carlsson 1983 a and b). An important spillover generator is the competence development needed to launch Saab Aircraft's civilian aircraft project. In addition, Volvo Aero has been able to increase its share of the civilian aircraft engine market from 5 percent in 1978 to about 60 percent in 1995. Ericsson has been able to acquire world leadership in microwave technology, advanced signal transmission, and special forms of computing. In fact, Ericsson would not have advanced so rapidly during the 1990s as a leading developer and producer of digital mobile telephone systems equipment, were it not for the earlier in-house development of certain critical military aircraft technologies (Eliasson 1995a). In addition, new composites developed within the JAS project have found uses elsewhere in Swedish industry. FFV Aerotech has been able to develop a strong market position both in the U.S. and in Europe in civilian aircraft maintenance. There are also several examples of firm start-ups on technology from the Saab military aircraft division. Perhaps even more important: the JAS project, contrary to earlier fighter development programs at Saab, has involved the development and use of advanced simulation models of materials strength, surface material tensions, etc., that will certainly become important competence elements in later advanced product design and in other engineering industries.[23]

EFFECTS OF PHYSICAL PUBLIC INFRASTRUCTURE AND R&D CAPITAL ON UNITED STATES MANUFACTURING PERFORMANCE

The choice between private and public infrastructure investment has been a topic of considerable controversy. It is here discussed from the point of view of generating positive, technological spillover effects.

---

[23] For more details on this see Eliasson (1995a).

Estimated elasticities of output and labor productivity with respect to changes in public infrastructure capacity differ significantly. Output elasticities in the range 0.3 to 0.4 have been reported from studies on national income time series data. Total factor productivity elasticities have been estimated to be 0.5 using cross-sectional country data. Berndt and Hansson (1992) and Morrison and Schwartz (1992) found that additions to public physical infrastructure capital significantly lowered labor requirements or costs, citing scale effects of the physical infrastructure investments. At a lower level of aggregation, however, estimated elasticities are smaller and Hulten and Schwab (1991) report no significant elasticities.

Nadiri and Mamuneas (1991) disaggregate further and distinguish between (1) public *infrastructure* and (2) publicly financed *R&D*, treating category (2) as direct 'unpaid inputs' in the private production process. Public infrastructure capital is composed of public investments in education, public research laboratories, etc. They pool time series and cross-sectional data from U.S. manufacturing and find significant direct effects of both kinds of publicly funded R&D on the cost structure and productivity growth of U.S. manufacturing. However, the second category, which is closer to the market than the first category, generated significantly higher marginal products.[24] In a later study, Mamuneas and Nadiri (1995) find that while publicly financed R&D increases cost savings, it crowds out privately financed R&D investment. They also find that publicly financed R&D carried out inside a particular industry has a stronger effect on cost savings than that of publicly financed R&D carried out outside of an industry, i.e. further away from applications and the market. Even though not phrased in those terms, closeness to industrial applications and markets appears to be a significant factor behind the documented productivity effects of R&D spending (see e.g. Zucker, Darby and Armstrong 1994;[25] Rivera-Batiz and Romer 1991). Henderson, Jaffe and Trajtenberg (1995) find that the increase in university patenting in the U.S. over the last 25 years is associated with an overall increase in university attention to commercial applications of research, perhaps due to increased industry funding of university research.

One possible definition of infrastructure capital is generality of its appropriability, access to the services of infrastructure capital being limited by the competence of the user to appropriate it with a minimum disbursement of resources (*receiver competence*). There is a large body of literature claiming that inadequate accumulation of public

---

24 The authors are currently working on a comparison of the same spillovers from pure private investments in R&D and the two categories of publicly financed R&D.

25 Zucker, Darby and Armstrong (1994) are very clear on this point. They conclude that "most production scientists generally are either employees or collaborators with the enterprise." These results refer to the U.S. or rather the Southern California biotech industry. Z-D-A go on to conclude that "intellectual capital when linked to the firm through market exchange mechanisms, including ownership, significantly accelerates firm production and growth" (p. 28). Z-D-A even emphasize the need in some advanced (biotech) industries for the 'learner' to work together with the creator of the new technology in order to
be capable of learning.

capital has contributed to substandard U.S. economic growth. From that perspective, Holtz-Eckin and Schwartz (1994) quite logically find it difficult to understand that there is practically no relation between infrastructure research and the literature on economic growth. They go on to study the effects on production growth of state investments (in the U.S.) in physical infrastructure capital. They find no evidence supporting the idea of a strong productivity boost from increased investments in physical infrastructure capital. Between 1971 and 1986 the effects would rather have been negligible. As mentioned, Hansen and Berndt (1992) also had difficulties finding significantly positive relationships between public infrastructure capital and economic growth in Sweden.

The negative results on physical public infrastructure capital contrast sharply with the spillover effects of intangible R&D capital. We know, of course, that physical private capital accumulation contributes to productivity growth. It would therefore be interesting to see whether the above difference has anything to do with the differences in source and origin of spillovers from infrastructure capital, private or public.

This result raises the question: how should R&D be organized to yield maximum economic growth? How much does the regime (public or private) matter? This question, however, does not lend itself to easy answers.

The stock of knowledge accumulation in the past is the fundamental input in the capacity of the current generation to generate new technology. Such knowledge is largely public, whether privately or publicly produced. Caballero and Jaffe (1993) confirm Mansfield's (1985) results that the rate of diffusion of new technology (as evidenced by patents granted and used) is quite fast and has become even faster. This increased speed is, however, also matched by a more rapid decline (over the last century) in the capacity of accumulated knowledge to generate new knowledge (measured by the patent citation rate).[26] It has therefore become more costly in terms of additional R&D needed to maintain past rates of economic growth.

The standard discourse on this matter has been about, on the one hand, basic versus applied research and, on the other hand, private versus public research. Trajtenberg, Henderson and Jaffe (1992) show that these distinctions are not empirically clear, and they study the effects of different "unmeasurable characteristics of the process that leads to innovations." *First*, they observe that "private, profit-making firms perform basic research and that it is profitable." *Second*, there are clearly visible success technology growth regions around universities (Route 128, Silicon Valley).

While the economist tends to define basic research by its difficulty of being appropriated, basically in terms of difficulties of establishing ownership or property rights to the results (profits), scientists tend to define basic research by its relation to *prior* research and in contrast to research oriented towards profit making.

All of those definitions abound with inconsistencies. By the first definition, software development would constitute basic research. The first definition also fails to

---

26 This is not surprising. The faster the diffusion, the faster new knowledge is used by others, and the less profitable it is for the inventor on the margin to develop it. The interesting thing is that this decline in the U.S. has been due to the rapid appropriation of U.S.-created knowledge by firms outside the U.S.

recognize the other side of appropriability (perhaps implicit in the second definition), namely that of being far from the application stage, where rents can be captured.

Trajtenberg, Henderson and Jaffe (1992) claim that "the basicness of research underlying an invention can be characterized by the nature of the previous patents cited by an invention" and that the "function of patents that are assigned to the same organization as the original invention is a measure of appropriability." Using such data they find that university research and research outcomes are more basic and harder to appropriate than those of corporations.

The authors, however, fail to recognize the importance of *entrepreneurship* for the appropriability process in and around university and corporate R&D facilities, which may explain the same differences. Corporate laboratories are closest to the market and should be able to generate more easily appropriable research. Researchers in corporate laboratories should also have a better sense than university-based laboratories of which innovations will carry the market in large volumes and which will not. (It is probably no coincidence that so many successful new start-ups in the information industry that have made it big over a decade or two have, in fact, often been based on research teams that have left other electronic firms' research laboratories). The important circumstance appears to be closeness to the market. In mature industries the technology is well established, tradable and transformable through technical universities. Innovative development, however, is in the form of new process solutions or product development with a strong commercial content. It thus occurs in and diffuses from firms.

The information technology (IT) industry development has been based on particular technological discoveries such as the transistor, etc., some of them in university labs, some of them in semi-public labs like Bell Laboratories. But most downstream IT development of a highly innovative nature has taken place in new businesses founded by groups of research people that have left corporate R&D labs in companies like Texas Instruments, Fairchild Camera, Hewlett Packard and above all Xerox and its PARC laboratory in Palo Alto (see Eliasson 1996a). Again the source of creation of new industry has been industry itself, and new technologies in the IT industry have been extremely advanced by any academic standard. Here private firms have functioned as advanced research institutes, generating large positive spillover effects in the form of establishment of new firms.

The *biotechnology sector* may be a different story, to judge from the development so far. Above all, the scientific foundation appears more profound than in IT.

Considerable controversy has surrounded the science-technology-industry links. While the political debate often assumes a direct positive relationship between higher education and research and productivity growth, those who have studied these relationships are not so sure (Eliasson 1992a, 1993b; Rosenberg 1994). It is rather typical that countries like Japan and Germany do not particularly encourage the more advanced forms of theoretical academic inquiry but rather encourage applied technical research. Historically, academia has been a quite hostile environment for industry, and applied research has much too often been looked at as the most low-brow form of academic inquiry. There are exceptions, however. The German chemical industry is said

(Locke 1989, p. 62) to have a university foundation, and the German Technical Institutes (Technische Hochschulen) that developed independently of, and outside, the university tradition are generally considered to have contributed to German economic development during the 19th century. Some of the basic inventions in information technology (not most, and few in later years, however) originated in the laboratories of technical universities and/or semi-public research institutions such as Bell Labs. By contrast, the new IT industries mostly originated in corporate development labs. Some changes may, however, be under way. More academically oriented research close to large firms may be turning academic research in a more entrepreneurial direction, and while IT has one foot in academia and one in industry, the new microbiology, biochemistry, biotechnical, and pharmaceutical fields certainly breed on research coming out of science labs in university environments in the United States. Zucker, Darby and Brewer (1994) argue that the more than seven hundred active biotechnology firms established in the U.S. in the last two decades[27] are based on the new developments in basic science. The timing, location and birth of biotech companies, they find, are primarily explained by intellectual capital measures, notably "the local number of highly productive 'star' scientists." The establishment and growth of new biotech firms was closely related to the presence of active star scientists and their collaborators. The stars did not have to be entrepreneurs themselves, but the intellectual capital that grew around them appeared to be the key input in the high tech new entry process. Life has, however, been tough for these firms and the market is crowded with competitors (*Business Week*, Sept. 26, 1994, p. 66 ff.).

Zucker, Darby and Brewer (1994) conclude that status is not necessarily associated with an academic institution. Scientists in private industry laboratories write as many scientific articles as do university scientists. Henderson and Cockburn (1993) also found that productivity effects in industrial (pharmaceutical) R&D improved significantly if participation in research programs could be run within the context of large and more diversified R&D efforts. Opportunities then increased to exploit knowledge across programs within the firm and internal and external spillovers could be more efficiently absorbed. This is much more typical of industry labs than academic labs, where fragmentation prevails.

There is still considerable uncertainty about the relative efficiency of public R&D and private R&D investment as a spillover source for general industrial productivity advance. However, evidence supports the hypothesis that the more commercial and applications-oriented private, firm-based research is more efficient in terms of yielding high technological rents for the R&D investing firm as well as generating commercially useful spillovers. This contradicts the early suggestions by Arrow (1962b) that spillover-intensive research should be moved to publicly financed research labs, since incentives in private industries would be too low. With capital spending on R&D rapidly increasing, the organization and efficiency of R&D work matter increasingly for

---

27 Or over 1,000 according to *Business Week*, Sept. 26, 1994, p. 66 ff.

overall performance. Hence, the choice of management regime (public, university or private laboratory), is becoming a much more serious concern than before both for top management and for public policy makers. Unfortunately, I have found no study that sets these three choices clearly against each other.[28] Summing up, Nadiri and Mamuneas (1994) conclude that there is irrefutable evidence that own investments in R&D benefits the investing firm, and that R&D investments also contribute to total factor productivity growth across firms in the same industry. The spillover effects, however, vary considerably. Privately financed infrastructure and R&D capital seems to have stronger productivity effects than publicly financed infrastructure and R&D capital.

POLICY IMPLICATIONS

The conventional way of looking at business competence development is to restrict attention to technical know-how, to understand learning as an investment in *more information*, and to identify such knowledge with measured R&D. If a firm is less informed than its foreign competitors, it needs time and protection and possibly financial support to gather more costly information in order to catch up. Thus went the old *infant industry* argument in international trade theory.

During the 1980s the infant industry argument from international trade theory was dusted off and marketed by well-known economists such as Paul Krugman under the name of *industrial targeting*. Industrial targeting became popular as a protectionist argument with the U.S. Congress of the 1980s.

The 1980s also saw the birth of the so-called 'new' growth theory, first formulated by Romer (1986) and Lucas (1988), making R&D a critical variable in the theoretical growth process and high R&D intensity identical to high-tech industry. As a consequence, the last few years have witnessed a rapid growth in both theoretical and empirical literature on *industrial research as the engine of economic growth*. A natural extension of this discussion is to look at the *technological systems* of firms and nations and to consider the possibility of national technological management. This paper includes a critical survey of this literature and its policy conclusions.

Again we see misunderstanding brew from too much focus on what can be seen and measured, and it is much too easy to lose sight of the fact that even though technology matters for firm performance, the main business competence is to be in the right market at the right time with the right product. R&D investment has become the focus of empirical applications of new growth theory, mostly because statistics of R&D investment have been collected for many years. The last few years have seen a burst of research on the productivity effects of R&D spending. Most of this recent literature has been reviewed in this paper. R&D investments are important, but they capture only a

---

[28] For instance, Nadiri and Mamuneas (1994) study the spillover effect of public R&D investments, having no variable for competing private R&D investments.

part of the total intangible capital stock residing in a modern firm that in large measure defines its competence base (Eliasson 1990a). Such statistical analyses tend to give too much importance to what is measured, i.e., R&D, compared to other competence characteristics. By current definitions and categories, for instance, a group of highly competent and entrepreneurial people that form a company but that do not set up an R&D lab will not be operating a high-tech business (Laestadius 1994). Even if they do and succeed, the success of the operation will be statistically attributed to the R&D lab rather than to the competence of the people establishing the firm. Such biases in the interpretation of results depend on the predominance of macroeconometric research. *To understand and to give advice on policy, a micro-based organizational interpretation is needed.*

To understand the relationships between technology and economic growth, three facts of industrial life have to be kept in mind: (1) learning in firms is organizational; (2) industrial competence is as much economic and commercial as it is technical; and most important, (3) learning does not take place by acquiring new information, but by watching, imitating and challenging the best competitors in the market. Thus, *efficient learning by firms can only occur in competitive markets*. Firms do not learn to improve by being offered cheap funding and protection. *Firms improve their performance in response to challenges from the best competitors in the market, and they do it predominantly through organizational change. The first and most important policy action for growth, hence, is to make sure that dynamically competitive markets be organized.*

The importance of private firms, notably international firms, as receivers, repositories and diffusers of technology has been more or less overlooked in the policy discussion of the last couple of decades. The policy focus has rather been on government-operated laboratories, universities, and technical institutes. The discussion has been intellectually shaped around the early writings on the matter, notably Arrow (1962b), and the presumed absence of incentives for private firms to engage in easily appropriable research. However, the last few years have witnessed a burst of research on private R&D spending, looking at private research as a provider of infrastructure spillover effects to the rest of industry. This research shows (*first*) that private firms are both good exploiters and diffusers of new technology, compared to public research institutions. New firms in new industries, notably in the information industry, rather than being initiated in university and public institutions, appear to have originated in and around the research laboratories of large, advanced firms in the U.S. The *second* important result on national technology policy is that *the organization of R&D is fundamental for research efficiency*, that industrial research does not stop at the laboratory gate but extends all the way to industrial scale production and marketing, and that, as in all other walks of life, incentives play a central role in the organizational solution. Incentives for entrepreneurship within large firms and for new innovative start-ups are critical for the economic benefits of technological improvements to be realized 'nationally.' Differences in the entrepreneurial culture go a long way towards explaining the superior performance of new firm growth in new technologies in the U.S. compared to Europe and (in particular) Sweden.

REFERENCES

Abramovitz, Moses, 1986. "Catching Up, Forging Ahead and Falling Behind," *Journal of Economic History*, 46 (2), 385-406.

Albrecht, James W., Pontus Braunerhjelm, Gunnar Eliasson, Jörgen Nilsson, Tomas Nordström and Erol Taymaz, 1992. *MOSES Database*. Stockholm: IUI.

Aoki, Reiko and Thomas J. Prusa, 1995. "Product Development and the Timing of Information Disclosure under U.S. and Japanese Patent Systems," NBER Working Paper No. 5063, NBER, Cambridge, MA.

Arrow, Kenneth J., 1962a. "The Economic Implications of Learning by Doing," *Review of Economic Studies*, 29 (3), 155-173.

--------, 1962b. "Economic Welfare and the Allocation of Resources for Invention," in Richard R. Nelson (ed.), *The Rate and Direction of Inventive Activity: Economic and Social Factors*. Princeton: Princeton University Press, 609-626.

Audretsch, David B. and Paula E. Stephan, 1995. "Company-Scientist Locational Links: The Case of Biotechnology," mimeo, Wissenschaftszentrum Berlin für Sozialforschung, Center for Economic Policy Research (CEPR) and Department of Economics and Policy Research Center, Georgia State University.

Bentzel, Ragnar, 1980. "Svensk ekonomisk tilväxt 1870-1975" (Swedish Economic Growth 1870-1975), in Erik Dahmén and Gunnar Eliasson (ed.), *op. cit.*, 157-200.

Berndt, Ernst R. and Bengt Hansson, 1992. "Measuring the Contribution of Public Infrastructure Capital in Sweden," IUI Working Paper No. 365, Stockholm.

Berndt, Ernst R., Catherine J. Morrison and Larry S. Rosenblum, 1992. "High-Tech Capital Formation and Labor Composition in U.S. Manufacturing Industries: An Exploratory Analysis," NBER Working Paper No. 4010, Cambridge, MA.

Bernstein, Jeffrey I. and P. Mohnen, 1994. "International R&D Spillovers between U.S. and Japanese R&D Intensive Sectors," NBER Working Paper No. 4682, Cambridge, MA.

Bernstein, Jeffrey I. and M. Ishaq Nadiri, 1988. "Interindustry R&D Spillover, Rates of Return, and Production in High-Tech Industries," *American Economic Review, Papers and Proceedings*, 78 (2), 429-434.

Bernstein, Jeffrey I. and M. Ishaq Nadiri, 1989a. "Rates of Return on Physical and R&D Capital and Structure of the Production Process – cross section and time series evidence," in Baldev Raj (ed.), *Advances in Econometrics and Modelling*. Boston and Dordrecht: Kluwer Academic Publishers, 169-187.

Bernstein, Jeffrey I. and M. Ishaq Nadiri, 1989b. "Research and Development and Intra-industry Spillovers: An Empirical Application of Dynamic Duality," *Review of Economic Studies*, 56 (2), 249-269.

Bernstein, Jeffrey I. and M. Ishaq Nadiri, 1991. "Product Demand, Cost of Production, Spillovers, and the Social Rate of Return to R&D," NBER Working Paper No. 3625, Cambridge, MA.

Bishop, John H., 1989. "Is the Test Score Decline Responsible for the Productivity Growth Decline?," *American Economic Review*, 79 (1), 178-197.

Braunerhjelm, Pontus, 1991. *Svenska underleverantörer och småföretag i det nya Europa – Struktur, kompetens och internationalisering* (Swedish Subcontractors and Small Business in the New Europe — Structure, Competence, and Internationalization). Stockholm: IUI.

Bresnahan, Timothy F. and Manuel Trajtenberg, 1992. "General Purpose Technologies. Engines of Growth?," NBER Working Paper No. 4148, Cambridge, MA.

Caballero, Ricardo I. and Adam B. Jaffe, 1993. "How High Are the Giants' Shoulders? - An Empirical Assessment of Knowledge Spillovers and Creative Destruction in a Model of Economic Growth," NBER Working Paper No. 4370, Cambridge, MA.

Carlsson, Bo, 1983a. "Industrial Subsidies in Sweden: Macroeconomic Effects and an International Comparison," *Journal of Industrial Economics*, 32 (1), 1-23.

--------, 1983b. "Industrial Subsidies in Sweden: Simulations on a Micro-to-Macro Model," in *Microeconometrics, IUI Yearbook 1982-83*. Stockholm: IUI, 61-72.

--------, 1992. "The Rise of Small Business: Causes and Consequences," in William James Adams (ed.), *Singular Europe: Economy and Polity of the European Community after 1992*. Ann Arbor: The University of Michigan Press, 145-169.

--------, 1996. "Small Business, Flexible Technology and Industrial Dymamics," in Zoltan J. Acs, Bo Carlsson and Roy Thurik, *Small Business in the Modern Economy*. Oxford and Cambridge, MA: Basil Blackwell Publishers, 63-125.

--------, (ed.),1995a. *Technological Systems and Economic Performance: The Case of Factory Automation*. Boston, Dordrecht and London: Kluwer Academic Publishers.

--------, 1995b. "Introduction," ch. 1 in Bo Carlsson (ed.), 1-19.

Carlsson, Bo, Erik Dahmén, Anders Grufman, Märtha Josefsson and Johan Örtengren, 1979. *Teknik och industristruktur – 70-talets ekonomiska kris i historisk belysning* (Technology and Industrial Structure — The Economic Crisis of the 70s in Historical Perspective). Stockholm: IUI and IVA.

Carlsson, Bo, Gunnar Eliasson and M. Ishaq Nadiri (eds.), 1978. *The Importance of Technology and the Permanence of Structure in Industrial Growth*, IUI Conference Reports 1978:2. Stockholm: IUI.

Carlsson, Bo and Rolf G.H. Henriksson (eds.), 1991. *Development Blocks and Industrial Transformation*. Stockholm: IUI.

Carlsson, Bo and Staffan Jacobsson, 1991. "What Makes the Automation Industry Strategic?" *Economics of Innovation and New Technology*, 1 (4), 257-269.

Carlsson, Bo and Rikard Stankiewicz, 1991. "On the Nature, Function and Composition of Technological Systems," *Journal of Evolutionary Economics*, 1 (2), 93-118.

Coe, David T, Elhanan Helpman and Alexander W. Hoffmaister, 1995. "North-South R&D Spillovers," NBER Working Paper No. 5048, Cambridge, MA.

Coe, David T. and Elhanan Helpman, 1993. "International R&D Spillovers," NBER Working Paper No. 4444, Cambridge, MA.

Cohen, Wesley M. and Daniel A. Levinthal, 1989. "Innovation and Learning: The Two Faces of R&D," *The Economic Journal*, 99 (397), 569-596.

Dahmén, Erik, 1950. *Svensk industriell företagarverksamhet* (Swedish Industrial Entrepreneurial Activity), Vol. 1 and 2. Stockholm: IUI.

Dahmén, Erik and Gunnar Eliasson (eds.), 1980. "Industriell utveckling i Sverige - teori och verklighet under ett sekel" (Industrial Development in Sweden - Theory and Reality during a Century). Stockholm: IUI.

Day, Richard H., Gunnar Eliasson and Clas Wihlborg (eds.), 1993. *The Markets for Innovation, Ownership and Control*. Stockholm: IUI and Amsterdam: North-Holland.

De Bondt, Raymond, 1995. "Spillovers and Innovative Activities," Paper presented at the 22nd Annual Conference of EARIE, Juan les Pins, September 3-5.

Eaton, Jonathan and Samuel Kortum, 1994. "International Patenting and Technology Diffusion," NBER Working Paper No. 4931, Cambridge, MA.

Eaton, Jonathan and Samuel Kortum, 1995a. "Trade in Ideas: Patenting and Productivity in the OECD," NBER Working Paper No. 5049, Cambridge, MA.

Eaton, Jonathan and Samuel Kortum, 1995b. "Engines of Growth: Domestic and Foreign Sources of Innovation," NBER Working Paper No. 5207, Cambridge, MA.

Eliasson, Gunnar, 1971, "Forskning, investeringar och ekonomisk utveckling" (Research, Investment and Economic Growth), *Affärsvärlden/ Finanstidningen*, No. 47, 1-3.

Eliasson, Gunnar, 1972. *International Competitiveness — An Empirical Analysis of Swedish Manufacturing*, Economic Research Report B13. Stockholm: The Federation of Swedish Industries.

--------, 1978, *A Micro-to-Macro Model of the Swedish Economy*, IUI Conference Report 1978:1. Stockholm: IUI.

--------, 1980. "Elektronik, teknisk förändring och ekonomisk utveckling" (Electronics, Technical Change and Economic Development), in *Datateknik, ekonomisk tillväxt och sysselsättning (DEK)*. Stockholm: Liber Förlag, 191-315.

--------, 1984, "Micro Heterogeneity of Firms and Stability of Growth," *Journal of Economic Behavior and Organization*, 5 (3-4), 249-274.

--------, 1986a. "Företaget som skapare och förvaltare av mänskligt kapital" (The Firm as an Innovator and Manager of Human Capital), in Gunnar Eliasson *et al.* (1986), 39-106.

--------, 1986b. "Innovative Change, Dynamic Market Allocation and Long-Term Stability of Economic Growth," Working Paper No. 156, IUI, Stockholm. Forthcoming in Paul David and Giovanni Dosi (eds.), *Innovation and the Diffusion of Technology*. Oxford: Oxford University Press.

--------, 1987. *Technological Competition and Trade in the Experimentally Organized Economy*. Research Report No. 32. Stockholm: IUI.

--------, 1989. "The Dynamics of Supply and Economic Growth - How Industrial Knowledge Accumulation Drives a Path-dependent Economic Process," in Bo Carlsson (ed.), *Industrial Dynamics: Technological, Organizational and Structural Changes in Industries and Firms*. Boston, Dordrecht and London: Kluwer Academic Publishers, 21-54.
--------, 1990a. "The Firm as a Competent Team," *Journal of Economic Behavior and Organization*, 13 (3), 273-298.
--------, 1990b. "The Knowledge-Based Information Economy," Chapter I in Gunnar Eliasson, Stefan Fölster, Thomas Lindberg, Tomas Pousette and Erol Taymaz, *The Knowledge Based Information Economy*. Stockholm: IUI, 9-87.
--------, 1991a. "The International Firm: A Vehicle for Overcoming Barriers to Trade and a Global Intelligence Organization Diffusing the Notion of a Nation," in Lars-Gunnar Mattsson and Bengt Stymne (eds.), *Corporate and Industry Strategies to Europe*. Amsterdam: Elsevier Science Publishers, 139-170.
--------, 1991b." Modeling the Experimentally Organized Economy - Complex Dynamics in an Empirical Micro-Macro Model of Endogenous Economic Growth," *Journal of Economic Behavior and Organization*, 16 (1-2), 153-182.
--------, 1991c. "Deregulation, Innovative Entry and Structural Diversity as a Source of Stable and Rapid Economic Growth," *Journal of Evolutionary Economics*, 1 (1), 49-63.
--------, 1991d. Financial Institutions in a European Market for Executive Competence; Chapter 7 in Clas Wihlborg, Michele Fratianni, and Thomas A. Willet (eds.), *Financial Regulation and Monetary Arrangements after 1992* Amsterdam: Elsevier Science Publishers B.V., 149-165.
--------, 1993a. Företagens, institutionernas och marknadernas roll i Sverige (The Role of Firms, Institutions, and Markets in Sweden), Appendix 6, Appendix volume 1 in SOU (Statens Offentliga Utredningar), *Nya villkor för ekonomi och politik* (New Conditions for Economy and Politics), SOU 1993:16, Stockholm.
--------, 1993b. *Den högre utbildningens ekonomiska villkor och betydelse* (The Economic Conditions and Importance of Higher Education), Report No. 2, DS 1993:94, Ministry of Education, Stockholm. Also published by IUI as Research Report No. 41.
--------, 1993c. "Endogenous Economic Growth through Selection," paper presented to the IARIW conference on "Microsimulation and Public Policy," Canberra, December 6-9. Also available as IUI Working Paper No. 397, Stockholm.
--------, 1994a. *Användningen av högre utbildade i företag - en studie grundad på intervjuer av företag*, (The Use of Highly Educated People in Firms — A Study Based on Firm Interviews). Report No. 14, DS 1994:119, Agenda 2000, Ministry of Education, Stockholm.
--------, 1994b. "The Theory of the Firm and the Theory of Economic Growth; an Essay on the Economics of Institutions, Competition, and the Capacity of the Political System to Cope with Unexpected Change," in Lars Magnusson (ed.), *Evolutionary*

and Neo-Schumpeterian Approaches to Economics. Boston and Dordrecht: Kluwer Academic Publishers, 173-201.

--------, 1994c. "Kunskap och framtida kompetensbehov" (Knowledge and Future Competence Needs), in Gunnar Eliasson and Eugenia Kazamaki Ottersten, *Om förlängd skolgång* (Concerning Extended Obligatory Education). Stockholm: IUI.

--------, 1994d. "Technology, Economic Competence and the Theory of the Firm - Discussing the Economic Forces behind Long-term Economic Growth" in Ove Granstrand (ed.), 125-164.

--------, 1994e. *Markets for Learning and Educational Services - a Micro Explanation of the Rôle of Education and Development of Competence*. DEELSA/ED/-CERI/CD, (94) 9 (04-Nov.). Paris: OECD.

--------, 1995a. *En teknologigenerator eller ett nationellt prestigeprojekt? — svensk flygindustri* (A Technology Generator or a National Prestige Project? — Swedish Aircraft Industry). Stockholm: City University Press.

--------, 1995b. "The Economics of Technical Change - The Macroeconomic Consequences of Business Competence in an Experimentally Organized Economy," *Revue d'Économie Industrielle*, Numero Exceptionnel, 53-82.

--------, 1995c. "Economic Growth through Competitive Selection." Paper presented at the 22nd Annual Conference of EARIE, Juan les Pins, September 3-5.

--------, 1996. *Firm Objectives, Controls and Organization*. Boston and Dordrecht: Kluwer Academic Publishers.

Eliasson, Gunnar, Fredrik Bergholm, Eva Christina Horwitz and Lars Jagrén, 1985. *De svenska storföretagen - en studie av internationaliseringens konsekvenser för den svenska ekonomin* (The Swedish Giants — A Study of the Consequences of Internationalization for the Swedish Economy). Stockholm: IUI.

Eliasson, Gunnar, Bo Carlsson, Enrico Deiaco, Thomas Lindberg and Tomas Pousette, 1986. *Kunskap, information och tjänster. En studie av svenska industriföretag* (Knowledge, Information and Services - A Study of Swedish Industrial Enterprises). Stockholm: IUI.

Eliasson, Gunnar and Bengt-Christer Ysander, 1983. "Sweden: Problems of Maintaining Efficiency under Political Pressure"; in Brian Hindley (ed.), *State Investment Companies in Western Europe*. London: Trade Policy Research Centre, 156-191.

Ethier, Wilfred J. and James R. Markusen, 1991. "Multinational Firms, Technology Diffusion and Trade," NBER Working Paper No. 3825, Cambridge, MA.

Feldman, Maryann P. and David B. Audretsch, 1995. "Science-Based Diversity, Specialization, Localized Competition and Innovation," (Johns Hopkins University and Wissenschaftszentrum Berlin für Sozialforschung and CEPR), paper presented to the 22nd Annual Conference of EARIA, Juan les Pins, September 3-5.

Fors, Gunnar, 1993. "Technology Transfer to Foreign Manufacturing Affiliates by Multinational Firms," IUI Working Paper No. 370, Stockholm.

Fredriksson, Billy, 1994. "Systems Engineering - A Holistic Approach to Product Development," *Griffin*, 94/95 (Nov.), 23-31.

Fölster, Stefan, 1991. *The Art of Encouraging Invention: A New Approach to Government Innovation Policy*. Stockholm: IUI.

--------, 1993a. "De teknologiska spridningseffekterna av JAS – en ny empirisk metod" (The Technological Spillovers of JAS — A New Empirical Method), mimeo, IUI, Stockholm.

--------, 1993b. "Do Subsidies to Cooperative R&D Actually Stimulate R&D Investment and Cooperation?," mimeo, IUI, Stockholm.

Gerschenkron, Alexander, 1962. *Economic Backwardness in Historical Perspective*. Cambridge, MA: Belknap Press.

Glete, Jan, 1993. "Long-Term Firm Growth and Ownership Organization. A Study of Business Histories," in Richard H. Day, Gunnar Eliasson and Clas G. Wihlborg (eds.), *The Markets for Innovation, Ownershop and Control*. Stockholm: IUI and Amsterdam: North-Holland, 155-177.

--------, 1994. *Nätverket och näringslivet* (Network and Industry). Stockholm: SNS.

Granstrand, Ove (ed.), 1994. *Economics of Technology*, Amsterdam: Elsevier Science B.V.

Granstrand, Ove and Sören Sjölander, 1993. "The Acquisition of Technology and Small Firms by Large Firms," in Richard H. Day, Gunnar Eliasson and Clas Wihlborg (1993), 301-320.

Greenstein, Shane M., 1994a. "Did Computer Technology Diffuse Quickly?: Best and Average Practice in Mainframe Computers, 1968-1983," NBER Working Paper No. 4647, Cambridge, MA.

Greenstein, Shane M., 1994b. "From Superminis to Supercomputers: Estimating Surplus in the Computing Market," NBER Working Paper No. 4899, Cambridge, MA.

Griliches, Zvi, 1986. "Productivity, R&D and Basic Research at the Firm Level in the 1970s," *American Economic Review*, 76 (1), 141-154.

Griliches, Zvi, 1994. "Productivity, R&D, and the Data Constraint," *American Economic Review*, 84 (1), 1-23.

Grossman, Gene and Elhanan Helpman, 1989a. "Endogenous Product Cycles," NBER Working Paper No. 2913, Cambridge, MA.

Grossman, Gene and Elhanan Helpman, 1989b. "Quality Ladders and Product Cycles," NBER Working Paper No. 3201, Cambridge, MA.

Gustafsson, Siv and J. Lindvall, 1978. *Kunskapsöverföring genom personalomsättning från flygindustriell till annan verksamhet* (Knowledge Transfer through Personnel Changes from Aircraft to Other Activity). Stockholm: Flygindustrikommittén (Fö 1978:01).

Harabi, Najib, 1995. "Channels of R&D Spillovers: An Empirical Investigation," Working Paper, Series D, Wirtschaftswissenschaftlisches Institut der Universität Zürich.

Helpman, Elhanan and Manuel Trajtenberg, 1994. "A Time to Grow and a Time to Reap: Growth Based on General Purpose Technologies," NBER Working Paper No. 4854, Cambridge, MA.

Henderson, Rebecca and Iain Cockburn, 1993. "Scale, Scope and Spillovers: The Determinants of Research Productivity in the Pharmaceutical Industry," NBER Working Paper No. 4466, Cambridge, MA.

Henderson, Rebecca, Adam B. Jaffe, and Manuel Trajtenberg, 1995. "Universities as a Source of Commercial Technology: A Detailed Analysis of University Patenting 1965-1988," NBER Working Paper No. 5068, Cambridge, MA.

Henderson, Vernon, 1994. "Externalities and Industrial Development," NBER Working Paper No. 4730, Cambridge, MA.

Hicks, Donald A. and Steven R. Nivin, 1995. "Spillover and Appropriation: Economic Returns to Regional Investments and Endowments," Political Economy Working Paper 95-23, University of Texas at Dallas (July).

Hines, James R, Jr., 1994. "Taxes, Technology Transfer, and the R&D Activities of Multinational Firms," NBER Working Paper No. 4932, Cambridge, MA.

Holtz-Eckin, Douglas and Amy Ellen Schwartz, 1994. "Infrastructure in a Structural Model of Economic Growth," NBER Working Paper No. 4824, Cambridge, MA.

Hulten, Charles R. and Robert M. Schwab, 1991. "Public Capital Formation and the Growth of Regional Manufacturing," *National Tax Journal*, 44 (4), 121-134.

Irwin, Douglas A. and Peter J. Klenow, 1994a. "Learning-by-Doing Spillovers in the Semiconductor Industry," *Journal of Political Economy*, 102 (6), 1200-1227.

Irwin, Douglas A. and Peter J. Klenow, 1994b. "High Tech R&D Subsidies: Estimating the Effects of Sematech," NBER Working Paper No. 4974, Cambridge, MA.

Jaffe, Adam B., 1989. "Real Effects of Academic Research," *American Economic Review*, 79 (5), 957-970.

Jaffe, Adam B., Manuel Trajtenberg and Rebecca Henderson, 1992. "Geographic Localization of Knowledge Spillovers as Evidenced by Patent Citations," NBER Working Paper No. 3993, Cambridge, MA.

Jovanovic, Boyan and Glenn MacDonald, 1993. "Competitive Diffusion," NBER Working Paper No. 4463, Cambridge, MA.

Jovanovic, Boyan and Rafael Rob, 1989. "The Growth and Diffusion of Knowledge," *Review of Economic Studies*, 56 (4), 569-582.

Jörberg, Lennart, 1961. "Growth and Fluctuations of Swedish Industry 1869-1912," *Studies in the Process of Industrialization*, Lund.

Jörberg, Lennart, 1983. "Teknikspridning och industriell förändring" (Technology Diffusion and Industrial Change) in *Festskrift till Kristof Glamann*, Odense Universitetsförlag, Odense, 235-248.(A revised version has been published in English as "The Diffusion of Technology and Industrial Change during the Nineteenth Century" in Kristine Bruland (ed.), *Technology Transfer and Scandinavian Industrialization*. London, 1991.)

Kamien, Morton I. and Nancy L. Schwartz, 1976. "On the Degree of Rivalry for Maximum Innovative Activity," *Quarterly Journal of Economics*, 90 (2), 245-260.

Katz, Michelle L., 1986. "An Analysis of Cooperative Research and Development," *Rand Journal of Economics*, 17 (4), 527-543.

Laestadius, Staffan., 1994. "Competence Requirements in Low-Tech Firms," KTH-index. Mimeo.

Lucas, Robert E., Jr., 1988. "On the Mechanics of Economic Development," *Journal of Monetary Economics*, 22 (1), 3-41.

Lundberg, Erik, 1961. *Produktivitet och räntabilitet* (Productivity and Profitability). Stockholm: SNS.

Mamuneas, Theofanis, P. and M. Ishaq Nadiri, 1995. "Public R&D Policies and Cost Behavior of the US Manufacturing Industries," NBER Working Paper No. 5059, Cambridge, MA.

Mansfield, Edwin, 1965. "Rates of Return from Industrial Research and Development," *American Economic Review*, 55 (2), 310-322.

Mansfield, Edwin, 1968. *The Economics of Technical Change*. New York: Norton.

Mansfield, Edwin *et al.*, 1977. "Social and Private Rates of Return from Industrial Innovations," *Quarterly Journal of Economics*, 91 (2), 221-240.

Mansfield, Edwin, 1985. "How Rapidly Does New Industrial Technology Leak out," *Journal of Industrial Economics*, 34 (2), 217-223.

Morrison, Catherine J. and Amy Ellen Schwartz, 1992. "State Infrastructure and Productive Peformance," NBER Working Paper No. 3981, Cambridge, MA.

Nabseth, Lars and George F. Ray, 1974. *The Diffusion of New Industrial Processes — An International study*. Cambridge: Cambridge University Press.

Nadiri, M. Ishaq, 1978. "A Dynamic Model of Research and Development Expenditure," in Bo Carlsson, Gunnar Eliasson and M. Ishaq Nadiri (eds.), 51-71.

--------, 1980a. "Sectoral Productivity Slow Down," *American Economic Review. Papers and Proceedings*, 70 (2), 349-352.

--------, 1980b. "Contributions and Determinants of Research and Development Expenditures in US Manufacturing Industris," in George M. von Fürstenberg (ed.), *Capital, Efficiency and Growth*. Cambridge, MA: Ballinger, 361-392.

------, 1991. "R&D Investments and the Catching Up Process in Major OECD Countries," mimeo, New York University.

--------, 1993. "Innovations and Technological Spillovers," NBER Working Paper No. 4423, Cambridge, MA.

Nadiri, M. Ishaq and Theophanis P. Mamuneas, 1991. "The Effects of Public Infrastructure and R&D Capital on the Cost Structure and Performance of U.S. Manufacturing Industries," NBER Working Paper No. 3887, Cambridge, MA. Forthcoming in *Journal of Public Economics*.

Nadiri, M. Ishaq and Theophanis P. Mamuneas, 1994. "Infrastructure and Public R&D Investments, and the Growth of Factor Productivity in US Manufacturing Industries," NBER Working Paper No. 4845, Cambridge, MA.

Nadiri, M. Ishaq and I. Prucha, 1993. "Sources of Growth of Output and Convergence of Productivity in Major OECD Countries," paper presented to the World Productivity Congress (WPC), Stockholm, May.

Nelson, Richard R., 1986. "Institutions Supporting Technical Advance in Industry," *American Economic Review*, 76 (2), 186-189.

Pavitt, Keith, 1979. "Technical Innovation and Industry Development," *Futures*, 11 (6), 458-470.

Phillips, Almarin, Paul Phillips and Thomas R. Phillips, 1994. "The Formation of the U.S. Market for Business Jets: A Study in Schumpeterian Rivalry," in Ove Granstrand (ed.), 289-322.

Phillips, Almarin, Paul Phillips and Thomas R. Phillips, 1994. *Biz Jets: Technology and Market Structure in the Corporate Jet Aircraft Industry*. Boston and Dordrecht: Kluwer Academic Publishers.

Ricardo, David., 1821. *The Principles of Political Economy and Taxation* (third ed.), London.

Rivera-Batiz, Luis, A. and Paul M. Romer, 1991. "International Trade with Endogenous Technological Change," NBER Working Paper No. 3594, Cambridge, MA.

Romer, Paul M., 1986. "Increasing Returns and Long-Run Growth," *Journal of Political Economy*, 94 (5), 1002-1037.

Rosenberg, Nathan, 1994. "Science - Technology - Economy Interaction," in Ove Granstrand (ed.), 323-337.

Salter, W., 1960. *Productivity and Technical Change*. Cambridge: Cambridge University Press.

Schumpeter, Joseph A., 1911 (English edition 1934). *The Theory of Economic Development*, Harvard Economic Studies, XLVI. Cambridge, MA: Harvard University Press.

Schumpeter, Joseph A., 1942. *Capitalism, Socialism and Democracy*. New York: Harper & Row.

Svensson, Roger and Gunnar Fors, 1994. "R&D and Foreign Sales: Evidence from Swedish Multinationals," IUI Working Paper No. 423, Stockholm.

Teplitz, Paul V., 1965. "Spin-Off Enterprises from a Large Government-Sponsored Laboratory," unpublished doctoral dissertation, Sloan School, MIT, Cambridge, MA.

Trajtenberg, Manuel, Rebecca Henderson and Adam B. Jaffe, 1992. "Ivory Tower Versus Corporate Labs: An Empirical Study of Basic Research and Appropriability," NBER Working Paper No. 4146, Cambridge, MA.

Tryggestad, Kjell, 1995. "Sourcing of Advanced Manufacturing Technology: The Role of Customer-Supplier Interaction," in Bo Carlsson (ed.), 211-239.

von Hippel, Eric, 1988, *The Sources of Innovation*, New York: Oxford University Press.

Vuori, Synnöve, 1992. "R&D, Technology Diffusion and Productivity Performance in Finnish Manufacturing Industries," in Synnöve Vuori and Pekka Ylä-Anttila (eds.), *Mastering Technology Diffusion – The Finnish Experience*, ETLA, Series B No. 82, Helsinki, 125-147.

Vuori, Synnöve, 1995. *Technology Sources in Finnish Manufacturing*, ETLA, Series B 108, Helsinki.

Vuori, Synnöve and Pekka Ylä-Anttila (eds.), 1992. *Mastering Technology Diffusion – The Finnish Experience*. Helsinki: ETLA.

Westerman, Johan, 1768. *Svenska Näringarnes Undervigt emot de Utländske, förmedelst en trögare Arbets-drift* (On the Inferiority of Swedish Manufacturers Compared to Foreign Ones due to a Slower Work Organization). Stockholm.

Zucker, Lynne G., Michael R. Darby, and Jeff Armstrong, 1994. "Intellectual Capital and the Firm: The Technology of Geographically Localized Knowledge Spill-overs." NBER Working Paper No. 4946, Cambridge, MA.

Zucker, Lynne G., Michael R. Darby, and Marilynn B. Brewer, 1994. "Intellectual Capital and the Birth of U.S. Biotechnology Enterprises," NBER Working Paper No. 4653, Cambridge, MA.

Zucker, Lynne G., Michael R. Darby, Marilynn B. Brewer and Yusheng Peng, 1995. "Collaboration Structure and Information Dilemmas in Biotechnology: Organizational Boundaries as Trust Production," NBER Working Paper No. 5199, Cambridge, MA.

9

MICRO-MACRO SIMULATION OF TECHNOLOGICAL SYSTEMS: ECONOMIC EFFECTS OF SPILLOVERS

Bo Carlsson, Gunnar Eliasson and Erol Taymaz

INTRODUCTION

As shown in Chapter 8, one of the most important and pervasive features of technological systems as they evolve over time is the interdependence of the various actors (through hierarchies and markets) within the system. A variety of terms have been used in the literature to describe basically the same phenomenon — networks, development blocs, technological spillovers, etc.

Given our definition of technological systems as "networks of agents interacting... to generate, diffuse and implement technology," we will now model spillovers in technological systems in such a way that their macroeconomic impact can be quantified and understood. We need to specify the mechanisms through which such networks are built and how innovation takes place in them and in the economy generally. This analysis is carried out in the context of a micro-based macroeconomic simulation model (MOSES). Technological change is viewed as taking place at the micro (firm) level. Firms vary in their choice of technology and in the degree to which they make successful use of a given technology. This is a result of their having different abilities to learn, to imitate others, and to generate new technology on their own. Firms form networks to increase both their connectivity and their effectiveness in exploiting the stock of knowledge and experience within the network (i.e., to increase their receiver competence or absorptive capacity). The mechanisms through which spillovers take place are (1) movement of people between firms, (2) imitation of other firms, (3) competent purchasing (subcontracting), and (4) entry of new firms (cf. p. 205 in Chapter 8).

The chapter is organized as follows. In the next section we present an overview of the model and a brief specification of the features of the model which are essential in our simulations. The model presentation is followed by a description of the simulation experiments and an analysis of the results. The concluding section summarizes the results and discusses the implications.

THE MODEL

Overview of the MOSES Model

MOSES (Model of the Swedish Economic System) is a micro-based macroeconomic simulation model developed at the Industrial Institute for Economic and Social Research (IUI) in Stockholm. It is a very large model, consisting of more than 10,000 equations. Therefore, a full presentation of the model is not practical here; only a few features pertinent to the present analysis will be described. For an overview of the model, see Carlsson and Taymaz (1995); for a more complete description, see Eliasson (1978, 1985, 1991), Albrecht *et al.* (1989), Bergholm (1989), and Taymaz (1991).

The purpose of the model is to explain the microeconomic foundations of economic growth. Growth is endogenized through four mechanisms: entry, re-organization, rationalization, and exit (Eliasson 1994). The primary focus of the model is on the individual firms of the manufacturing sector and how they interact and compete in markets. This sector is therefore modeled in greater detail than other sectors: a government sector, a household sector, and a foreign trade sector. There are also sectors for agriculture/forestry/fishing, construction, oil, electricity, services, and finance, although these are not explicitly modeled at the micro level.

The manufacturing sector is divided into four industries (raw material processing, intermediate goods, investment goods, and consumer non-durables). Each industry consists of a number of firms, some of which are real (with data supplied mainly through an annual survey), and some of which are synthetic. Together, the synthetic firms in each industry make up the difference between real firms and the industry totals in the national accounts. There are approximately 150 real decision-making units covering about 30 % of industrial employment and output, and about 75 synthetic units at the beginning of the simulation.[1] This number increases during the simulations as a result of new entry.

Firms in the model constitute short- and long-run planning systems for production and investment. Each quarter, each firm begins by forming price, wage, and sales expectations that they confront, through their internal accounting system, with a profit margin target. These expectations and targets are then used as inputs into the production planning process in which each firm sets a preliminary production/employment plan. The basic inputs to this planning process are (1) the firm's initial position (level of employment, inventories, etc.), (2) a specification of the feasible production/-

---

[1] The 150 real decision-making units represent divisions within the 40 largest manufacturing companies plus several medium-sized firms.

employment combinations (determined by past investments), i.e., the firm's production function, and (3) a set of satisfactory production/employment combinations.

The firm's initial (*ex ante*) production and employment plans need not be consistent with those of other firms. If, for example, the aggregated employment plans for all the firms exceed the number of workers available at the wage levels the firms intend to offer, an adjustment mechanism is invoked to ensure *ex post* consistency. In the case of labor, the adjustment takes place in a stylized labor market, where the firms' employment plans confront those of other firms as well as labor supply. The labor supply is treated as homogeneous, i.e., labor is recruited from a common "pool" but can also be recruited from other firms. However, the productivity of labor depends on where it is employed. This process determines the wage level, which is thus endogenous in the model. In a similar manner, firms' production plans are revised after a market confrontation in the domestic product market, and domestic prices are set.[2]

There is also a capital market where firms compete each quarter for investment resources and where the rate of interest is determined. Given this interest rate, firms invest as much as they find it profitable to invest, in view of their profit targets.

The exogenous variables which determine the potentials attainable in the model are the rate of technical change (which is specific to each sector and raises the labor productivity associated with new, best-practice investment — see further below) and the rate of change of prices in export markets. What drives the model is the incentive system implicit in the feedback mechanisms (particularly in the labor and product markets). It should be noted further that firms which are unable to reach their profit targets or whose net worth becomes negative, exit from the industry.

Learning, training and R&D activities were not explicitly modeled originally in MOSES, although there is a principal design of an R&D module not programmed into the model in Eliasson (1985). Ballot and Taymaz (1993, 1994) have developed new modules for those activities. Empirical support for the modeling of internal learning is found in the survey of spillover literature in chapter 8 and in the study on the biotechnology competence bloc in chapter 6. In the next section, we will describe the specification of technology. We will then explain how (general) knowledge and (specific) skills affect the performance of the firm. The specifications of the production function and of how investment decisions are made, as well as a short summary of the markets in Moses, are provided in Appendix 9-1.

Technology[3]

The technological level of the firm can be represented by a set of "techniques" as follows.

---

2 There is also an export market whose specification need not concern us here.

3 This section is based on Ballot and Taymaz (1994).

$$F^P = \{f_1^P, f_2^P, \dots f_n^P\} \quad [1]$$

where F^P is the technology used by the firm, and f_i^P is the i th technique, I=1,2,...,n. Superscript P denotes the relevant technological regime. For simplicity (and without any lack of generality), a technique can be assumed to have only two values/alternatives, $f_i^P \in \{f_i^{1P}, f_i^{2P}\}$.

The best-practice technology of a technological regime (the "global technology"), which is defined similarly, describes the best *combination* of techniques. The firm which uses *all* techniques in the set of global technology reaches the highest technological "level" defined by that regime. A firm can know only *a part* of the global technology (or the opportunity set, see Eliasson 1990) and its technological level is determined by the correspondence between the global technology and the technology applied by the firm.

In our experiments, we use a 40-element vector for the global technology. The technological level of the firm depends on the degree of correspondence (DC) between the global technology and the technology employed by the firm as follows.

$$DCP = \sum\nolimits^{n}_{i=1} a_i w_i \quad [2]$$

$$a_i = \begin{cases} 0 \text{ if } t_i^P \neq f_i^P \\ 1 \text{ if } t_i^P = f_i^P \end{cases}$$

where w_i is the weight for the ith technique, T^P the global technology of the pth regime, and F^P the technology used by the firm. t_i^P and f_i^P denote the ith technique of T^P and F^P, respectively.

The technological level of the firm (MTEC) is computed by an exponential function of the DC value.

$$MTEC^P = \alpha^P \, exp(\beta^P DC^P), \quad [3]$$

where α^P and β^P are industry- and regime-specific parameters ($\alpha > 0$, $\beta > 0$) and *exp*(..) the exponential function. In our model, we use the same DC function for all global technologies. Differences in α and β values create a hierarchy between the different global technologies. Thus, the value of $\alpha^P exp(\beta^P)$ defines the absolute limit for the Pth regime, because DC = 1 if $F^P = T^P$. Although there may be many alternative specifications for the DC and MTEC functions, the one used in our model is quite flexible and sufficient for our purpose.

In our simulations, there are three technological regimes in each sector: 1, 2, and 3. We may think of them as historically developed alternative technologies. For example, in the case of factory automation they may represent conventional machining, mass production, and flexible production, respectively. At the beginning of the simulation, all firms use Tech 1. Then they innovate and imitate Tech 2 and/or Tech 3. The technological level of Tech 3 is higher than that of 2, etc.

Learning and Incremental Innovations

Firms use "genetic algorithms" to discover the global technology of a certain technological regime. A firm has a memory to retain k alternative technology sets at a time (in our experiments, 3 sets), and uses the set that has the highest degree of correspondence. Firms "learn" about the global technology by recombining their own technologies (*experimentation*), by recombining their sets with other firms' sets (*imitation*), or by *mutations*. One of the most important processes of evolutionary dynamics, *selection*, takes place at the sectoral level through the selection of firms. Badly performing firms (and technologies used by them) will be forced to exit by the competition process in the market. This is the way learning takes place at the national economic system level (see Eliasson 1995).

The learning process which takes place at the beginning of each year is executed in four steps for all technologies in each firm's memory. *First,* the firm decides whether it will try experimentation (in-house search) or imitation (external search). The higher their stock of general knowledge, the more more likely firms are to try imitation. The probability also depends on the number of firms in the sector using the same type of technology. For example, firms using flexible production technology are more likely to imitate other users of the same technology if there are many of them. But if there are few other users, the firm has to experiment.

Second, if the firm decides to experiment, it will select a technology *from its memory* for recombination. If it decides to imitate instead, it will select a technology for recombination *from another firm* in the same market using the same type of technology. Advanced firms are more likely than others to be imitated.

Third, the firm selects randomly a number of elements from the technology set to be used in recombination. The number of elements to be replaced is determined by the firm's R&D expenditures aimed for incremental innovation/imitation. The values of those elements (i.e., techniques) are replaced by the corresponding elements from the selected technology vector. If the degree of correspondence increases, the firm keeps the modified technology in its memory. Otherwise, the existing technology remains in the memory.

Finally, the technology of the firm may mutate. The probability of mutation depends on the stock of general knowledge and the firm's R&D expenditures aimed for incremental innovation. In the case of mutation, the element which is mutated is replaced by its opposite value ($0 \to 1$, and $1 \to 0$). Thus, an advanced firm which spends a lot on R&D for improvement of its current technology has a higher probability of success than a less advanced firm which spends less on within-regime R&D.

We assume that outside search and mutation probabilities depend on firms' stock of general knowledge. A firm with a large stock of general knowledge will be able to experiment more with other firms' techniques than will a firm with a small stock and, hence, will achieve a higher rate of learning. Intuitively, external search should be better than internal search, since the set of externally available technologies is both broader and larger than internally available technology (because the number of firms in the sector is greater than the number of technologies that reside in a firm's memory). Similarly, forcing firms to compete in wider markets will not only subject them to possibly devastating competition and exit. It will also expose the firm to a richer learning experience. Moreover, except the most advanced firm, at least one firm's technological level is higher than that of the imitating firm. Our simulation experiments with the learning module supports this intuition.[4]

The number of elements to be changed in experimentation, imitation, and mutation is determined by real (ex post) R&D expenditures. In a sense, the firm buys experiments from its R&D unit, and the quality (the probability of success) of those activities depends on the stock of general knowledge.

A firm can improve its technological level by learning and incremental innovations only within the limits of its global technology (technological regime). When the firm gets closer to the limit, it starts to allocate more funds for radical innovation because of difficulties in improving its technological level within the existing regime. Through radical innovation, the firm jumps to another technological trajectory. This is illustrated in figure 9-1.

Figure 9-1 Incremental and Radical Innovation

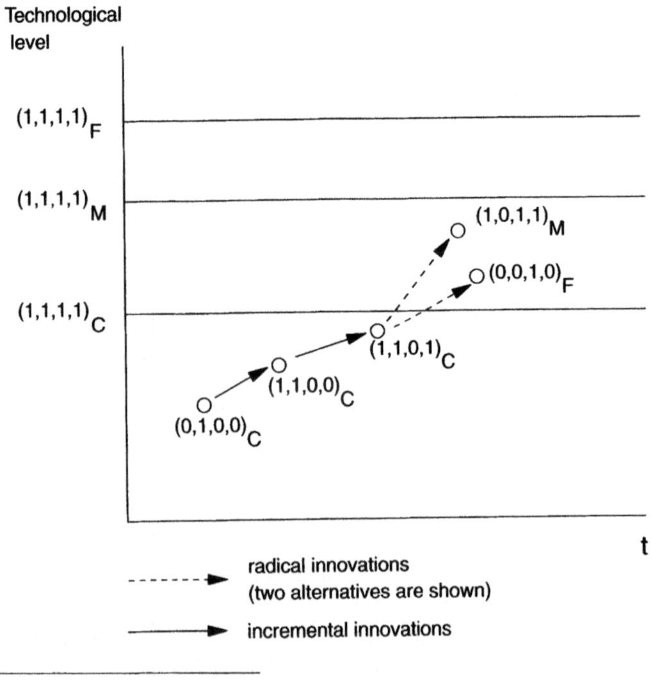

---

[4] In developing the learning module, we made some simulations with only that module to see the relative effects of changes in internal vs. external searches. The results showed that the rate of learning (i.e., the rate of discovery of elements of the global technology) was higher in the case of external searches than in the case of internal ones.

Let us assume that there are three technological regimes: conventional machining (C), mass production (M), and flexible production (F). These technologies consist of four techniques each. For example, the conventional technology consists of the following elements (techniques):

1. Type of cutting tool (1 if high-speed steel, 0 otherwise)
2. Cutting speed (1 if 3600 rpm or greater, 0 otherwise)
3. Cutting feed (1 if 10 cm/min, 0 otherwise)
4. Type of coolant (1 if water, 0 otherwise).

We assume that the highest technological performance (i.e., the highest technological level) is obtained when the firm is using a high-speed steel tool with 3600 rpm speed, 10 cm/min feed, and water coolant. Then the best practice of the conventional technology is represented by the following set:

$\{1,1,1,1\}_C$

where the first element of the set shows the type of cutting tool, the second element cutting speed, etc.

If a certain firm uses $\{1,0,1,1\}_C$, it shows that the firm does not use the appropriate cutting speed. Mass production and flexible production technologies are defined similarly. 'Incremental innovation' means discovering the elements (techniques) of the best practice technology. If the firm discovers all elements, i.e., if the firm uses the best practice technology, it will achieve the highest performance level defined by that technology.

Let us assume that we can rank these technologies in terms of their highest performance levels as follows:

$F \succ M \succ C$

In other words, the firms using the best-practice flexible production technology achieve a better performance than those using the best-practice mass production technology. 'Radical innovation' means changing the technology itself (from C to M or F). When the firm achieves a radical innovation, it will try to learn (discover) the appropriate elements of that technology to improve its performance.

The figure shows these three technologies and the highest performance levels that can be achieved in each (indicated by the horizontal lines). The firm uses $\{0,1,0,0\}_C$ at the beginning. In other words, the firm uses the conventional technology but it uses correctly only the cutting speed. All other elements are wrong. Since the firm is far below the $\{1,1,1,1\}_C$ line (i.e., since there is scope to improve the technology), it can improve its technological performance by incremental innovations. In our example, it first innovates the correct cutting tools (the first element is changed from 0 to 1), then innovates the correct cooling type. When the firm gets closer to the limits of conventional technology, it becomes difficult to improve the performance by incremental innovations because most technological opportunities are already exhausted. Then the firm will try radical innovations. It can either innovate mass production or flexible production. After radical innovation, the firm starts with randomly selected elements. In the figure, two hypothetical radical innovations are shown (from $\{1,1,0,1\}_C$ to $\{1,0,1,1\}_M$, and from $\{1,1,0,1\}_C$ to $\{0,0,1,0\}_F$). Note that the firm will not adopt the flexible technology, because $\{0,0,1,0\}_F$ has a lower performance than $\{1,1,0,1\}_C$ even though it has a much better potential. We assume that the firm does not know the growth potential of new technologies. It compares different technologies only on the basis of available (limited) information.

Radical Innovations and User-Producer Interaction

Firms, especially those close to the technological frontier, may try to achieve a radical innovation (a new type of technology) or to imitate a radical innovation from other firms, as shown in Figure 9-1.

Whether a firm engages in radical innovation or imitation is determined in three steps. The first step is a probabilistic draw for innovation/imitation. The probability of a radical innovation depends on real R&D expenditures aimed for radical innovation, the stock of general knowledge, and knowledge spillovers from other firms in the same sector which is also a positive function of the stock of general knowledge. If the firm turns out to be an innovator, then the new technology PTN to be innovated is determined in the second step; if not, the firm imitates.

In the second step, the new technology to be innovated PTN is randomly drawn from an exponential distribution. The selection probability is inversely related to exp(PTN), which means that new technologies with high potential are less probable than those with lower potential. Finally, in the third step, the techniques of the new technology are determined randomly. The new technology should be seen as a process prototype. Then, besides learning-by doing, the firm tries to improve the technology by experimenting and/or imitating other firms. Even if no other firm is using the new technology, it can learn by experimenting. It estimates the rate of learning over five future years and compares the estimates of the technological efficiencies of the new and the old technology. The learning rate depends on the number of users. The new technology is adopted if it looks better than the old. The old technology is then forgotten, and the firm would have to re-innovate or imitate a technology to go back to it, and this would be costly.

The treatment of radical imitations (adopting a new technology through imitation) introduces an additional variable, the distance between the current technology used by the firm and the technology to be imitated (PTM). The probability of imitation depends on the same variables: R&D expenditures for radical innovation, the general stock of knowledge, and knowledge spillovers from other firms. It also increases with the level of the technology currently used by the firm, since a firm in a high technology has a higher level of technological knowledge. Then the probability that the Mth technology will be imitated depends positively on the number of firms using PTM, the index number of PTM, and the distance between PTM and the firm's current technology. Finally, the techniques of the new technology are selected randomly, since an imitator does not know all the techniques of the imitated firm (there is no patent licensing in the present version). The same trial procedure as for radical innovations takes place.

User-producer interaction also plays a major role. The diffusion of radical innovations requires that capital goods producers and users can imitate technologies from one another. In other words, capital goods producer firms can imitate a radical innovation from *all* firms; other firms (users) can imitate from capital goods producers *and* the firms in their own sector. Thus, we recognize that producers can "learn" from users. In that way, the capital goods sector becomes a "nodal" industry that facilitates the diffusion of radical innovations. Once a radical innovation is adopted by most of the firms in all sectors, the economy will move into a new technological trajectory defined by the radically new technology.

Training and R&D Activities

The stock of specific skills is enhanced by investment in specific training and the volume of past production (learning by doing), but the efficiency of learning by doing is determined by the stock of general knowledge. Competent workers learn faster in the production process. The general stock of knowledge is enhanced by investment in general training. The stocks are depreciated at different rates. The depreciation rate (or the obsolescence rate) of the stock of specific skills, ρ_s, is a function of the rate of improvement in the case of incremental innovations. A different (and much bigger) value is used for radical innovations. (For a more detailed specification of training and R&D, see Appendix 9-1.)

General knowledge, once created, is applicable in all firms and, therefore, transferable. If employees with a high level of general human capital move to another firm, they will increase the stock of general knowledge of the new firm. Firm-specific skills, as the name implies, cannot be transferred from one firm to another. Therefore, firms can increase the stock of specific skills only by specific training and learning-by-doing, whereas the stock of general knowledge can be increased by general training and hiring highly educated workers from other firms. (Sometimes firm-specific skills can be increased only if the firm's general knowledge is increased, but that has not been modeled.)

In our model, it is assumed that the stock of knowledge of a firm affects its problem-solving capabilities. Firm-specific skills play a critical role in the application of what is learned about the global technology. There are two aspects of the "application" process. First, a firm learning more about the global technology can update/improve its current stock of productive equipment. The improvement of existing fixed capital stock as a result of learning global technology depends on the stock of firm-specific skills. Second, the actual use of existing fixed capital depends on the stock of firm-specific skills. A firm endowed with the most productive equipment cannot produce any output at all if employees are not trained to use that machinery. Thus, firm-specific skills can be used both in updating existing equipment (of the old vintage) by favoring learning-by-doing, and in effectively operating the (updated) equipment.

The specification of investment in training and the specification of markets are provided in Appendix 9-1.

Network Interaction: Spillovers and Imitation

There are two types of network interactions among firms: spillovers and imitation. First, knowledge and R&D spillovers are the inputs into innovative activity. Connectivity (the CONNECT parameter) determines the amount of knowledge and R&D capital that a firm emits to other firms. The receptivity parameter determines how much a firm uses the information emitted by other (network and non-network) firms. Connectivity depends on the place of a firm. If it belongs to a network, it gives off more information to those within the network. Receptivity depends on the stock of general human capital within the firm. Therefore, if a firm does not have general human capital, it cannot benefit from being a member of a network.

Second, a firm can imitate techniques (incremental) or the technology (radical) used by other firms. This is somewhat different from the spillover case. In that case, they share *inputs*. In the imitation case, they share the *results* of the innovative process. This is like the strategic cooperation between firms. When two firms (e.g., IBM and Microsoft) come together to innovate a new product/process, they combine their stocks of human capital (and R&D). When Microsoft imitates Apple by developing Windows, we have the second type of interaction.

Thus, firms benefit from belonging to a network mainly for two reasons:

(1) **Increased receptivity.** The effective stock of knowledge (receiver competence) in the firm increases because it can draw on the knowledge and R&D in other firms (especially firms within the same network), not just its own. As a result, the probability of engaging in external search increases, the probability of successful mutations as well as radical imitation and innovation increases, and the learning efficiency is improved — learning by doing is more effective because production rises and experience accumulates more quickly.

(2) **Increased connectivity.** Doors are opened to other firms in the network. Firms therefore become more aware of what others are doing; in effect, the opportunity set increases. For example, if a firm realizes that another firm is using a particular technology, it is more likely to adopt it. This enhances both incremental and radical innovation. Firms can use part of the knowledge stock of other firms in the same network. Spillovers from non-network (network) firms are determined by $CONNECT_{other}$ ($CONNECT_{net}$). In our simulations, $CONNECT_{other}$ is equal to 1 and $CONNECT_{net}$ is equal to 10. (Note that this is a firm-specific variable. The firm can determine the level of knowledge it transmits. This level is fixed in the current simulations, but we plan to endogenize it.)

SIMULATIONS

Specification of Experiments

We have run nine 50-year simulations which may be grouped in the following way. In the **first group**, the main thrust is to explore the implications of different make-up of network membership. This group consists of four simulations in which the main differences are the criteria for membership in a network:

NETRANDOM: There is one network in each sector. At the beginning of the simulation, some firms enter the network. The probability of joining the network is the same for all firms and is equal to 1/3. As new firms are born and enter the industry, they also join the network with 1/3 probability, so that the member/nonmember ratio remains constant. Thus, around 1/3 of the firms in each sector belong to a sector-specific network. This simulation may be thought of as the reference case with which the other experiments are compared.

NETGENHK: 1/3 of the firms in each sector form a more selective sector-specific network. Network members are the firms that have the highest stock of general human capital. Firms are ranked by their stock of general human capital, and the top 1/3 of these firms form the network. A new firm joins the network only if it is among the top 1/3 of firms. In all other experiments, if a firm joins a network, it will stay in it. In this experiment, a firm stays in a network only if it keeps a sufficiently large stock of general human capital (general knowledge). Ceteris paribus, one would expect this case to yield more favorable results than the NETRANDOM case, but that may not always occur if some of the advanced firms are close to the global technology frontier. In that case, technical progress may be temporarily faster in less advanced firms.

NETINTER: There are again four networks, but in this case the networks are intersectoral, i.e., they are made up of members of all sectors. At the beginning of the simulation, firms enter randomly into the networks. The probability of being in a network is again 1/3. New firms entering business randomly select a network (among the four alternatives). New firms join a network with 1/12 probability. Since there are 4 networks, the probability of a new firm joining any network is 1/3. Thus, around 1/3 of all firms belong to one of the four intersectoral networks.[5]

NETLINK: In this experiment, the five technologically most advanced firms (close to the global technology in regime 1) in each sector are first selected. Next, 7-8 more firms from the same sector are randomly selected. Finally, 7-8 firms from the capital goods sector are also randomly selected. Together, the selected firms form a network. Therefore, there are 3 types of firms in each network: advanced users, capital goods

---

[5] There are two types of learning from other firms. First, firms can learn "elements" of the technology from other firms. If Firm A uses {1,0,0,1} and Firm B {1,0,1,1}, Firm A can learn the third element from Firm B. Therefore, the rate of learning is correlated with the number of firms using the same technology. Second, the firm can learn the technology of other firms. If Firm A uses conventional technology and Firm B flexible technology, Firm A can adopt the flexible technology.

producers, and less sophisticated users (followers). Altogether, there are four networks. The total number of network members is around 1/3 of all firms (as in the case of all other experiments). In this experiment, it is assumed that if a user firm has made a radical innovation, capital goods producers in the same network will immediately learn that the user innovated something radical, and try to imitate/adopt that innovation. The producer, however, knows only that there is a change in the user's technology but does not know the technology itself. Therefore, it can imitate/adopt it only if it also invests in R&D aimed for radical innovation.

The **second group** of simulations consists of five experiments in which the magnitude or **intensity of receptivity and/or connectivity** is reduced in comparison with the first group. In each case, a random network is formed, similar to the NETRANDOM case:

LOWRECEPT: In this experiment, the receptivity parameter is reduced to 0.025 from 0.5. That means that the firm cannot benefit as much from information emitted by other firms.

LOWCONNECT: Low connectivity experiment. In other experiments, the connectivity is equal to 10 for the firms that form a network. In this experiment, the connectivity parameter is equal to 2 for firms forming a network. For firms that do not belong to any network, the connectivity parameter is equal to 1. In other words, in the LOWCONNECT experiment, the information exchange between network member firms is two (rather than ten) times higher than that between non-members.

NORECEPT: In this experiment the receptivity parameter is set equal to 0. That means that there is no knowledge spillover. In other words, firms do not *indirectly* benefit from the knowledge stock of other firms. (However, they can still imitate techniques or technologies of others.)

NOCONNECT: In this experiment, the CONNECT parameter is equal to 0. There is neither spillover nor imitation. Firms are "information tight:" they do no emit any information. Their processes are not observed by other firms. In other words, if a firm wants to change its techniques or technology, it will have to do so on its own.

NONET: The combination of NORECEPT and NOCONNECT.

All experiments are run for 55 years, but the first 5 years are not analyzed here because the model adjusts itself to some data inconsistencies in the initial state description at the beginning of the simulation.

Simulation Results

The simulation results for each decade (after the initial 5-year period) are summarized in table 9-1 and are shown graphically in Appendix 9-2. It is hardly surprising that the simulations in which the technological spillovers are the most intensive (i.e., the first group) are the ones generally exhibiting the highest overall economic growth rates and rates of return. What is perhaps surprising is the magnitude of the reduction in the economy's performance when the receptivity and connectivity effects are reduced or eliminated. The average GNP growth rates in the first group of simulations range from 3.3 to 3.5 percent per year, compared with only 0.7 to 3.3 percent in the second group. This suggests that firms learn more from each other and from their environment than they do from their own in-house research and development in isolation. This result, of course, depends on the micro-micro (within-firm) assumptions made, but it seems reasonable; there was no way of anticipating this macroeconomic result purely from understanding the principles of the model. This outcome should be taken as indicative of strong systems effects due to viable combinations of innovation, learning, and spillovers.

Comparing first the two groups of experiments, we find that the performance is higher in the first group in terms of all the variables reported in table 9-1 (except investment whose magnitude has no intrinsic value but rather must be evaluated in terms of the growth it generates). The only simulation in the second group which comes close to the GNP and manufacturing growth rates in the first group is the LOWRECEPT case. The LOWCONNECT case performs well in terms of rate of return during the last two decades — largely, it turns out, because firms fail to invest — but performs dismally in output growth. The LOWRECEPT case exhibits a decent learning rate and generates reasonable stocks of knowledge, skills, and R&D, but the same is true for the LOWCONNECT case only during the first three decades. The learning rate in this case falls throughout the whole simulation. Apparently, firms get discouraged when they are prevented from learning from other firms. Their investments in physical capital and R&D decline, and their stocks of both general knowledge and specific skills fall precipitously.

The NOCONNECT case shows what happens when firms cannot imitate each other's innovations: they invest a lot in both physical capital and R&D but still cannot keep up; the rate of return and learning rate are dismally low. The NORECEPT and NONET cases show absolute decline in both GNP and manufacturing towards the end of the period; the stocks of knowledge and R&D capital also fall in these cases.

Looking next at the various specifications of who the members of the networks are (as distinct from the intensity of contacts between them), i.e., comparing the first group of simulations (NETRANDOM, NETGENHK, NETINTER, and NETLINK), there are no major differences in the overall economic growth rates. The average GNP and manufacturing growth rates for the whole 50-year simulation range from 3.3 to 3.6 and from 3.6 to 4.3 percent per annum, respectively. The rates of return in manufacturing are generally high, ranging between 7 and 10 percent on average over the period as a whole. The main differences that occur are with regard to the magnitude and timing of investment in capital and R&D which, in turn, lead to differences in labor productivity growth rates and in the build-up of both general and specific skills. Also, something apparently goes wrong in the NETINTER case in the last ten years of the simulation, causing a collapse in labor productivity growth, a sharp reduction in the GNP and manufacturing growth rates, and a much more modest increase in investment than in most of the other cases. But in the first four decades the NETINTER case shows the highest rates of accumulation of R&D and skills.

Table 9-1 Summary of Simulation Results

PE-RI-OD	NET-RAN-DOM	NET-GEN-HK	NET-INTER	NET-LINK	LOW-RE-CEPT	LOW-CON-NECT	NO-RE-CEPT	NO-CON-NECT	NO-NET
GNP growth rate (average annual growth rate per decade, percent)									
1	3.38	3.24	3.60	3.57	2.88	2.92	2.60	1.99	1.36
2	3.05	3.04	2.95	2.80	3.04	2.84	2.35	2.11	1.59
3	3.57	3.44	3.73	3.71	2.46	2.65	1.55	2.71	1.08
4	3.91	3.89	3.72	3.82	4.16	1.14	-0.48	3.14	0.26
5	3.64	3.35	2.39	3.87	3.83	-1.61	-1.31	3.04	-0.64
Avg.	3.51	3.39	3.28	3.55	3.27	1.59	0.94	2.60	0.73
Manufacturing growth rate (average annual growth rate per decade, percent)									
1	4.55	4.10	4.76	4.75	3.70	3.57	3.38	2.72	1.65
2	3.95	3.93	3.61	3.36	4.21	3.57	3.08	2.94	2.37
3	4.39	4.20	4.38	4.40	2.98	3.43	1.84	3.93	1.10
4	4.32	4.17	4.39	4.25	4.46	0.19	-2.17	3.53	-0.48
5	4.04	3.51	1.10	4.22	4.08	-3.28	-2.58	3.36	-1.95
Avg.	4.25	3.98	3.65	4.20	3.89	1.50	0.71	3.30	0.54
GNP (relative to NETRANDOM, end of period)									
1	1.00	1.00	1.02	1.02	0.95	0.98	0.90	0.87	0.81
2	1.00	0.99	1.04	1.01	0.96	0.98	0.84	0.80	0.71
3	1.00	0.94	1.01	0.99	0.82	0.90	0.66	0.70	0.55
4	1.00	0.98	0.98	1.00	0.88	0.65	0.42	0.68	0.38
5	1.00	0.96	0.88	1.01	0.85	0.38	0.26	0.64	0.25
Manufacturing output (relative to NETRANDOM, end of period)									
1	1.00	0.98	1.02	1.03	0.91	0.96	0.85	0.84	0.74
2	1.00	1.00	1.05	1.01	0.98	0.95	0.80	0.78	0.65
3	1.00	0.93	0.97	0.96	0.78	0.84	0.58	0.68	0.48
4	1.00	0.99	0.92	0.97	0.86	0.55	0.30	0.66	0.29
5	1.00	0.96	0.78	1.00	0.79	0.26	0.15	0.64	0.16
Rate of return (average annual return per decade, percent)									
1	-1.82	-1.13	-0.90	-0.90	-1.81	-0.81	-0.88	-0.59	-1.22
2	-6.72	-6.62	-4.35	-5.92	-7.44	-7.20	-5.71	-6.75	-6.80
3	4.71	4.89	11.10	3.97	-0.21	2.04	1.01	0.31	-5.40
4	16.76	20.49	22.93	17.75	7.44	21.29	10.24	4.68	0.61
5	22.77	30.53	21.40	29.78	14.28	26.10	9.52	6.23	1.40
Avg.	7.14	9.63	10.04	8.94	2.45	8.28	2.84	0.78	-2.28

Table 9-1 (continued)

PE-RI-OD	NET-RAN-DOM	NET-GEN-HK	NET-INTER	NET-LINK	LOW-RE-CEPT	LOW-CON-NECT	NO-RE-CEPT	NO-CON-NECT	NO-NET
Stock of special skills (end of period, billion SEK)									
1	102	97	101	112	86	98	78	93	80
2	143	147	156	122	137	125	91	140	89
3	215	125	184	175	171	125	90	141	99
4	239	202	322	349	221	68	60	157	68
5	278	348	352	466	212	55	51	191	56
Stock of general knowledge (end of period, billion SEK)									
1	111	104	110	119	107	103	104	104	106
2	178	162	168	170	150	158	114	140	111
3	261	237	286	228	215	201	119	190	121
4	379	314	470	377	285	127	77	237	87
5	521	549	528	683	383	81	56	330	66
R&D stock (end of period, billion SEK)									
1	79	70	72	92	65	67	64	54	55
2	161	155	159	153	131	140	95	114	83
3	260	205	300	208	187	189	109	135	98
4	355	304	469	401	245	125	77	169	74
5	472	546	526	826	325	94	62	241	59
Learning rate (average annual growth rate per decade, percent)									
1	6.52	5.91	5.51	6.15	5.18	5.58	4.14	0.16	0.16
2	3.69	4.19	4.04	3.59	3.26	4.95	2.36	0.02	0.02
3	4.29	2.45	3.42	3.78	3.10	3.45	2.20	0.02	0.02
4	3.50	3.38	4.90	4.74	3.88	1.98	0.49	0.02	0.02
5	3.50	3.20	3.67	4.10	3.00	0.47	0.25	0.04	0.02
Avg.	4.30	3.83	4.31	4.47	3.68	3.29	1.89	0.05	0.05
Investment (average annual investment per decade, MSEK)									
1	16,347	16,022	16,006	16,818	16,014	15,791	16,374	19,200	18,183
2	25,999	24,512	27,722	23,004	26,564	25,190	21,747	33,701	31,146
3	38,202	37,798	32,279	16,612	51,617	23,878	23,514	95,112	53,449
4	125,990	101,932	54,828	31,533	100,864	11,657	22,117	245,300	59,344
5	348,524	215,376	91,689	213,401	239,100	11,704	25,017	812,504	93,739
Labor productivity growth rate (average annual growth rate per decade, percent)									
1	2.32	2.46	2.29	3.14	0.95	1.79	0.93	0.40	-0.50
2	4.37	4.62	4.41	2.82	4.40	3.71	2.75	3.27	1.52
3	4.95	2.64	5.82	5.47	2.05	3.97	1.80	1.69	0.87
4	3.23	4.67	3.86	6.09	3.74	0.29	-0.43	1.77	-0.41
5	3.57	4.80	0.02	5.37	2.95	-1.00	-0.77	2.22	-0.87
Avg.	3.69	3.84	3.28	4.58	2.82	1.75	0.86	1.87	0.12

An examination of the stocks of special skills, general knowledge, and R&D shows that the NETLINK case outperforms all the others. It also has the highest learning rate and labor productivity growth rate. Thus, the spillover effects are slightly more positive when there are vertical as well as horizontal linkages among the participants than in the case of absence of such linkages (NETRANDOM and NETINTER). The spillovers (as measured by the resulting stocks of special and general knowledge as well as R&D) are greater when only the firms with the most human capital (NETGENHK) form the network than when the membership is random (NETRANDOM and NETINTER). The learning and spillover effects are impressive in the NETINTER case until the collapse in the last decade of the simulation. But the results prior to that suggest that having network membership drawn from a wide variety of actors yields positive spillover and learning effects.

The differences among the simulations in the second group are also illuminating. The LOWRECEPT case shows steadily increasing stocks of general knowledge, special skills, and R&D, as well as a fairly high learning rate — just not as high rates of growth as in the first group. Thus, reduced spillover effects yield generally worse performance. But the results get much worse when the connectivity within the network is reduced (the LOWCONNECT case); the stocks of knowledge and R&D actually fall. Firms reduce their investment, and both GNP and manufacturing output eventually fall as well. A comparison between the LOWRECEPT and NORECEPT cases shows that even a reduced intensity of spillovers is better than none: the stocks of knowledge and R&D actually fall in the last half of the simulation in the NORECEPT case. In the NO-CONNECT case, firms are unable to imitate each others' innovations even though they are still able to reap benefits from spillovers. They invest more than they do in any of the other simulations and achieve steady, albeit not impressive, economic growth — at the expense, however, of low rates of return. The NONET case, finally, shows the dismal results of slow or negative knowledge accumulation.

CONCLUSION

The simulations suggest that it is important to be connected to a network. The basic reason is the limited knowledge and experience in each firm compared to that which exists elsewhere. Thus, the yield on each investment, whether in physical capital or knowledge, rises as a result of spillovers, and the economy grows faster. It does not seem to matter much who the other network members are; the important thing is that networks expand the range of options available to each participant and increases the experience base upon which decisions can be made. The performance of the economy clearly declines when both connectivity and receptivity are reduced.

REFERENCES

Albrecht, James, F. Bergholm, G. Eliasson, K.A. Hanson, C. Hartler, M. Heiman, T. Lindberg and G. Olavi, 1989. *MOSES Code*. Research Report No. 36. Stockholm: IUI.

Ballot, Gérard and Erol Taymaz, 1993. "Firm Sponsored Training, Technical Change and Aggregate Performance in a Micro-Macro Model," paper presented at the IARIW special conference on micro simulation and public policy, Canberra, Australia, December 6-8.

Ballot, Gérard and Erol Taymaz, 1994. "Training, Learning and Innovation. A Micro-to-Macro Model of Evolutionary Growth," paper presented to the International J. A. Schumpeter Society Conference, Münster, Germany, August 17-20.

Bergholm, Fredrik, 1989. *MOSES Handbook*. Research Report No. 35. Stockholm: IUI.

Carlsson, Bo (ed.), 1995. *Technological Systems and Economic Performance: The Case of Factory Automation*. Boston and Dordrecht: Kluwer Academic Publishers.

Carlsson, Bo and Erol Taymaz, 1995. "The Importance of Economic Competence in Economic Growth: A Micro-to-Macro Analysis," in Bo Carlsson (ed.), *op. cit.*, 359-389.

Eliasson, Gunnar, (ed.), 1978. *A Micro-to-Macro Model of the Swedish Economy*. IUI Conference Reports 1978:1. Stockholm: IUI.

Eliasson, Gunnar, 1985. *The Firm and Financial Markets in the Swedish Micro-to-Macro Model — Theory, Model and Verification*. Stockholm: IUI.

Eliasson, Gunnar, 1990. "The Firm as a Competent Team," *Journal of Economic Behavior and Organization*, 13 (3), 273-298.

Eliasson, Gunnar, 1991. "Modeling the Experimentally Organized Economy: Complex Dynamics in an Empirical Micro-Macro Model of Endogenous Economic Growth," *Journal of Economic Behavior and Organization*, 16 (1-2), 153-182.

Eliasson, Gunnar, 1994. "Technology, Economic Competence and the Theory of the Firm — Discussing the Economic Forces behind Long-Term Economic Growth," in Ove Granstrand (ed.), *Economics of Technology*. Amsterdam: North Holland, 125-164.

Eliasson, Gunnar, 1995. "The Economics of Technical Change—The Macroeconomic Consequences of Business Competence in an Experimentally Organized Economy," *Revue d'Economie Industrielle*, numéro exceptionnelle, 53-82.

Taymaz, Erol, 1991. *MOSES on PC: Manual, Initialization, Calibration*. Stockholm: IUI.

APPENDIX 9-1

The Production Function

The production function for each firm in MOSES is of the following form:

$$Q_t = QTOP_t*(1\text{-}exp(\text{-}TEC_t*L_t/QTOP_t)) \quad [1]$$

where Q is the potential output (in physical units) for a given employment level in number of hours (L_t), QTOP is the maximum level of output which is approached asymptotically when infinite amounts of labor are used, given a certain level of capital stock, and TEC is the productivity of the first unit of labor (to be more precise, the slope of the production function at the origin), and *exp*(.) the exponential function. Subscript t denotes time.

The maximum output, QTOP, depends on the (real) stock of physical capital, the stock of specific skills, and the efficiency of the stock of physical capital as follows.

$$QTOP_t = QTOPFR_t*\{MINRT+[(1\text{-}MINRT)*ST_1*(1\text{-}exp(\text{-}SPECTR_t/ST_2))]\} \quad [2]$$

where QTOPFR is the level of productive capacity (the asymptotic limit of the QTOP variable) and SPECTR is the stock of specific skills. ST_1 and ST_2 are industry-specific parameters. MINRT is the minimum (percentage) level of output that can be produced with no stock of specific skills. It shows the productivity level of completely unskilled workers. Thus, if SPECTR=0, QTOP=QTOPFR*MINRT. On the other hand, as SPECTR $\rightarrow \infty$, QTOP $\rightarrow$ QTOPFR. The productive capacity, QTOPFR, is determined by the efficiency and amount of physical capital (QTOPFR = EFF*PK, where EFF is the output/capital ratio, and PK the stock of physical capital).

As shown in the specification of the production function (equations 1 and 2), there are two critical *technology* variables that determine the performance of the firm: EFF and TEC. There are two methods to upgrade these variables. In the case of *embodied* change, the technological level is upgraded by investment, since new equipment embodies the stock of knowledge possessed by the firm. In the *disembodied* case, the EFF and TEC variables are increased by implementing what is known by the firm (organizational changes, rationalization, etc.).

Investment is the first mechanism to increase the EFF and TEC variables, since newly installed capital embodies MEFF and MTEC levels of technology. The technological level of the capital stock is upgraded by investment as follows:

$$TEC_t^\tau = [(TEC_{t-1}*QTOPFR_{t-1})+(MTEC_t*\Delta QTOPFR_{t-1})]/(QTOPFR_{t-1}+\Delta QTOPFR_{t-1}) \quad [3]$$

$$EFF_t^\tau = [(EFF_{t-1}*PK_{t-1})+(MEFF_t*INV_{t-1})]/(PK_{t-1}+INV_{t-1}) \quad [4]$$

where INV is the amount of investment and ΔQTOPFR is the addition to the productive capacity by investment ($\Delta QTOPFR_{t-1}=INV_{t-1}*MEFF_{t-1}$). Superscript τ indicates that the TEC_t^τ and EFF_t^τ variables have temporary values within each time period (quarter). Their final values for the period t are calculated in equations 5 and 6 below.

MEFF is the efficiency of newly installed capital and MTEC is the level of labor productivity associated with new capital. As shown in equations 3 and 4, the TEC and EFF variables are calculated as weighted averages of the technological levels of different vintages of capital. In a sense, the MEFF and MTEC variables reflect the stock of knowledge possessed by the firm. The technological level of the productive equipment actually used (measured by the TEC and EFF variables) is lower than the level known by the firm because of the vintage effect.

We assume that the firm can also gradually update its existing equipment without any investment in physical capital stock by applying what is learned as follows:

$$TEC_t = TEC_t^{\tau} + (MTEC_t - TEC_t^{\tau}) * ST_3 * [1 - exp(-SPECTR_t/ST_4)] \quad [5]$$

$$EFF_t = EFF_t^{\tau} + (MEFF_t - EFF_t^{\tau}) * ST_5 * [1 - exp(-SPECTR_t/ST_6)] \quad [6]$$

where ST_3, ST_4, ST_5 and ST_6 are industry-specific parameters. Thus, as $SPECTR \rightarrow \infty$, the TEC and EFF variables are updated to the amount equal to ST_3 and ST_5 percent of the difference between known and applied knowledge. Thus, we assume that the stock of *specific* skills (the skills specific to the physical capital used in the firm) determines the pace of disembodied change for a *given* level of the MTEC and MEFF variables. The values of the MTEC and MEFF variables depend on the efficiency of learning which is determined by the stock of *general* knowledge.

Equations 1 and 2 define the production function. Equations 3-6 show the way the technology variables are changed in the model. The following capital accumulation equation closes the system of production.

$$QTOPFR_t = [QTOPFR_{t-1} * (1 - \delta_q)] + \Delta QTOPFR_{t-1} \quad [7]$$

where δ_q is the (constant) rate of depreciation.

Training and R&D Activities

Training is essential in generating general knowledge and firm-specific skills. We also allow learning-by-doing to take place. The stocks of general knowledge and specific skills are accumulated as follows:

$$SPECTR_t = (SPECTR_{t-1} * (1 - \rho_s)) + LEARNEFF_{t-1} * f(Q_{t-1}/L_{t-1}, INVST_{t-1}) \quad [8]$$

$$GENTRSTOCK_t = (GENTRSTOCK_{t-1} * (1 - \rho_g)) + INVGT_{t-1} \quad [9]$$

where ρ_s and ρ_g are depreciation parameters, and INVST and INVGT are (real) specific and general training expenditures per employee, respectively. f(..) is an exponential function, and LEARNEFF is the efficiency of learning which depends on the stock of general knowledge.

Investment in Training

The level of desired investment in training depends on three variables: existing stocks of knowledge and specific skills, (the inverse of) the rate of utilization of potential capacity (QTOP/QTOPFR), and sales revenue. Firms tend to increase stock at a certain rate and to spend a part of their sales revenue on training. If the QTOP/QTOPFR ratio is low, the firm will spend more on training since a low value of that ratio indicates that the firm is not able to use efficiently its productive capacity because of the lack of specific skills.

The desired level of investment in R&D depends on the stock of general knowledge, sales revenue, and the emphasis on radical innovations. If the firm spends a large part of its R&D funds for generating radical innovations, the level of R&D funds will be increased.

In addition to training and R&D, the firm calculates its desired level of investment in physical capital and liquid assets. Then, given the level of net cash flow, the firm decides the level of desired borrowing (desired total investment *minus* net cash flow). The actual level of borrowing depends on the resources of the bank and total demand for borrowing. Finally, after the level of borrowing is set in the credit market, the firm allocates its resources (net cash flow *plus* net borrowing) among four different assets (training, R&D, physical capital and liquid assets) in proportion to their desired levels.

Total investment in training, INVTR, is allocated to general and specific training. The allocation between general and specific training depends on a distribution parameter and the QTOP/QTOPFRONT ratio. Similarly, total investment in R&D, INVRD, is also allocated to two different types of R&D investments: RDRADICAL (for radical innovations) and RDINC (for incremental innovations). The allocation depends on recent improvements in the MTEC and MEFF variables and the number of firms adopting an advanced technology. If the firm is able to improve the MTEC and MEFF variables, then it will spend more on incremental innovations. On the other hand, if many firms are using a new global technology, the firm will spend more on radical innovations to adopt the new technology.

Markets

Firms confront other firms on the various markets, on a quarterly basis. First they make plans, based on adaptive expectations. The top management sets a target rate of return as a minimum profit margin based on past experience (satisficing rule). This target translates into a minimum labor productivity level. Then the division head examines whether this level is compatible with the potential production Q allowed by the present employment level. If not, the firm will search (through trial and error) for a level of production that satisfies both the minimum productivity level and the (unknown) production frontier. The employment plan is then set.

The labor market is decentralized. Each firm sets its wage rate. When it wants to hire, it tries both the unemployment pool and the employees from the other firms. The latter quit if they are offered a wage rate at least 10% higher than their present wage rate. If the wage offer does not attract these employees, the firm may raise the wage a limited number of times before the contracts are signed. Then firms hire or lay off. At the end of the quarter, there remain unemployed workers and vacant jobs simultaneously. The resulting macroeconomic effect is characterized by a Beveridge curve (not always stable) and a wage curve (unemployment depresses wages or rather wage growth).

The average stock of general human capital of the transferred workers is equal to the average stock of the attacked firm. The new stock of general human capital in the attacking firm is a weighted average of the existing and the transferred workers.

Simultaneously production plans are confronted on the product markets (one per industry) and may undergo revisions, since plans may not be compatible, due either to the consumers' purchasing power or the market for inputs. The prices are set endogenously in the process. Presently there is only one price per industry, since the products of the different firms are not differentiated.

Finally, there is a capital market where firms compete each quarter for investment resources. The rate of interest is set endogenously within a band. Firms may get rationed on their borrowing.

APPENDIX 9-2

Figure A9-2-1 Manufacturing Output (index, NETRANDOM - 100)

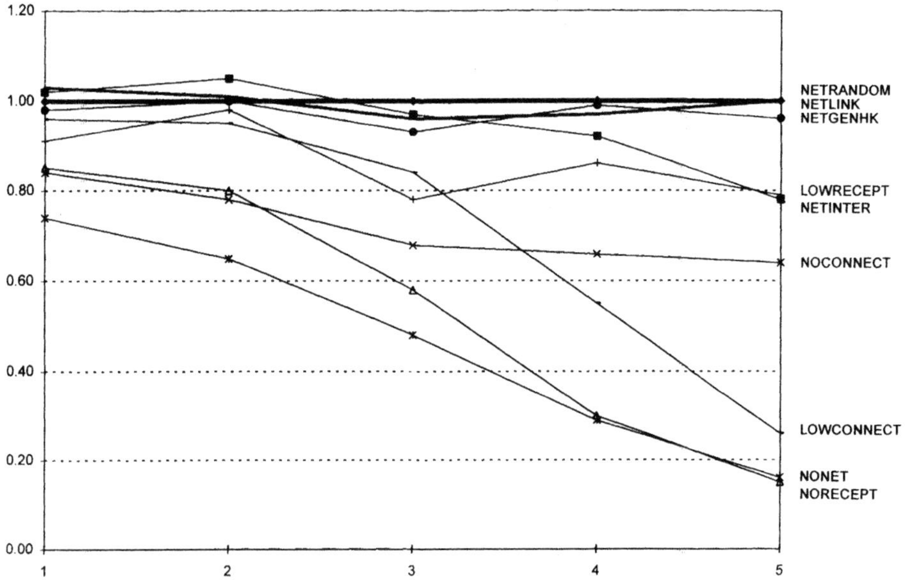

Figure A9-2-2 Learning Rate (average annual growth rate per decade, percent)

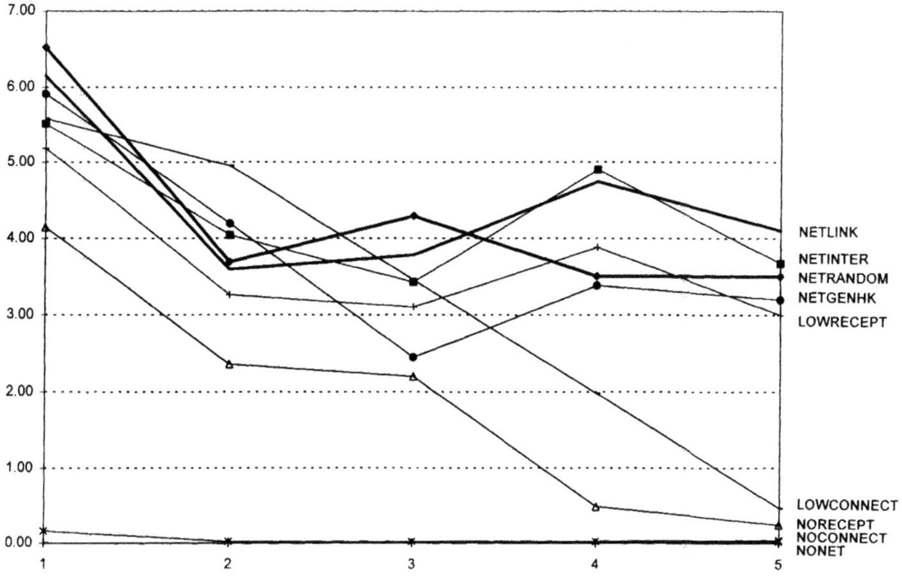

Figure A9-2-3 R&D Stock (end of period, billion SEK)

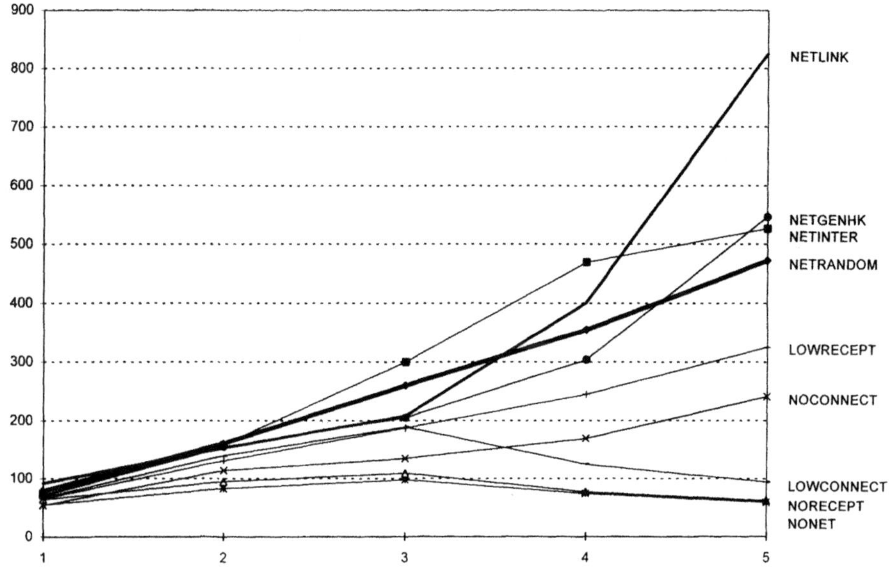

Figure A9-2-4 Rate of Return (average annual return per decade, percent)

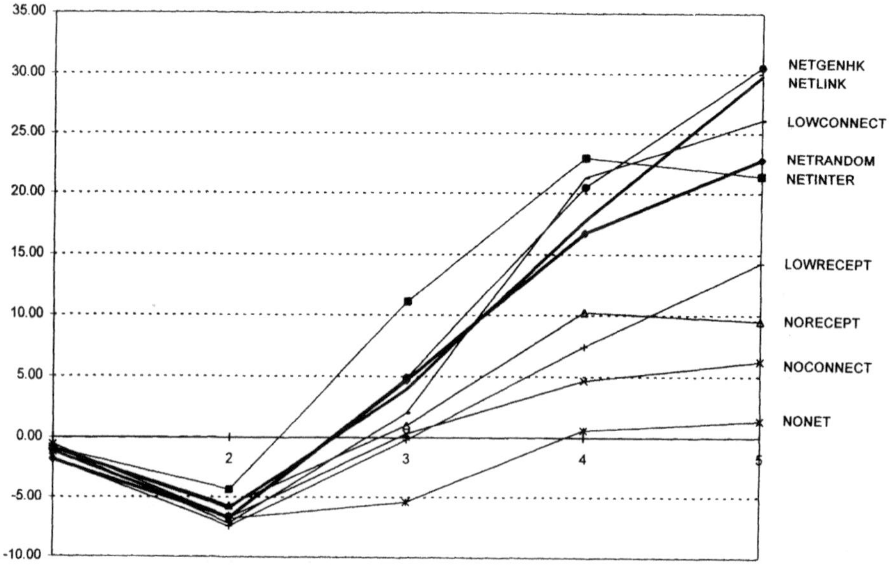

Figure A9-2.5 Stock of General Knowledge (end of period, billion SEK)

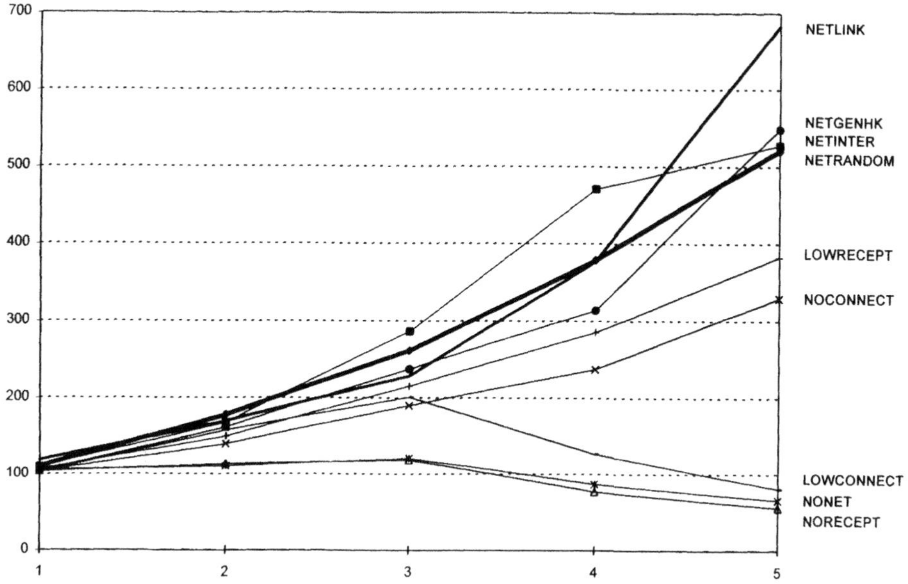

Figure A9-2.6 Stock of Special Skills (end of period, billion SEK)

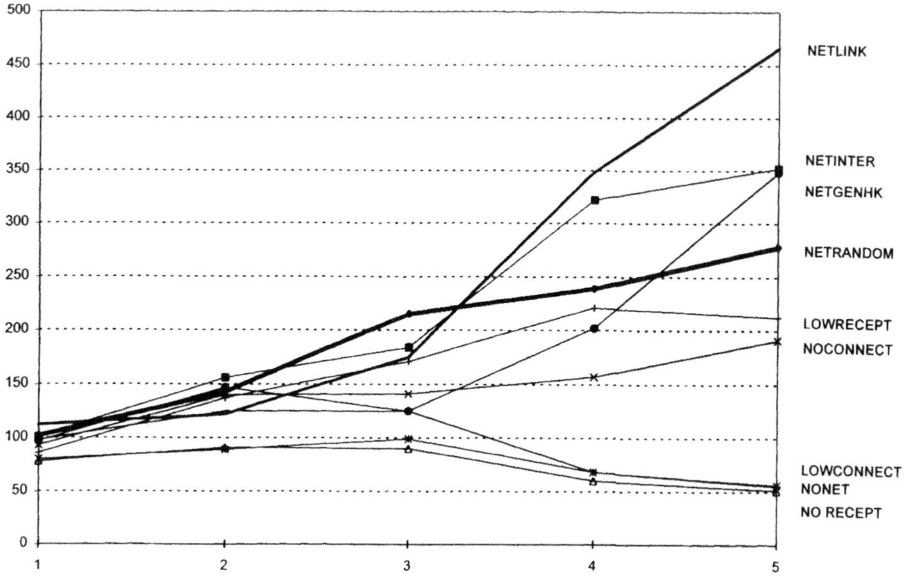

Figure A9-2-7 Labor Productivity Growth Rate (percent per year)

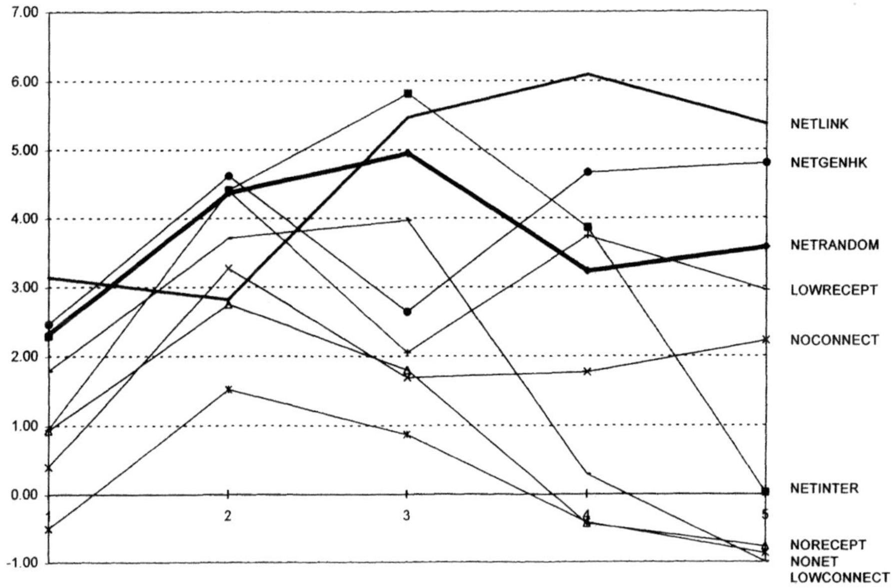

10

FOUR TECHNOLOGICAL SYSTEMS: WHAT HAVE WE LEARNED?

Bo Carlsson

INTRODUCTION

New insight is the result of creative thinking and innovative research. Systematic empirical observations, creatively interpreted, form the foundation of path-breaking new theory. When formulated and tested, such theory extends the frontier of knowledge and opens up new avenues for research.

The preceding chapters have summarized the work carried out by our research team during several years in each of the four areas of technology covered thus far. We have utilized the theoretical framework specified at the beginning of our study, namely the technological systems approach (see Carlsson and Stankiewicz 1991). It is not surprising that we have made new discoveries in the course of our work — more precisely, we have learned how better and more systematically to organize observations about technological systems — so that we are now in a position to extend and improve the theoretical framework for future studies. Therefore, rather than simply summarizing our findings in the preceding chapters, we will systematize and reformulate our observations according to a more formal and fully elaborated scheme than employed up to now. In particular, reflecting upon our research experience thus far, we have found that there are several aspects of technological systems which were left out in our initial approach but which can be incorporated if we go more deeply into our material and include additional sources as well. The most important aspect has to do with dynamics: the evolution over several decades of technological systems as distinct from a static view of each system at a moment in time. Asking time-related questions forces us to examine the time frame and initial conditions of each system, to analyze who the various actors have been, what role they have played, how these roles have shifted over time, and how the institutional framework has evolved. Other aspects which are also given more emphasis below than in the preceding chapters are the role of capital (especially the sources of venture capital), international linkages, and geographic proximity.

The chapter is organized as follows. The general framework is presented in the next section. This is followed by an analysis of each of the four main features of technological systems. Some further reflections on similarities among the systems and on questions for future research conclude the chapter.

THE GENERAL FRAMEWORK

One of the basic assumptions underlying our work is that the technological opportunities facing any economic agent are virtually unlimited; the pool of global possibilities has practically no boundaries. The first problem for each economic agent is therefore to identify and successfully exploit a small subset of opportunities, one of the main constraints being the boundedness of the agent's own economic competence (defined as the ability to identify, expand and exploit business opportunities — see Carlsson and Eliasson 1994).

Some areas of the global opportunity set have been fairly well explored and mapped out, others not at all. Some areas are highly accessible, with well-built access routes and well-established rules and procedures. Here knowledge is highly codified. Provided potential users have the proper equipment, they can access the technology. This results in a short distance between the science frontier and best industrial practice. Other areas are barely touched by human inquiry and not yet applied in economic activity. Here access is much more haphazard and experimental. In some areas, the knowledge required for successful exploitation is highly specific and acquired by only a few actors, whereas in others general knowledge in the form of a broad range of skills is necessary and may be acquired by many actors. Sometimes new discoveries are made which radically alter the economic potential and accessibility of a technology, generating vast changes in one or several systems.

The nature of knowledge in each area determines the ways in which new discoveries are made and new combinations of previous knowledge are successful. This means that the potential for technological spillovers (intended as well as unintended learning from others) varies between areas and that the prerequisites to absorb spillovers also vary. Identifying new opportunities and finding ways (i.e., technologies) to exploit them may require a combination of global search or monitoring and in-house research and development.

Our work on four technological systems in Sweden suggests that there are four features of such systems which determine their performance, i.e., the degree of successful economic exploitation: (1) the nature of knowledge and spillover mechanisms; (2) the receiver competence (or absorptive capacity, to use the terminology of Cohen and Levinthal 1990) on the part of various actors, (3) the connectivity between various parts of the system, and (4) the vigor of variety creation and selection mechanisms. These features are summarized in Table 10-1.

Table 10-1 General Framework for the Analysis of Technological Systems

Nature of Knowledge and Spillover Mechanisms
System definition
The nature of knowledge:
- tacit vs. codified or articulated
- component vs. architectural
- embodied vs. disembodied

Spillover mechanisms: who and how?

Receiver Competence
Prime movers:
- Who created/raised awareness?
- Who took the early initiatives?
- Entrepreneurship

How was critical mass created?
Market failures/obstacles?
How were the obstacles overcome?
Role of institutions and public policy
- Source and role of venture capital
- Role of finance and financial markets
- Role of academia
- Role of educational policy: proactive vs. reactive
- International links

Connectivity
Importance of proximity/clustering and in what sense?
User-supplier relations
Problem-solving networks
- Network characteristics
- Who built the networks?
- Bridging institutions
- Role of business groups
- Role of public policy

Informal/personal networks

Variety Creation Mechanisms
"Bounded vision" - how characteristic?
Path dependency; lock-in effects
Entry and exit
International impulses
Role of public policy

As already indicated, **the nature of knowledge** in each field determines both **the potential and the mechanisms for spillovers**. If knowledge in a particular system is tacit by nature (as distinct from codified or articulated), spillovers are likely to take place primarily via personnel transfers. Thus, physical proximity and high receiver competence are likely to be important. If the crucial knowledge is tied to individual components of the system rather than more architectural (over-arching) in character, the carriers of knowledge are likely to be more numerous and dispersed. If knowledge is embodied in artifacts rather than disembodied, the receiver competence required for successful transfer may be much lower.

Receiver competence refers to the ability to tap into the global technology set. In each field studied thus far, a few leading edge firms — prime movers — who are also major (potential) users have formed the link to the global opportunity set. They have done so by raising their own receiver competence to a high level. This has been achieved via investment in R&D, most of which is oriented towards locating the technology frontier and identifying who is on it, the remainder of the R&D effort being devoted to raising in-house competence and (occasionally) pushing out the global technology frontier. Through this activity, leading-edge firms raise not only their own competence; they also generate significant spillovers, (a) by setting high standards in the form of specific technical requirements for their suppliers, and (b) by training their own labor force, thus supplementing the public education system. These highly trained workers, mainly engineers, raise the receiver competence of the firms they work for, but some of their knowledge spills over to other firms as well, either directly via personnel transfers, or indirectly via business transactions. Suppliers as well as customers can learn via business transactions, and competitors can learn by imitation.

The key questions therefore are: who are the prime movers, and what has been their specific role in creating awareness, taking initiatives, and overcoming obstacles? What role have various institutions played? In particular, what have been the sources of finance (esp. venture capital) and the role of academia and international linkages, and what has been the role of public policy?

The nature and importance of spillovers are determined by the degree of **connectivity**. A technological system which exhibits a dense network of links among its participants is more robust and benefits more from spillovers than one which has weak and few links. The higher the density in a given geographical area, the greater the potential for spillovers.

Each technological system exhibits a variety of overlapping/interrelated networks. There are three main types: (1) the buyer-supplier links, (2) the technical problem-solving network, and (3) the informal national community network. In the systems we have studied in Sweden, the buyer-supplier relations are essentially international in scope, reflecting the characteristics of a small, open economy.

In the technical problem-solving network, which frequently overlaps the first, various actors addressing a shared R&D task are inter-linked via collaboration and exchange. It is the problem-solving network that provides the core links of the technological system.

The 'national community network' is informal and links, by way of personal contacts, individuals active in the various sub-fields and specialties of the technology. The common interest makes it both extensive and loosely knit. This relatively stable network cuts across the various sectors and levels of the technological system.

In order to describe the connectivity within each system, the key questions are what types of networking relationships exist, who are the participants, and what is the role of clustering, geographic or otherwise?

The vigor (or survival, for that matter) of the system over the long haul is determined also by the extent to which it is continually rejuvenated by being exposed to new challenges in the form of new competitors (both players and products). The market is often an exceedingly tough selection environment; failure and exit are inherent in the competitive process. The more path-dependent and resistant to change the actors are, the greater are the risks of a gradual reduction in the number of players and hence the number of links into the global opportunity set. **Mechanisms to create more variety** are needed in order to avoid the gradual erosion and eventual collapse of the system.

The main idea here is to examine how closed or open the system is, the extent to which the main players have a broad or narrow vision of their role, and the extent to which they benefit from or are impeded by their past history. To what extent is entry of new players encouraged? What is the role of institutions and of public policy?

Each of these features operates within an institutional framework in the form of both 'rules of the game' and certain organizations. Institutions may evolve independently or they may co-evolve with the technological system of which they are a part. They may certainly be influenced through public policies which enhance or hinder each feature. For example, receiver competence can be enhanced through improved public education; connectivity can be increased by building or encouraging "bridging institutions," and variety can be created by promoting entrepreneurship via tax reform and promotion of venture capital formation.

The use of the systems concept varies somewhat in our analysis of the four systems. Factory automation and electronics are both viewed as consisting of a cluster of closely related technologies with many common (or at least highly similar) components (institutions, buyers, suppliers, networks, etc.). Powder technology is viewed as consisting of two main clusters (powder metallurgy and ceramics), while the technological system for pharmaceuticals and biotechnology is viewed mainly from the point of view of the pharmaceutical industry whose technology base has shifted over time. Factory automation and pharmaceuticals have a long history in Sweden, while electronics and powder technology have a shorter history.

In principle, it is possible to apply the framework suggested here to an analysis of the evolution of each system. However, a truly dynamic analysis is more hinted at than fully carried out as it would require more information along the time dimension than is currently available, given our resource constraints and the original design of the study.

NATURE OF KNOWLEDGE AND SPILLOVER MECHANISMS

The characteristics of knowledge in each area of technology determine both the likely carriers of such knowledge and the mechanisms through which it can be transferred and diffused. This section defines the boundaries of the four technological systems studied and the types of spillover which occur in each area.

Factory automation is defined as "the cluster of technologies... focused on solving certain technical problems involving the enhancement, assistance, and... substitution of human labor through the use of certain techniques or machines in manufacturing operations" (pp. 37-38 above). Further, the technology consists of both disembodied knowledge (constituent disciplines and specialties) and knowledge embodied in artifacts such as numerically controlled machine tools, industrial robots, etc.

The academic disciplines which constitute the knowledge base are well established internationally, particularly in Germany, but they are not especially strong in Sweden (Granberg 1995). The Swedish strength has traditionally been in the practical application of this knowledge in artifacts and techniques. Such applications are highly task- or machinery-specific and involve mostly tacit knowledge which incorporates and builds on long engineering traditions, i.e., it is strongly path-dependent. It is also strongly component-oriented and embodied in equipment, but the equipment itself fits into complex systems. This means that high receiver competence is needed, particularly in the form of systems know-how. It is therefore not surprising that there are several highly competent leading-edge users, each with its own systems know-how, and that these prime movers have played a dominant role.

The technological system for **electronics and computers** consists of three main clusters: radio and telecommunications, military electronics, and power electronics (power generation and distribution). There are also two lesser clusters focused on factory automation and medical electronics. In all of these clusters, our primary interest is in the application of digital information technology. Thus, the time frame is from the 1970s onward.

The nature of knowledge is difficult to characterize, particularly in view of the fragmentation of the field. It is useful to distinguish between product design, production, and application of electronic hardware. The three main clusters considered in the technological system as we have defined it here represent different applications of microelectronics. While the design and production of semiconductors and other microelectronic hardware require different configurations of knowledge, the application side is driven by detailed knowledge of each specific application. This means that the relevant knowledge is often tacit in character, embodied in software, and therefore integrated with other aspects of the hardware. This also implies a high degree of path dependence.

Our examination of the **pharmaceutical sector** consists of two studies, both involving Astra, the largest Swedish pharmaceutical company. One of the studies (chapter 5) focuses on the development of so-called beta blockers, and the other (chapter 6) on the competence bloc supporting biotechnology, especially as exemplified in the development of the gastrointestinal drug Losec. Both studies emphasize the experimental nature of the discovery process (i.e., the importance of serendipity and the unpredictability of outcomes), the dependence on academic research and on close working relationships between academia and industry. The importance of a decentralized organization of R&D (i.e., multiple decision makers) is also stressed. Interestingly, the technology base in the pharmaceutical industry shifted during the period studied from chemistry to biology, and Swedish firms happened to be particularly well positioned to take advantage of this shift.

Powder technology is defined as "the field of technical knowledge and competence underlying the development and production of powders and powder-based materials for engineering applications requiring advanced structural materials properties." Powder technology is, actually and potentially, a generic ('enabling') technology of great pervasiveness, being useful in a broad range of applications and products.

Powder technology comprises two major sub-fields, powder metallurgy (PM) and engineering ceramics. The powder metallurgy side has evolved as an outgrowth from the metallic materials and metals-based designs which have dominated Swedish industry for well over a century. Engineering ceramics, by contrast, is still, by and large, a field of research and experimental development, of prototype trials and minor niche products. Commercially, the field is still in an embryonic stage.

RECEIVER COMPETENCE

Having defined each technological system and the type of spillovers specific to each, we turn now to an analysis of the prime movers and the institutional environment in which they operate.

The knowledge component in **factory automation** has a long history going back to the late 19th century in Sweden, while the particular artifacts which constitute factory automation products and techniques go back no further than the 1960s. Sweden's traditional strength in engineering industries goes back hundreds of years in mining and metalworking and about 100 years in mechanical and electrical engineering (associated with the founding of such companies as Asea, L.M. Ericsson, SKF, and Alfa Laval, to mention but a few). A machine tool industry grew up alongside and in close technical collaboration with these metalworking and engineering firms.

The strongly international orientation of Swedish companies generally and, in particular, of the large Sweden-based multinational firms which are also major users (potential and actual) of factory automation has played a crucial role in creating early awareness of the potential for each type of factory automation equipment and/or technique. The high technical competence of several major Swedish firms such as Asea (now ABB) and Volvo enabled them to serve as receptacles connected to the global opportunity set by quickly identifying and absorbing new technology and transmitting new technical requirements to the rest of the technological system. This is shown by the fact that Swedish suppliers of automation equipment were among the pioneers in

Europe in their respective fields. The leading-edge users did more than buy equipment, however; in several cases they also started new companies.

Thus, advanced domestic users of all types of factory automation equipment and techniques created a market which previously did not exist and therefore created first mover advantages for Swedish suppliers. Being early, the Swedish suppliers of automation equipment entered the market while the critical mass requirements were still quite modest, thus offsetting the disadvantage of a small domestic market. Financially strong and technically advanced customers absorbed much of the financial and technical risks. Despite the formation of a few new companies, not much venture capital has been required. The few new ventures that have been formed (ABB Robotics and BT Carrier Systems) have been spin-offs from large companies. The risk capital involved in R&D is modest and largely supplied by the buyers. Internal finance (within existing companies and groups of companies) has been the most important source of finance. Some of the major Swedish business groups (especially Wallenberg and Volvo) have been directly involved. Statsföretag (the state-owned holding company) was involved in two machine tool firms, SMT and Sajo (without commercial success, however), and some small private groups have taken over several small machine tool companies in the last decade. But the overall impression is that not much external capital has been put into this system, except for industrial robots (Asea).

The nature of knowledge in this technological system is such that academic research does not play an important role (except to the extent that it is application-oriented), while basic engineering training is important. The Swedish academic system has performed well in both of these aspects. It has contributed to receiver competence all along. Since the new technology in this field builds largely on traditional mechanical engineering, the academic institutions responded well and in timely fashion in mechanical engineering but somewhat sluggishly in electrical engineering and electronics.

In **electronics**, the Swedish receiver competence can be broadly characterized by

- traditional strength in mechanical engineering;
- a fairly strong position in parts of the electro-mechanical field, primarily in power generation and transmission as well as in telecommunications;
- close user-supplier relations precisely in the areas of traditional strength in electro-mechanical engineering; and
- a strong position in military research along with substantial and competent procurement by the Swedish military from the domestic military industry, including both aircraft and artillery, and later also missiles.

There is an interesting pattern of interaction between public and private prime movers in this field. The dominant actors in each cluster are public sector agencies representing buyer competence and private companies, supplier competence. In radio and telecommunications, the main user is Televerket (the Swedish Telecommunications Agency, now renamed Telia) and the main supplier is Ericsson. In military electronics, FMV (the Defense Procurement Agency) is the buyer and Saab, Bofors, and Ericsson are the main suppliers. In power electronics and other generation and transmission equipment, Vattenfall (the State Power Board) is the user and Asea the main supplier. In medical electronics, the regional health care providers are the major customers and a variety of small private companies (e.g. Siemens-Elema) are the suppliers. In each cluster, other actors such as consultancy firms and parts of academia also play important roles.

The private companies and public sector agencies have jointly played a particularly important role in raising awareness and providing finance. The public agencies have acted as competent buyers of electronics products. They have also played a major part in shaping the institutional setting for the emergence of electronics-based products. For example, the decision to create a common standard for mobile telephony in the Nordic countries created a large market which aided Ericsson's successful mobile telephony venture, and which also prompted other firms such as Nokia in Finland to enter the market. This helped these firms attain economies of scale and to accumulate experience which gave them an advantage in the international markets.

At the same time, the private companies have played a corresponding role on the supply side. Especially noteworthy is the contribution of the Wallenberg group of companies. All of the dominant private firms in each field (Asea, Ericsson, Saab, and Bofors) are Wallenberg companies. The prime mechanism for new growth has been diversification of these and other firms into electronics-based products to supplement or replace the functions previously served by mechanical products. The internal capital market within the group has been a major source of finance.

The role of academia has been multifaceted. On the one hand, there has been a set of areas of strength in academic research which has proved to be helpful to industry (e.g. in power electronics and radio technology). On the other hand, the response of the technical universities has, overall, been slow in terms of increasing the supply of engineers in this field. The same applies to the process of expanding the number of chairs in electronics and computer science.

The Swedish **pharmaceutical industry** emerged between the wars predominantly as a series of off-shoots from pharmacy businesses. It consisted of a number of small firms catering almost exclusively to the local market. There was little effort to develop original products. What little R&D did occur was mostly directed towards acquiring foreign technologies (e.g., sulfa drugs and penicillin).

Since the 1960s, and particularly the last two decades (the period covered in this study), the Swedish pharmaceutical industry has enjoyed extraordinary success. This success is based on a happy confluence of several factors which originated much earlier. Due to the absence of a strong chemical industry in Sweden, the Swedish pharmaceutical industry developed along different lines from those of its main competitors, particularly in Germany and Switzerland. The technical/scientific culture of the Swedish pharmaceutical industry is rooted in pharmacology instead of chemistry and has developed strong links to both academia and the health care sector. Together with other Scandinavian countries and Britain, Sweden has pioneered in clinical research. There is high user competence in the health care sector, in the regional health care providing agencies, and particularly in about 10 teaching hospitals. The close collaboration between these customers and the pharmaceutical companies has led to a high competence level of clinical testing in the country and contributed to the emergence of many new product ideas. Taken together, these features of the technological system constituted an excellent position for the Swedish pharmaceutical companies to take advantage of the new opportunities which subsequently opened up in biomedicine.

The rapid growth in the last twenty years has been accompanied by consolidation of the Swedish companies. The dozen or so Swedish companies which existed at the beginning of the 1970s have merged into two major companies, Astra and Kabi-Pharmacia. The latter has recently merged with Upjohn in the United States.

The traditional source of capital in the industry has been retained earnings. But when Astra scored major successes, e.g. in developing the first beta blocker to be approved by the U.S. Federal Drug Administration for the U.S. market, it was forced to form a joint venture with Merck to market the drugs. Kabi as a subsidiary of the government-owned Statsföretag secured government finance for the joint venture with Genentech (KabiGen). This financing was vital for the speed at which Kabi moved into genetic engineering, as the company did not have enough internal resources, nor were there any venture capital firms capable of taking the risk. The state funding was not restricted to setting up the new company but financed, at least partly, the research contract with Genentech. Some of the money was supplied by a government risk capital fund, Industrifonden.

The pharmaceutical industry needs people with research degrees. An increase in their supply is therefore a function of a prior expansion in academic research. The expansion that did take place was clearly not sufficient to satisfy the demand from industry; in the mid-1980s there was a large shortage of such competence in Sweden. The shortage of competent staff affected the academic units also, thus perpetuating the problem of lack of genetic engineering competence. STU initiated a program to foster genetic engineering capability at Swedish universities in 1979, after such work had already been started at the universities in the mid-1970s. Given that the first cloning of a gene took place in 1973, that the recombinant DNA experiments were first discussed in a public forum in 1974, and that the first firm to exploit rDNA research (Genentech) was formed in 1976 (OTA 1991, p. 5), the response appears to have been early in Sweden.

The strong biomedical research base in Swedish universities makes for strong international links. Kabi (through its contacts with Genentech) and Astra (through its research unit in Bangalore and its research conducted in various hospitals around world) have been strongly internationally oriented all along and are well plugged into the international networks.

In **powder metallurgy (PM)**, some materials firms (e.g. Sandvik), drawing on their existing competence in iron and steel metallurgy, recognized the possibilities arising from the breakthroughs in carbide alloys by German and American firms in 1928 (Carlsson 1984, p. 99). Thus, previously existing iron and steel metallurgy firms became the prime movers and provided significant entrepreneurship, mostly by forming subsidiaries. Most of the companies involved are members of the dominant business groups. The required capital has been obtained internally within these firms. Critical mass was created by integrating PM with on-going metallurgic activities and products in the companies that ventured into the area.

In **engineering ceramics**, there are no major prime movers; entrepreneurial activity is yet to be seen. The technology is still in an R&D phase. The few commercial ventures that exist are basically subsidiaries of large firms. Several Swedish companies have manifested an interest in the possibilities of engineering ceramics by engaging in monitoring, materials testing, or exploratory development. However, active industrial involvement has been limited mainly to some potential leading-edge users and a few materials- and process-oriented firms. Government policies in the 1970s were designed to broaden and reinforce the scientific and technological infrastructure of ceramics. A series of national R&D programs has also been implemented subsequently.

In ceramics, critical mass has not yet been achieved. The main market failure is missing information (i.e., technological shortcomings or weaknesses in the technology base) and missing product markets. These obstacles remain.

As far as the role of academia is concerned, the overall PT effort undertaken by Swedish university- and institute-based groups is quite limited. A man-year estimate for the PM field (prior to the launch of the PT-90 program in 1991) gives a maximum of 20, while the total number of researchers working in the ceramics field is put at approximately 40. In view of the severe resource constraints, the activities engaged in are remarkably diversified, with small-scale projects (single-investigator projects are common) addressing a variety of problems. In the overall profile, materials engineering and applications are more prominent than materials science and basic technology.

On the PM side, the educational system has been largely reactive. The principles and standard processes of PM have been introduced into engineering education as a natural response to the industrial success. On the ceramic side, however, the system may be said to have been proactive to some degree. Not that there has been a large engineering education program; rather, the increase in the training of researchers has led to a greater emphasis on ceramic materials in the materials science programs. This is a response to the international development in ceramics, not due to a dramatic increase in the demand for trained personnel by Swedish industry.

CONNECTIVITY

The third main component in technological systems is the networking/clustering which binds together the various actors and makes technological spillovers possible.

In **factory automation** as well as in the engineering industries more generally, the nature of knowledge (tacit, component-oriented, and embodied in equipment) gives the users a dominant role. A long tradition of close user-supplier relationships in the engineering industries in Sweden makes it difficult to distinguish between the buyer-supplier and 'problem-solving' networks. Leading users, by virtue of their high technical and economic competence, have imposed high standards on parts suppliers (subcontractors) as well as suppliers of factory automation equipment. The system is closely interconnected; it is also highly international. Even if Swedish users would prefer to work with domestic suppliers, they often have to go to foreign ones because of a limited product range and competence among domestic suppliers. As a result, some 75 percent or so of factory automation technology is supplied by foreign companies. Geographic proximity plays a role but is overshadowed by considerations of technical competence. "Cultural" proximity in terms of technical approach and manufacturing philosophy is also important.

There are also well-developed informal (primarily domestic) 'community' networks, some having simply evolved over a long period, others formed more recently with the assistance of bridging institutions. Among such institutions are IVF (The Institute for Production Engineering Research, a cooperative institute established in 1964), STU/Nutek (originally established in 1968, the major technology policy agency in the country), Mekanförbundet (the Engineering Industry Association), and IVA (The Royal Swedish Academy of Engineering Sciences), both pre-existing. The combination of strong bridging institutions and informal networks has resulted in a remarkably well-integrated system.

In **electronics**, the main actors in each of the three clusters existed prior to the era of electronics. As in the case of factory automation, the buyer-supplier networks are largely international, i.e., most suppliers of electronic components are foreign, as are most of the customers of the final products. Over a long period of time, intense problem-solving networks have also evolved in each area. Five collective research institutes operate within the field of electronics and computer technology. These institutions, each with its own specialty, were designed as bridging institutions between the universities and industry and have been largely financed by Nutek. They were initiated as a result of close interaction between industry, academia, and Nutek.

In **the pharmaceutical industry**, Sweden has a long and well-established tradition of excellence in biomedicine and related basic disciplines such as biochemistry. The fact that the leading Swedish firms were rooted in pharmacy businesses rather than the chemical industry meant good links to the medical profession and through it to biomedical research.

This particular technical/scientific culture of the Swedish companies and the strong links between these firms and the health care sector, coupled with the high degree of centralization of that sector, have clearly shaped the pharmaceutical and biotechnology sector in Sweden. The close collaboration between the leading teaching hospitals and the pharmaceutical companies has led to a high competence level of clinical testing in the country and contributed to the emergence of not only many new product ideas but also products that have passed through the clinical tests. The existence of a strong scientific base in clinical medicine and biomedicine has also contributed. The private companies, in collaboration with the public health care providers, have built these institutions, over many years. The strong biomedical research has been state-funded all along.

The relatively small scale of the Swedish firms has forced them to link up with academic research through consultancies, R&D collaboration, funding of academic chairs, recruitment, and location of R&D facilities near academic research centers. Because of the limited size of the domestic resource base, many of these linkages are international.

Perhaps the most remarkable feature (particularly in view of the small size of the country) of the biomedical research complex in Sweden is its completeness and integration, both between the academic side and clinical practice side and between basic, preclinical, and clinical fields. This has been achieved through several institutional means, including the development of excellent academic hospitals; more than 40 percent of academic medical research in Sweden is conducted in such hospitals. Furthermore, until recently there has been a strong link between medical education and research. As a result, the Swedish health care system has been to a high degree staffed by doctors who were, if not practicing researchers, at least research-minded. Such persons play an important role in creating linkages between the learning processes which are rooted in medical practice and the more formal R&D in its various forms.

An additional characteristic of the Swedish R&D system is that it is very well connected internationally. Especially after World War II, intensive links have been formed with the United States which clearly dominates the biomedical field. For example, Kabi formed a joint venture at an early stage with Genentech, the leading American firm in genetic engineering. By forming links with the local academic community, the Swedish companies have acquired an excellent observation post to watch the international biomedical scene. Because of insufficient domestic capabilities in the academic system with regard to biomedicine, Astra decided to build a research unit in Bangalore, India, home of the Indian Institute of Science (Sind 1991:5). Thus, this is a case of plugging into the global system; the part of the technological system which supports genetic engineering applications in Swedish firms is only partially located there; a large part of it is located abroad.

Academic researchers took the initiative in building bridging institutions, working with pre-existing organizations. In addition, Kabi and Astra have each formed their own bridging institutions to link with researchers outside Sweden. A few of the academic spin-offs in the 1980s had genetic engineering capabilities. Some of these biotechnology firms have acted as a bridge between academic research and industry. This seems to be an area in which specific technical competence, often tied to specific individuals or organizations, is more important than geographic proximity.

The Swedish **powder technology** system comprises a number of networks involving a broad spectrum of actors, from powder makers and equipment suppliers, through producers of powder-based materials and parts, to user firms, mainly on the PM side. The buyer-supplier networks are, to a large extent, international. The materials industry is heavily dependent on foreign users: overall, the export share of the materials and parts supplied amount to more than 90% of the sales value. At the same time, the Swedish engineering-industry users appear to be similarly dependent on foreign suppliers of finished PT-based parts and components. Only a minor portion of the finished PT parts used in mechanical and electrical engineering applications are in fact purchased at home.

These conditions limit the scope for the development of strong user-producer links within the Swedish PT system. This is a natural result of the limited resource base of a small country and the specialized input requirements. But even though Sweden is a small country, surprisingly many parts of the production chain are represented domestically in some areas (e.g. hot pressed materials and cemented carbides). The strong dependence on international links may be viewed as both a potential strength and a potential weakness. Continual monitoring of global markets reduces the risk of insularity, while the absence of domestic links significantly limits the scope for the development of strong user-producer links within the Swedish PT system and hence the achievement of critical mass. Geographic proximity is important to the extent that users engage in joint R&D projects with the materials industry and the research infrastructure.

In the problem-solving networks, the dominant pattern is that of joint problem solving by pairs or triplets of institutional actors. The PT system can thus be seen as a loosely connected aggregate of 'mini-networks.' These are characterized, first, by a large vertical span, extending from policy agencies (as promoters and organizers of collective R&D) and academic science departments to leading-edge users. Second, the problem-solving networks are, to a far greater extent than the user-supplier networks, nationally based and circumscribed. Finally, these networks are also strongly influenced by government action. In many cases, they are the direct results of programs or major projects initiated and organized by national policy agencies.

As far as problem-solving in the PM field is concerned, *Jernkontoret*, the cooperative organ of the Swedish iron and steel industry, has played an important role as a supporter and coordinator of industrial R&D projects.

The Swedish PT system is supported by a broadly inclusive ("everybody knows everybody") community network, the nodes of which are formed by the leading practitioners — such as chief engineers, senior researchers and R&D managers.

In the 'national community network,' representatives of the *institute sector* have often performed a bridging or mediating role vis-à-vis industry and academia. Academics are, to a considerable extent, active in the applied areas of the field, in many instances in collaboration with a relevant institute.

The community network is also internationally connected through numerous informal links between the community members and their foreign colleagues. The establishment and maintenance of such links takes place partly within the framework of international associations, conferences, symposia, etc.

VARIETY CREATION MECHANISMS

The fourth major component of technological systems is the ability of the system to create or at least sustain a variety of technical solutions, actors, and institutions in order to counteract the continual pressure of selection mechanisms, i.e., the constant weeding out of the less successful, which otherwise reduces the variety.

Factory automation represents a case of high receptivity to new technology on the part of both suppliers and users (particularly the latter, especially via their strong international exposure), and large amounts of high quality information being exchanged via existing networks.

In numerically controlled machine tools (NCMTs), the large number of actors (until recently) has ensured a variety of technical approaches. In FMS (flexible manufacturing systems), the differentiation strategies chosen by SMT and Sajo failed in the face of strong competition from cost leaders (especially Japanese companies). In industrial robots, ABB Robotics' strategy of cost leadership (in conjunction with technical leadership) has proved viable thus far, with other companies (both domestic and foreign) supplying application hardware and software. There has been consolidation of the robotics industry (both worldwide and in Sweden), but new companies are also still being formed. In automatic materials handling there is now only one major firm, but there were more during the 1980s.

The main selection mechanism has been the international market. This has worked predictably well, but to the point of near extinction of some supplier segments in Sweden, especially in NCMTs and FMS. But this has been the experience in other countries as well, not least the United States. This raises a difficult question: how many major suppliers can be expected in each field in a small country?

Due to their international exposure, Swedish companies have had a rather broad vision. Prime examples are ABB's introduction of microprocessors in robots already in 1974 and SMT's early moves into both NCMTs and FMS. The factory automation technologies have basically extended the life span and dominance of the existing engineering industries; the path dependency has been reinforced. This has been good for traditional industries but has probably had less beneficial effects for other parts of the Swedish economy. Given the fact that most of the user industries are relatively mature, it is not surprising that the number of exits far exceeds the number of new entries.

In factory automation, the role of public policy, especially technology policy, has been largely supportive but not prominent, more reactive than proactive. The main role has been to build bridging institutions, support networks, and train a sufficient number of engineers, not that of creating and maintaining a variety of options.

In **electronics**, the awareness of the opportunities offered by new technology has been high in the system as a whole, although more so in some parts than in others. The small number of suppliers may, however, create a problem as far as variety creation is concerned, even though the main expansion of the system (at least in commercial terms thus far) has been due to Ericsson's diversification into a new field, mobile telephony. This represents variety creation by the leading supplier. The paucity of actors is caused by both a strong selection environment resulting in exit of many firms in areas outside the three main clusters and by a lack of diversification into the field by the large, and strongly path-dependent, mechanical engineering firms. A hopeful sign, though, is an increase in the number of smaller electronics and computer science-based firms which may lead to a process in which the path dependency of the larger firms is counterbalanced and variety is increased. But we have not yet seen many new technology-based firms as a source of expansion in the electronics industry.

The major thrust of public policy in shaping the technological system has been via procurement. The procurement policies by state-owned bodies has had a strongly positive effect on the system in that it has induced and helped large firms to switch to new technologies and products.

The vast programs in the 1980s in the field of microelectronics and system design represent a second aspect of policy which has been beneficial to the development of the technological system, even though the semiconductor efforts have not met with a strong industry response. Those programs could have been implemented earlier and the procurement programs could have been better coordinated with the educational and research policies. These policies were, if not failures, at least not very successful in terms of the timing of their expansion in this field. Finally, policies (primarily tax-related) influencing the supply of risk capital to new technology-based firms, although not specific to electronics, have not been conducive to growth of new firms in electronics and computer science.

In **pharmaceuticals** there are only two major pharmaceutical companies remaining in Sweden (or one-and-a-half, after the recent merger of Pharmacia and Upjohn), but they have taken very different approaches and each has its own networks. Given the risks involved, not much more can be expected in a small country.

This does not appear to be a case of "bounded vision" constraining the development potential. Kabi got involved very early in biotechnology. Astra was at least 5 years behind but appears to have been hampered much less by its past than many competitors in Germany and Switzerland. The strong pharmacological heritage seems to have helped. Thus, path dependency appears to have been relatively mild.

The case studies of Astra and its subsidiary Hässle in chapters 5 and 6 demonstrate the importance of the organization of corporate R&D. Without the combination of strong commitment to R&D of top management and a decentralized R&D organization with a high degree of autonomy of subsidiary R&D units, it is doubtful that either the beta blockers or the peptic ulcer drug would have been developed successfully. "Tight" management of processes characterized to a high degree by serendipity and experimentation as well as complexity and cross-disciplinarity is not appropriate. Instead, networking involving multiple units in diverse academic and industrial fields, both at home and abroad, and technological entrepreneurship are necessary ingredients.

As far as entrepreneurship is concerned, there were about 80 new biotech companies formed in Sweden in the 1980s through academic spin-offs (Laage-Hellman and Axelsson 1986). This number compares favorably with that in the United States, where about 400 companies were started during the same period (OTA 1991, p. 3). But in contrast to the U.S., none of the Swedish companies has grown much beyond start-up size — a problem which seems to be endemic in the Swedish economy. Creating conditions favorable to firm formation and growth should therefore be a major focus of public policy. Many of the Swedish companies, similarly to those in the U.S., were financed by the infant venture capital market in the early 1980s, but some were also financed by the government bodies STU and Industrifonden.

In the pharmaceutical industry, STU/Nutek appears to have come in early and played a proactive role in shifting the technology base towards biotechnology; policy appears to have been oriented towards building capabilities and networks rather than supporting particular ventures. But in the more traditional (non-biotech) sectors of the industry, technology policy has not played much of a role, except insofar as the regulation which is characteristic of this industry in all countries has been relatively rigorous in Sweden with respect to clinical testing, apparently resulting in benefits to the Swedish companies in gaining approval for their products in foreign markets.

In **powder technology**, technical problem solving involves processes of trial-and-error learning where the experience gained guides the development of alternative concepts or variant designs. The generation of variety can be seen as taking place at all stages; there is thus an ever-growing diversity of powders and powder-based materials, forming and densification methods, processing methods, etc. In the search for optimal or satisfactory solutions, selection (or the reduction of variety) occurs both 'top-down', through the user needs or specifications that are to be met, and 'bottom-up', by virtue of the fact that the choice of raw materials and powders limits the options available higher up the chain.

The relative strength of the industrial/commercial part of the technological system in powder metallurgy is largely a reflection of the traditional strength of Swedish engineering industry in metallurgy and metal-based activities. But in itself, powder technology opens up new areas of application. The ceramic side of the technological system does not have much of an industrial tradition to fall back on in Sweden, nor in other countries. This reflects the absence of major leading-edge users, similar to those that have been so influential in the electronics field. This is a problem not only in Sweden which might be addressed at least partially through public procurement.

In powder metallurgy, some infrastructure support in the form of government programs has been provided but was reduced during the 1980s, with reference to the relative maturity of the technology and the strong or adequate, independent capabilities of industry. In ceramics, public policy has been much more proactive, with early efforts to promote the growth of the system.

The basic aim of the more recent PT-90 program is the comprehensive ('systemic' or 'holistic') one of strengthening the system as a whole by attending to all of its major components and the way in which they interact. The emphasis is on closing certain gaps or creating certain links among the actors/sectors. The breadth of the enterprise manifests itself in different dimensions: (i) horizontally, in the coverage and hoped-for cross-fertilization of powder metallurgy and engineering ceramics; (ii) vertically, in the ambition to achieve closer integration across the production chain and to engage the entire range of actors, from infrastructure units to user-industry firms; and (iii) with regard to the spectrum of activities involved which encompasses the basic categories of R&D, education, and technology diffusion. Given the breadth and systemic orientation of the program, it is hardly surprising that it is also pervaded by a strong concern with the bridging of internal gaps and with the development and effective use of actor networks, notably networks of the problem-solving type.

CONCLUSION

The complexity of each system makes for many differences between the systems. In fact, the differences seem to outnumber the regularities. But even though the number of systems studied here is too small for generalizations, a few concluding reflections may be in order. These observations are grouped under two headings: similarities among the four systems, and main strengths in each system.

Similarities among the Four Systems

A systems perspective inevitably emphasizes the interaction between actors and institutions. The focus is on the system as a whole rather than its individual actors; the interaction of its participating individuals and institutions determines the system's properties. Some of the interaction is the result of intentional acts of individual actors, while some occurs as a result of largely unintended spillovers. The main function of the technological system is to capture, diffuse, and magnify such spillovers. The most important features of the system are the *characteristics of knowledge and spillover mechanisms* which determine the potential spillovers, the *receiver competence* (absorptive capacity of each actor, particularly that of the prime movers in the system), the *connectivity* (the degree to which various components in the system are tied together), and the *mechanisms which create variety* within the system.

A common feature in all the systems studied thus far is their evolutionary character. The systems are continually changing, some slowly and more or less predictably, others dramatically and unpredictably. They emerge over several decades, but not out of thin air: they are strongly path dependent. Initial conditions in the form of pre-existing technology base, actors, and networks play an important role. The nature and locus of knowledge in each field place important requirements on the entire system and its components, the relationships among various actors, and the institutions which facilitate their interaction. Therefore, the traditional role and specialization of the main actors and the culture of which they are a part constitute important features which shape, and are shaped by, the system.

Our study has brought out the importance of certain types of interaction within the networks, particularly the identification of specific technical requirements. This takes place in the interface between competent customers and other actors within the system. Entrepreneurial activity is crucial but is successful only when the technical requirements are well understood. User-supplier interaction is often a good vehicle, but the important part is the technical problem-solving rather than the buying and selling of inputs.

Institutions play an important role in each system, but the make-up of the institutional framework varies from one system to another, and from one time period to another. Given the small size of the Swedish economy, it is perhaps inevitable that international linkages figure prominently in each system. The ability to tap into the global system may be the single most important institutional feature in each system studied thus far. Going abroad for talent, information, and capital is clearly an effective and quick way to bypass domestic constraints in the educational and financial systems. This is undoubtedly one of the contributing factors to the strong reliance on a few well-established and well-connected Swedish firms in each area. These firms function as channels through which the Swedish system is connected to the global system and through which domestic competence is acquired or enhanced. Perhaps the apparently modest role of new firms is partly a consequence of this phenomenon.

Main Strengths in Each System

Strength lies in different places in each system. The high degree of diffusion of factory automation in Sweden is attributable largely to strong traditions in mechanical engineering which have resulted in several highly competent users. These firms also happen to be internationally well connected. Another main source of strength resides in long-established networks and bridging institutions.

In electronics, the Swedish suppliers (which are also major multinational firms) excel in the application of new (electronic) technology to traditional (mechanical) areas (products and processes), with government agencies playing a significant role as competent buyers.

The extraordinary success of the Swedish pharmaceutical industry in the last couple of decades is attributable chiefly to three factors:

(1) access to an excellent biomedical knowledge base

(2) the ability to access that knowledge base; the historical roots, particularly in pharmacology, and biomedical and clinical research have played significant roles; and

(3) strategies and managerial culture allowing the companies effectively to exploit the technological opportunities, especially in making the transition from a chemical to a biological technology base.

Sweden's relatively good performance in powder metallurgy is strongly related to the country's historical traditions in metallurgy and its industrial applications, and to the development in some important niches of strong vertical actor networks. In ceramics, the main strength resides in hot isostatic pressing and to some extent academic research.

REFERENCES

Carlsson, Bo, 1984. "The Development and Use of Machine Tools in Historical Perspective," *Journal of Economic Behavior and Organization*, 5 (1994), 91-114.

Carlsson, Bo (ed.), 1995. *Technological Systems and Economic Performance: The Case of Factory Automation*. Boston and Dordrecht: Kluwer Academic Publishers.

Carlsson, Bo and Gunnar Eliasson, 1994. "The Nature and Importance of Economic Competence," *Industrial and Corporate Change*, 3 (1), 687-711. Also published as chapter 3 in Carlsson (1995).

Carlsson, Bo and Rikard Stankiewicz, 1991. "On the Nature, Function, and Composition of Technological Systems," *Journal of Evolutionary Economics*, 1 (2), 93-118. Also published as chapter 2 in Carlsson (1995).

Cohen, Wesley M. and Daniel A. Levinthal, 1990. "Absorptive Capacity: A New Perspective on Learning and Innovation," *Administrative Science Quarterly*, 35 (1), 128-152.

Granberg, Anders, 1995. "The Academic Infrastructure of Factory Automation," in Bo Carlsson (ed.), 1995, 119-164.

Laage-Hellman, Jens and Björn Axelsson, 1986. *Bioteknisk FoU i Sverige — en kartläggning* (Biotechnical R&D in Sweden — A Mapping). Uppsala: IM-Group.

OTA (Office of Technology Assessment), 1991. *Biotechnology in a Global Economy*, OTA-BA-494. Washington, D.C.: U.S. Government Printing Office, October.

SIND (Swedish National Board for Industrial Development), 1991. *Biotekniken i svensk industri — Bokslut och prognos inför 90-talet* (Biotechnology in Swedish Industry — Status and Prognosis for the 90s). SIND Reports 1991:5.

Stankiewicz, Rikard, 1986. "Utvecklingsdynamik inom genteknisk FoU i USA, Japan och Europa" (Genetic Engineering - R&D Dynamics in the USA, Japan and Europe), in Guno Haskå and Rikard Stankiewicz, (eds.), *Strategier för Bioteknisk Forskning och Utveckling* (Strategies for Biotechnology Research and Development). Lund: Stiftelsen Bioteknisk Forskning and FPI, 5-16.

Stankiewicz, Rikard, 1994. "Spin-off Companies from Universities," *Science and Public Policy*, 21 (2), 99-107.

11

IN SEARCH OF USEFUL PUBLIC POLICIES — KEY LESSONS AND ISSUES FOR POLICY MAKERS

Bo Carlsson and Staffan Jacobsson

INTRODUCTION

The purpose of this chapter is to discuss what lessons can be drawn for public policy from our work on technological systems. The analysis in the preceding chapters has shown that public policy with respect to technological systems needs to be broadly conceived; it should not be confined to the domain of technology policy alone but should also include aspects of educational policy, science policy, and industrial policy more broadly.

Our analysis rests on the basic assumptions[1] which were presented in chapter 1 and on the main characteristics[2] of technological systems described in chapter 10. Given our assumptions and our findings about system characteristics, the conventional approach to public policy — which is based on market failure — is no longer adequate. It is inadequate not only because market failures are ubiquitous but also because institutional and network failures are also common. In addition, the failures may interact and create failures at the level of an entire technological system. Thus, as shown in the next section, the policy discussion needs to be broadened to encompass system failures as well. In the subsequent section, we pose a set of questions for public policy, primarily

---

1 The assumptions are the following: (1) the technological opportunities facing any economic agent are virtually unlimited; the pool of global possibilities has practically no boundaries. (2) Bounded rationality: decision makers base their actions on limited knowledge. (3) Firms' technological competencies are, although differentiated, highly stable and path-dependent. The search for new ideas and new solutions is mostly 'local' (although not necessarily local in a geographic sense). (4) System perspective: the system as a whole is the unit of analysis (rather than its constituent elements). (5) Dynamics: the technological systems we have studied are constantly evolving. They are not given, once for all.

2 In addition to the nature of knowledge determining the spillover mechanisms in each system, the characteristics determining the performance are receiver competence, connectivity, and the presence and strength of variety-creating mechanisms.

with respect to the creation of new technological systems. The final section contains our main conclusions.

FROM 'MARKET FAILURE' TO A SYSTEMS APPROACH FOR PUBLIC POLICY

The theoretical foundation of industrial, trade and technology policy normally rests on the notion of market failure.[3] Market failures — the failures of market mechanisms to reach an optimal solution to an economic problem — may be caused by the presence of increasing returns to scale and scope, externalities, missing markets, coordination problems, uncertainty, etc. All of these are present in abundance in each technological system. Indeed, it could well be argued that market failures in a neoclassical sense are at the heart of the innovation process.

In the context of evolutionary economics, the power of the market lies not just in allocating resources more or less well but also in serving as a powerful (although often blunt and heavy-handed) selection device and in providing incentives for learning and discovery. Even though it may lack allocative optimality at any and every point in time (Dosi 1993, Eliasson 1984, 1991a, 1996), the market has dynamic implications by constituting the selection environment and influencing the process of variety creation.

The selection and discovery processes can, of course, be shaped or strongly influenced by market failures. The case of video recorders, described by Arthur (1990), indicates that market failure (n the presence of increasing returns and network externalities) may lead to the selection of an inferior technology. Lock-in effects of long duration in inferior technologies can be argued on theoretical grounds to be a normal consequence of boundedly rational agents' learning in vast and heterogeneous business opportunity sets (Ballot and Taymaz 1996). We would argue, therefore, that phenomena which are labeled market failures in a static or comparative static framework are important and often positive influences in a dynamic context. This is illustrated particularly well in the Losec case study (see chapter 6): the development of the peptic ulcer drug narrowly escaped cancellation by the top corporate management no less than five times, yet was allowed to proceed to successful completion because of a combination of public policy intervention and a sufficiently decentralized corporate structure to permit a variety of views to exist.

The processes of selection and variety creation are influenced not only by market-related phenomena but also by the character of the entire technological system which constitutes the innovation system. The market represents but a limited part of the context for innovation and diffusion. Institutions and networks also constitute building blocks in that system; the nature and functioning of institutions and networks influence the processes of selection and discovery. Therefore, in addition to market failures,

---

3 For clear statements of that approach, see Corden (1974) and (1980).

institutional and network failures also need to be considered as part of the domain of public policy. By combining market, institutional and network failures, we also open up for the possibility that there may be failures at the level of entire technological systems. We will therefore first discuss network and institutional failures and then proceed to system failures.

Network, Institutional, and System Failures

Network Failure

Technological knowledge is to a varying degree tacit (person-embodied). Each technology embodies elements which cannot be written down in blueprint form or are difficult to verbalize and can therefore not be diffused easily. As this tacit component in the knowledge base increases, technological accumulation is more experience-based, and communication as well as technology transfer take place primarily via inter-personal contacts. Consequently, formal or informal networks are important routes for the transfer of tacit knowledge (Metcalfe 1992). Moreover, networks can partly compensate for limitations in the firm's search space on account of both bounded rationality and bounded vision (Fransman 1990). The network may also improve the resource base and the degrees of freedom of the individual firm (Håkansson 1987, 1989).

Thus, the character of the networks to which the firm belongs has a bearing on the type of information and knowledge to which the firm has access. The connectivity of the constituent parts of the system matters, too, i.e., the amounts of information and knowledge dispersed may vary among networks. This means that a well-functioning technological system rests on strongly positive and reciprocal external economies which tie together users, suppliers, competitors and related firms (cf. Porter 1990)[4] — as well as academic and other institutions. The reciprocal flow of information may well result in a blending of visions of the future (technological expectations) among various actors. Sharing the same vision may lead to a reduction of perceived risk and a quasi-coordination of investments among formally independent actors.[5] Innovation and diffusion turns, in this way, into a collective activity, in addition to being an individual one. Networks are therefore central to the innovation process.

---

[4] Indeed, the importance of networks suggests that technological systems, and therefore the process of generating and diffusing new technology, rest heavily on reduction of transaction costs via internalization of transactions within networks.

[5] For instance, the interaction between the technologically leading firms (in terms of factory automation) and the local supplier industry resulted in a blending of their visions of the future production technology. The whole network then invested in technological capabilities and design developments which were interchanged within the network.

Networks do not necessarily grow spontaneously, however. As Saxenian (1994) argues, there may be institutional and organizational obstacles to the growth of a 'collective identity.' To the extent this is the case, *network failure* would arise. Such failures are either 'weak'— firms are not well connected to other firms with an overlapping technology base — or 'strong' — individual firms are guided by others (i.e., by the network) in the *wrong direction* and/or fail to supply each other with the required knowledge. The source of a strong network failure lies in different knowledge bases and different assumptions concerning present, but most importantly, future (missing) markets. When new technologies become available, providing expanded and different technological opportunities, the bounded (and different) visions (Fransman 1990) of managers imply that firms and their networks may differ greatly in their perception of and ability to seize these opportunities. Rationality, on these assumptions, could well be argued to be not only limited but quite different among both firms and networks (Eliasson 1990). This opens up for '*strong network failures*.'Thus, the other side of the coin of tight networks is that these can contain considerable inertia, suggesting that there may be slow adjustment to new technological opportunities (Ehrnberg and Jacobsson 1996).

Institutional Failure

Industrial dynamics depend not only on the behavior of individual firms and networks; institutions matter, too. These can be both 'hard' in the form of organizations such as educational and financial institutions and 'soft' such as the laws, culture, and customs of a country or region.

Among the 'hard' institutions, universities in particular have a key role to play by fostering both a high receiver competence and variety by pursuing high-risk research activities as well as through educating engineers and scientists in new scientific and technological fields. However, universities differ in their behavior with respect to the speed and strength of response to new scientific and technological opportunities.

Other 'hard' institution are those performing a 'bridging' role in terms of acting as information exchanges within the technological system. Examples in the factory automation case are the Institute for Production Engineering Research (IVF), the Swedish National Board for Industrial and Technical Development (STU/NUTEK), and the Engineering Industry Association (Mekanförbundet).

Other institutions which play a role within the technological system are the financial institutions (especially the suppliers of venture capital and other long-term capital for new types of business ventures and the rules under which such funds are allocated), the existence and nature of business groups and their ability to allocate financial, technical, and managerial resources, and government policy bodies.

Culture adds a 'softer' dimension to institutions but is nevertheless important in many ways. Perhaps one of the more influential is the willingness to share resources with other actors, a factor underlined by Saxenian (1994). Another way in which culture influences technological systems is via the entrepreneurial spirit in the country/region.

These institutions can either enhance or diminish the functioning of technological systems, in particular their ability to generate and exploit new ideas, and to create variety. Institutions can not be seen as neutral with respect to various technological systems — each may require a different institutional configuration — nor are they necessarily formed or reshaped endogenously or automatically, opening up for *institutional failure*. At the same time, they are malleable and can be influenced by many types of actors, including firms and government.

System Failure

Firms are located within highly specific regional (or national) parts of global technological systems with specific and unique competence, networks and institutions forming regional or national innovation contexts (Freeman 1987; Carlsson and Stankiewicz 1991, Patel and Pavitt 1994a). As increasing returns apply to the process of innovation, an initial specialization tends to be reproduced and strengthened, resulting in different, uneven, and divergent technological development among countries (Patel and Pavitt 1994b). While these local parts of the systems operate under increasing returns and in virtuous circles (positive feedbacks) in expansionary periods, there is a considerable risk that:

- firms, institutions and networks become 'locked in' to the 'old' technologies. Thus, the cumulativeness and path dependency of innovation lead to risks of lock-in into technological, institutional and networking cul-de-sacs. Hence, market, institutional and network failures may combine into systems failure.
- if a search is undertaken outside traditional areas, it is done in a highly localized fashion. The cases of factory automation (chapter 3) and powder technology (chapter 7) are illustrative. In the former, there was a highly localized search from mechanical engineering to 'mechatronics' and in the latter case, from iron and steel metallurgy to certain niches in the powder metallurgy field, particularly cemented carbides and other hard materials.
- they manage to actually *hinder the process of diversity creation* in the local economy. Electronics and computer technology in Sweden is a case in point where a virtuous circle in mechanical engineering may well have prevented the emergence of a strong technological system in electronics. In other words, there may be feedback mechanisms whereby "...a selection process may consume variety" (Metcalfe 1992, p. 103). On the other hand, the pharmaceutical industry case studies in chapters 5 and 6 show that it is possible to break out of path dependence and thus avoid failure. The mechanism there was a combination of organizational innovation, a particular corporate culture, serendipity, and occasional public policy intervention. The strong emphasis on

medical research in technology policy, the organization of health care in Sweden, and the strong research orientation of Swedish physicians also contributed.

In this dynamic *cum* systems perspective, the role of government differs from that in a static framework. First, it gives a different perspective on, and justification for, policy where actors external to industry may have a significant role to play in the long-term evolution of industry. Even though there are serious problems of accumulating firm-, technology- and industry-specific knowledge in the state bureaucracy in order to remedy market failures, and even though there are also obvious problems of 'government failure' (Krueger 1986), there is still a role for public policy.

Second, in thinking about policy, we need to go beyond remedying market failures and also address network, institutional and system failures. Third, the mechanisms for policy intervention need to be such that they also address institutional and network development, thereby preventing system failures.

ISSUES FOR PUBLIC POLICY

There are two possible directions of public policy resulting from our analysis, namely strengthening or preserving existing systems and creating new systems. Given our basic assumptions (virtually unlimited global technical opportunities, bounded rationality, stable and path-dependent technological competencies, and the need for a systems *cum* dynamic perspective), there is no doubt that by far the most important but also most difficult task for public policy is to promote the creation of new technological systems. This is the primary focus of the discussion below. Our purpose is to formulate questions for public policy concern, the answers to which usually depend on the circumstances in each case. To a degree, we will also identify some policy options.

We begin by discussing questions and options relating to influencing the receiver competence of the economy. This is a function of both the 'prime movers' and the academic sector. We proceed to the issue of connectivity and network formation. The final area of concern is a set of mechanisms that influence the creation and maintenance of variety. We will here focus on the role of the capital market and the importance of new technology-based firms.

Improve Receiver Competence

Receiver competence is a concern of both private and public actors. The receiver competence of private actors determines their ability to exploit new technological opportunities and is raised through research and development, hiring of competent personnel, training of personnel (via both formal education and on-the-job training), and accumulation of experience (learning by doing). The concern of public sector actors lies not only in their influence over the educational system and university R&D but also in the risks of 'lock-in' discussed above and in the opportunities of creating positive externalities through various forms of spillovers.

We will divide the discussion of receiver competence into two parts. The first deals with the role of prime movers, primarily private firms, and how policy may act towards these. The second part deals with the academic sector.

The Role of Prime Movers

Prime movers perform four important tasks with respect to a new technology. They raise awareness, undertake investment in the new technology, give it legitimacy, and diffuse it through various mechanisms to other actors. In the course of this, they may be instrumental in creating a process characterized by increasing returns and industrial transformation. Prime movers are therefore key players in industrial dynamics and in the creation of new technological systems. We will elaborate on these four tasks.

First, a fundamental function in the promotion of industrial dynamics is the identification of new technological opportunities at an early stage and *raising of the awareness* of these opportunities as broadly as possible. But awareness may lead to action only if *entrepreneurship* is present. There must be someone with the vision and daring to take action and bring all the other necessary ingredients together. The entrepreneur can arise either out of existing firms or from elsewhere. But without that visionary spark, nothing happens. In the U.S. electronics system, the military initially performed much of this function whilst in the Swedish case, several advanced customers, including the military, did the same. In the U.S. case of genetic engineering, it was a cooperative effort between academia, an entrepreneur and venture capital, whereas in Sweden the first effort was undertaken by a state-owned firm, partly financed by a government-run venture capital fund (McKelvey 1996; see also chapter 6 above).

Second, capital must be available, and it has to be *venture* capital, not just funding. Venture capital serves the dual function of supplying the necessary funds (along with absorbing financial risks) and making sure that the necessary business competence is in place. Few entrepreneurs have all the required management skills, in addition to technology-specific knowledge, and this becomes a major obstacle for the business to take off. Just as problematic, there may be few venture capitalists with a knowledge of the specific technology(ies), particularly if it is an emerging one. Prime movers can either exploit an internal capital market or supply capital along with guidance and competence to other firms, e.g., through procurement. A division of labor between research companies with specialized knowledge and skills devoted to discovery of new business opportunities and large, traditional companies with marketing and mass production capabilities may be a good way to organize the supply of venture capital (as illustrated in chapters 5 and 6).

Third, the technology must be given *legitimacy* in the eyes of the market and institutions. As Lundgren (1991) argues in the case of image processing technology in

Sweden, the early support of the state gave legitimacy to the technology in the eyes of the capital market. A similar process was noted in the U.S. electronics industry, for instance, in the case of the Apollo project.

Fourth, the technology must be diffused to other local actors. This may happen through a whole range of mechanisms, as demonstrated by Eliasson in the case of SAAB Aircraft in chapter 8 of this book.

With localized search procedures for new technology, path dependence, and the existence of self-reinforcing patterns of specialization of regional/national technological capabilities, the issues of who the prime movers are and under what conditions they perform these functions are crucial for policy makers. Thus, a *key issue for policy makers is to create conditions and institutions conducive to monitoring the international scene and to encourage the establishment of local prime movers, to act early (in an international perspective), and to promote strong links to the local economy.*

Government policy may influence the emergence of prime movers by a range of instruments. First, it can *foster industry's awareness* of the opportunities of new technology. The vastness of the global set of technical opportunities and the uncertainties associated with these opportunities may make it difficult for individual agents to venture out on their own. Monitoring technological developments around the world, identifying the technological frontier and who is on it, and disseminating the resulting information within the own organization are important functions already carried out by large (multinational) firms (often the prime movers in their respective fields) (Eliasson 1991b). The primary vehicle for these activities is corporate research and development, most of which is privately funded. However, these activities can be supplemented and enhanced through public agencies through various means.

Fransman (1990, p. 286) argues that the role played by MITI (the Ministry of International Trade and Industry) in the Japanese collaborative research program could, in part, be seen as doing this by helping firms to expand their perceived opportunity set, in particular in periods of rapidly increasing technological opportunities. Such a role is, of course, not restricted to MITI.[6] A similar case is the Swedish powder technology program which aims at establishing links between centers of excellence and prospective users, thereby raising the awareness of the latter to the opportunities of powder technology (see chapter 7).

Second, *advanced public procurement policies* can ensure that the necessary investments are undertaken in the new technology by prospective prime movers. A technologically competent buyer can specify not only the price and performance of a new product but can also contribute to its design. Moreover, by paying for the seller's R&D it reduces the risks of venturing into new technologies. Public procurement has been shown to be a useful instrument for this purpose, as the electronics case in both Sweden and the U.S. suggests (see chapter 4 and Carlsson and Jacobsson 1996b) and

---

6 In the literature on Korea's phenomenal industrial development, a key role of the state has been to enlarge the perceived opportunity set by 'field manipulation' (Jones and SaKong 1980, Jacobsson and Alam 1994).

is revealed in the case of SAAB Aircraft (see chapter 8). The prime movers have in these cases been a combination of public agencies and private firms.

A less orthodox policy is one which builds more on the strength of private procurement where the role of the state is more of a coordinating one. An interesting example of this is Nutek's present experiment as a coordinator of procurement of electric vehicles and energy-efficient refrigerators from a whole range of private and public actors. Through Nutek's coordination, a poorly articulated and fragmented demand is transformed into a visible and sizeable demand providing incentives for suppliers to venture into new fields.

Third, the public sector can have an *indirect effect on the private sector* such as in the pharmaceutical field where a decisive factor behind the success of Swedish firms has been publicly run hospitals with a significant international standing in clinical research (see chapter 5).

Fourth, policies which stimulate the formation of new technology-based firms (NTBFs) can supplement private initiatives in exploiting opportunities in new areas of technology. While most of these new firms will not become prime movers, the occasional one, or a cluster of related firms, may take on that role. As we have argued elsewhere (Carlsson and Jacobsson 1996b), in the case of the U.S. electronics industry, competently handled procurement policy had a major influence on the growth of small technology-based firms. Indeed, it may well be argued that the leverage (in terms of inducing new behavior) of any public resources spent on procurement may be much higher if they are focused on smaller firms since these are more apt to change strategy in response to a smaller sized 'carrot.'[7]

The Role of the Academic Sector

The main responsibility of public policy in the area of receiver competence lies in the building of strong educational institutions. The academic sector influences industrial dynamics in several ways: by identifying new emerging technologies, shaping an awareness of their potential, spinning off firms exploiting that technology, and increasing society's absorptive capacity by accelerating research and education in new technologies when they are judged to be on the verge of becoming economically interesting.

It is important that policy makers ensure that the educational institutions' behavior is characterized by an early expansion of research and education in new fields.[8] The

---

[7] A procurement policy which is limited to smaller firms may have the additional benefit of intensifying competition for the larger firms, thereby providing a 'stick' for them to change their behavior.

[8] The crucial role played by U.S. universities in both electronics and biotechnology illustrates the central role of the academic sector in the formation of a new technological system. The Swedish experience in both these technologies illustrates the same phenomenon. Unfortunately, in the Swedish case, the universities lacked the proactive behavior of some of the U.S. universities and, hence, failed to support the growth of

importance of good timing can be appreciated if we consider the lead-time from the point when a new technology shows a good economic potential to when 'volume production' of engineers and Ph.D.'s with an orientation towards the new technology is possible.

In those technologies which prove to be economically viable, society then has a competence base to rely on as demand for specialized personnel accelerates in that area. The lead time is therefore reduced significantly, albeit at a price of a possible 'over-investment' at any given point in time. But in an evolutionary world, this is the normal state of affairs; 'waste' is always present: "...static inefficiencies are the necessary cost which must be incurred if economic systems are to develop and evolve" (Metcalfe 1992). The opportunities foregone in the absence of such investments are likely to be even costlier (see also Eliasson 1992; Carlsson and Jacobsson 1996a).

At the point where the uncertainty of the economic importance of a new technology is significantly reduced, universities need to 'change gears' and build capacity for high-volume production of engineers and scientists at all levels, including Ph.D.'s. As was mentioned in the study of the Swedish pharmaceutical industry (chapter 5), the demand for Ph.D.'s is now rapidly outstripping the supply; this has been the case for many years in the case of engineers with an orientation towards electronics and computer science. In the latter case, the expansion of training of engineers in electronics and computer science was about 10 years behind that in the United States and lagging 15 years behind the key inventions in electronics (see chapter 4).

A pro-active policy of the educational units is therefore essential and central to any public policy in countries where the universities are either state-run or centrally funded. Equally important, it is necessary for the academic system to have the ability rapidly to 'change gears.' A central task for policy makers is therefore *to monitor the behavior of the educational institutions. An associated problem of first magnitude for policy makers is how to design decision-making systems and incentive structures for educational institutions which are compatible with a pro-active and 'gear-changing' behavior.*

A few lessons can be drawn from our studies in this regard. First, the uncertainties about future states and the bounded vision of each agent means that actions with respect to new technology often do not involve choosing between clearly defined alternatives with known outcomes but rather consist of experiments. As each actor has a unique competence base and vision of the future, a decentralized structure leads to different decisions by each actor. That is, decentralization means that society as a whole explores a larger set of options than in a centralized structure. There is, therefore, a greater chance that a proactive and 'gear changing' behavior is followed by at least some actors.

Second, a 'gear-changing' behavior implies that the educational institutions should satisfy the demand for education in a whole range of segments, including not only the

---

strong technological systems in not only genetic engineering but also in electronics (see chapter 4). Indeed, in both cases, an external actor to the universities, the Swedish Board for Technical Development, had to intervene and take on a role as an educational policy maker.

'elite' segment. Take the case of computer science in Sweden. In spite of the shortage of computer scientists (see chapter 4), the supply capacity is still quite limited and most applicants to the education are rejected. The rejections include a number of students with significantly higher than average grades. For instance, in one of the major universities, the minimum average grade for students accepted to that education was 4.4 in 1996 (on a 5-point scale). Thus, even though there is a shortage in Sweden of high school students with an engineering and natural science orientation, many of those who are interested in a technical education and have fairly good grades are rejected. In part, this is probably due to factors outside the universities (a centralized decision making system where each university has had to ask government for permission and funding to expand education in a particular field). In part, however, this can be explained by the existence of fairly homogeneous universities which follow the same strategy. A key element in that strategy is to keep the intake of students low in order to have a high average grade of the accepted students. This is a way not only to recruit the best students but also to obtain a high status for the university. This is probably true for major universities in all countries, but in the Swedish context these were oligopolists until recently.[9] In other words, if none of the six universities accepted an applicant for, say, computer science, the student had to refrain from studying that subject. The lack of 'gear-changing capacity' and the consequent excess demand for computer science education probably stems, in part, therefore from a combined effect of homogeneous universities attempting to obtain a high status.

The educational infrastructure must therefore include a varied set of actors, some of which are more 'elitist' and cater to the students with the best grades, whereas others cater to the demand for students with other profiles, which may include, for instance, a spatial preference. Simply put, the educational system must be aligned to the needs of the market. Taken jointly, this leads us to the conclusion that the educational system should have a structure which is both decentralized and characterized by diversity.[10]

Increase Connectivity

Increasing the degree to which various components in each technological system are tied together is a 'natural' area for public policy concern. For instance, our study has shown the important role played by bridging institutions, many of which are public or semi-public agencies. The main function is to increase the number and intensity of linkages among businesses, between business and academia, among these and policy agencies, and especially to stimulate international linkages of all agents. Recent

---

9 In turn, this may well have reduced the pressure from an expanded undergraduate training in a specific field on the renewal of the teaching staff and graduate education.

10 In the Swedish case, the emergence of a whole set of colleges and a couple of new universities is radically changing the educational scene.

examples of such linkage-building activities in Sweden are the formation of so-called 'competence centers' linking academia and business; interdisciplinary collaboration on complex technical systems among business, research institutes, and academic institutions; and increased links to the European Union at all levels. The role of public policy should be to promote linkages which do not come about without intervention, not least those between academic institutions and business, and to promote system integration. Of particular importance is the establishment of strong links of local networks to the international system. Considerations of physical proximity and high density (critical mass) may stimulate transfer of knowledge, through various mechanisms, between firms.

Policy may be directed towards remedying both 'weak' and 'strong' network failures (see above). In the former case, the policy objective is to strengthen the functioning of an already developed technological system. In the latter case, it is to reshape the network configuration for the purpose of contributing to the growth of a new system.[11] *A prime task for policy makers is therefore to monitor how networks adjust to changing technological opportunities. The associated problem is how to influence the process of network creation and change.*

A natural area for policy is a broker function, either directly or indirectly. That is, government policy may have the function of 'matching' firms (and other organizations) which, so far, have had little or no contact with each other (Alänge and Jacobsson 1994). This matching can be done in many ways. Policy may, for instance, involve matching university researchers with firms, as was done by Stanford University where the policy agent was not the state but leaders at the University (Saxenian 1994). It may also involve the creation of various types of bridging institutions.

Create Variety

Given the strong path dependence and the powerful market mechanisms which weed out most new initiatives, perhaps the most important area of all for public policy is to build institutions which sustain and increase variety. Indeed, variety is necessary for long-term stability of economic growth (Eliasson 1984). Variety requires multiple players, each acting under different assumptions and with different knowledge. The most important vehicle for variety creation is free entry into markets. Above, we emphasized the role of both prime movers and academia in sustaining and creating variety. In this section, we focus on two important facets of the process of variety creation via entry of new firms into the market: the functioning of the capital market, particularly with regard to stimulating entrepreneurship, and the emergence of new technology-based firms (NTBFs).

---

[11] In both cases, the creation of trust and collective identities is at the heart of the process.

Entrepreneurship and the Capital Market

The capital market needs to be broad and diversified enough to meet the requirements of each technological system. Because of the nature of innovative and entrepreneurial activity, the main source of capital is often the venture capital market, not the regular capital market. The most important function of venture capital is not so much to supply funds and absorb risk as to provide management competence (identification of needed skills, help in recruiting key personnel, networking contacts and skills, etc.). The lack of a well-functioning venture capital market in Sweden (as well as in other European countries) has often been noted. The short supply of private individual wealth, the concentration of stock ownership in relatively few hands (particularly in a few business groups), the absence of a market for initial public offerings, the lock-in effects resulting from double taxation of corporate profits, the centralization of retirement saving in government funds rather than in a multitude of private pension funds, and the resulting lack of opportunities for the general public to invest in new ventures are all likely contributing factors. The paucity of venture capital in Sweden is probably an important factor behind the country's poor performance in the 'high tech' industries, notably the electronics industry. In addition to these reasons, many of which can be changed by legislation, the small size of Sweden probably matters. Because of the highly specialized competence required for many high-tech ventures, it may not be possible in a small country to find competent and willing venture capitalists locally or domestically, making it necessary to turn to the international market instead. In contrast, it may well be argued that it was precisely the presence of venture capital (even though it came from inside the Astra company rather from the open capital market) which made possible the success in both beta blockers and peptic ulcer drugs.

This suggests that the attention of policy makers should be directed towards improving the functioning of the capital market (particularly the venture capital market) and towards changes in tax and other legislation to remove the obstacles to growth of a venture capital market and entrepreneurship in general. Given the high risks and need for experimentation, the optimal venture capital market is likely to be characterized by many small players and many venture capitalists, in addition to the existing large firms with their internal capital markets. The competence requirements are such that a significant part of the venture capital will have to come from the international market.

New Technology-Based Firms

A special form of spillover is the formation of new technology-based firms through spin-offs from either existing firms or universities, utilizing knowledge which otherwise may not have been commercially exploited. The frequency of spin-offs and the evolution of the firms depends on a range of factors outside of the individual starting the firm, factors which can be influenced by policy (Jacobsson and Rickne 1996). These include the functioning of the capital market (see above), the attitude towards and degree of support for academics commercializing research results, and procurement policies (see above). A vital role for public policy is to monitor the birth and evolution of new technology-based firms and to shape policies which strengthen this mechanism of industrial dynamics.

CONCLUSIONS

The objective of this final chapter is to search for useful public policies to support the creation and strengthening of technological systems, and to identify a set of key issues for policy makers. We began with the following assumptions: a) virtually unlimited technological opportunities facing each agent; b) the behavior of agents is characterized by bounded rationality; c) the technological competence of firms is stable and path-dependent in its evolution. Taken jointly, this means that firms should be seen as pursuing experiments (mainly localized) rather than optimizing technology and strategy choice. d) The experimentation is undertaken within a context defined by a set of technological systems; and e) these systems are constantly evolving and new systems emerge frequently, i.e. the experimentation occurs in a dynamic world.

With these assumptions as a starting point, we discussed the domain of policy by broadening the perspective from 'market failure' to a wider field. This broader domain includes failures in networks and institutions as well as at the level of the technological systems. That is, whole technological systems may fail to be properly developed due to a lack of interaction of its constituent elements.

The domain of policy is broad but policy makers need to be directed towards specifics. We suggested that three features of technological systems are particularly important and raised a set of such specific issues for public policy makers to consider.

Improve the receiver competence of the economy. In part, this relates to policy towards prime movers, which are key actors who raise awareness of a new technology, undertake investment in it, give it legitimacy, and diffuse it locally. Policies are needed to promote the creation of prime movers and they formation of strong linkages to the rest of the economy. In part, this refers to the academic sector which needs to be both proactive with respect to new technology and have the ability to 'change gears' in the supply of engineers and scientists in a new field. Policy must ensure that decision-making systems and incentive structures exist which are compatible with such behavior. Decentralization and a diverse set of actors are key elements of such a policy.

Increase connectivity. A system perspective emphasizes spillovers, inter alia mediated through the interaction between actors which takes place in various types of networks. The degree of connectivity influences the amount of information and knowledge which is diffused through the system. A high connectivity is, however, not automatically created (by market forces) but is based on the development of trust and a collective identity. This process should be the concern and object of policy. Moreover, with the emergence of new technology, networks may have to be reformed. A policy concern is to both monitor and influence the process of network creation and change.

Sustain and improve variety. Whereas both prime movers and academia influence variety, and new networks provide a vehicle for a magnification of that influence, variety is also influenced by the workings of the capital market and the process of entry of new technology-based firms. The capital market can not be expected to evolve spontaneously to suit the requirement of each emerging technological system. Policy needs to monitor the evolution and make sure that a mismatch is avoided. Another source of variety is the creation of new technology-based firms. The 'birth' and evolution of such firms can be expected to be highly influenced by the characteristics of the technological system of which they are a part. Policy needs to monitor the birth and development of such firms and make sure that their external environment (the technological system) is conducive to a dynamic behavior.

REFERENCES

Alänge, Sverker and Staffan Jacobsson, 1994. "Erfarenheter från det svenska teko-programmet — en syntes i ett näringspolitiskt perspektiv", report submitted to NUTEK, Department of Industrial Management and Economics, Chalmers University of Technology, Göteborg.

Arthur, W. Brian, 1990. "Positive Feedbacks in the Economy," *Scientific American*, February, 92-99.

Ballot, Gérard and Erol Taymaz, 1996. "Human Capital, Technological Lock-in and Evolutionary Dynamics." Paper presented to the Sixth Conference of the International Joseph A. Schumpeter Society, Stockholm, June 3-5.

Carlsson, Bo (ed.), 1995. *Technological Systems and Economic Performance: The Case of Factory Automation.* Boston and Dordrecht: Kluwer Academic Publishers.

Carlsson, Bo and Rikard Stankiewicz, 1991. "On the Nature, Function, and Composition of Technological Systems," *Journal of Evolutionary Economics*, 1 (2), 93-118. Also published as chapter 2 in Carlsson (1995).

Carlsson, Bo and Staffan Jacobsson, 1996a. "Technological Systems and Industrial Dynamics: Implications for Firms and Governments," in Ernst Helmstädter and Mark Perlman, *Behaviorial Norms, Technological Progress, and Economic Dynamics. Studies in Schumpeterian Economics.* Ann Arbor: The University of Michigan Press, 261-283.

Carlsson, Bo and Staffan Jacobsson, 1996b. "Diversity Creation and Technological Systems: A Technology Policy Perspective," in Charles Edquist (ed.), *Systems of Innovation: Technologies, Institutions and Organizations.* London: Pinter Publishers (forthcoming).

Corden, W. H., 1974. *Trade Policy and Economic Welfare.* Oxford: Oxford University Press.

Corden W.H., 1980. "Trade Policies," in J. Cody, H. Hughes and D. Wall, *Policies for Industrial Progress in Developing Countries.* Oxford: Oxford University Press.

Dosi, Giovanni, 1993. Comments by Giovanni Dosi in UNCTAD, Report of Ad Hoc Expert Group on Technology Policies in Open Developing Country Economies, UNCTAD/ITD/TEC/3.

Ehrnberg, Ellinor and Staffan Jacobsson, 1996. "Technological Discontinuities and Incumbents' Performance - An Analytical Framework" in Charles Edquist (ed.), *Systems of Innovation: Technology, Organizations and Institutions*. London: Pinter Publishers (forthcoming).

Eliasson, Gunnar, 1984. "Micro Heterogeneity of Firms and Stability of Growth," *Journal of Economic Behavior and Organization*, 5 (3-4), 249-274.

Eliasson, Gunnar, 1990. "The Firm as a Competent Team," *Journal of Economic Behavior and Organization*, 13 (3), 273-298.

Eliasson, Gunnar, 1991a." Modeling the Experimentally Organized Economy - Complex Dynamics in an Empirical Micro-Macro Model of Endogenous Economic Growth," *Journal of Economic Behavior and Organization*, 16 (1-2), 153-182.

Eliasson, Gunnar, 1991b. "The International Firm: A Vehicle for Overcoming Barriers to Trade and a Global Intelligence Organization Diffusing the Notion of a Nation," in Lars-Gunnar Mattsson and Bengt Stymne (eds.), *Corporate and Industry Strategies to Europe*. Amsterdam: Elsevier Science Publishers, 139-170.

Eliasson, Gunnar, 1992. "Business Competence, Organizational Learning, and Economic Growth: Establishing the Smith-Schumpeter-Wicksell (SSW) Connection," in Frederic M. Scherer and Mark Perlman (eds.), *Entrepreneurship, Technological Innovation, and Economic Growth. Studies in the Schumpeterian Tradition.* Ann Arbor: The University of Michigan Press, 251-275.

Fransman, Martin, 1990. *The Market and Beyond.* Cambridge: Cambridge University Press.

Freeman, Christopher, 1987. *Technology Policy and Economic Performance*. London: Pinter Publishers, Ltd.

Håkansson, Håkan, 1987. *Industrial Technological Development: A Network Approach.* London: Croom Helm.

Håkansson, Håkan, 1989. *Corporate Technological Behavior: Cooperation and Networks.* London: Routledge.

Jacobsson, Staffan and Ghayur Alam, 1994. *Liberalization and Industrial Development in the Third World. A Study of Government Policy and Performance of the Indian and Korean Engineering Industries*. New Delhi: SAGE Publications.

Jacobsson, Staffan and Annika Rickne, 1996. "New Technology-Based Firms and Industrial Renewal — a Literature Review," mimeo., Department of Industrial Management and Economics, Chalmers University of Technology, Göteborg.

Jones, Leroy P. and Il SaKong, 1980. *Government, Business, and Entrepreneurship in Economic Development: The Korean Case.* Cambridge, MA: Harvard University Press.

Krueger, Anne, 1986. "Changing Perspectives on Development Economics and World Bank Research," *Development Policy Review*, 4 (3).

Lundgren, Anders, 1991. *Technological Innovation and Industrial Evolution — the Emergence of Industrial Networks,* Doctoral Dissertation, The Economic Research Institute/Stockholm School of Economics.

McKelvey, Maureen, 1996. *Evolutionary Innovation.The Business of Biotechnology.* Oxford: Oxford University Press.

Metcalfe, J. Stanley, 1992. "The Economic Foundations of Technology Policy: Equilibrium and Evolutionary Perspectives," mimeo, University of Manchester.

Patel, Parimal and Keith Pavitt, 1994a. "National Innovation Systems: Why They Are Important and How They Might be Measured and Compared," Economics of Innovation and New Technology, 3 (1), 77-95.

Patel, Parimal and Keith Pavitt, 1994b. "Uneven (and Divergent) Technological Accumulation among Advanced Countries: Evidence and a Framework of Explanation," Industrial and Corporate Change, 3 (3), 759-787.

Porter, Michael E., 1990. "The Competitive Advantage of Nations", *Harvard Business Review*, 68 (2) (March-April), 73-93.

Saxenian, AnnaLee, 1994. *Regional Advantage. Culture and Competition in Silicon Valley and Route 128.* Cambridge and London: Harvard University Press.

AUTHOR INDEX

SUBJECT INDEX

Economics of Science, Technology and Innovation

1. A. Phillips, A.P. Phillips and T.R. Phillips: *Biz Jets. Technology and Market Structure in the Corporate Jet Aircraft Industry.* 1994 ISBN 0-7923-2660-1
2. M.P. Feldman: *The Geography of Innovation.* 1994 ISBN 0-7923-2698-9
3. C. Antonelli: *The Economics of Localized Technological Change and Industrial Dynamics.* 1995 ISBN 0-7923-2910-4
4. G. Becher and S. Kuhlmann (eds.): *Evaluation of Technology Policy Programmes in Germany.* 1995 ISBN 0-7923-3115-X
5. B. Carlsson (ed.): *Technological Systems and Economic Performance: The Case of Factory Automation.* 1995 ISBN 0-7923-3512-0
6. G.E. Flueckiger: *Control, Information, and Technological Change.* 1995 ISBN 0-7923-3667-4
7. M. Teubal, D. Foray, M. Justman and E. Zuscovitch (eds.): *Technological Infrastructure Policy. An International Perspective.* 1996 ISBN 0-7923-3835-9
8. G. Eliasson: *Firm Objectives, Controls and Organization. The Use of Information and the Transfer of Knowledge within the Firm.* 1996 ISBN 0-7923-3870-7
9. X. Vence-Deza and J.S. Metcalfe (eds.): *Wealth from Diversity. Innovation, Structural Change and Finance for Regional Development in Europe.* 1996 ISBN 0-7923-4115-5
10. B. Carlsson (ed.): *Technological Systems and Industrial Dynamics.* 1997 ISBN 0-7923-9940-4

KLUWER ACADEMIC PUBLISHERS — BOSTON / DORDRECHT / LONDON

GPSR Compliance

The European Union's (EU) General Product Safety Regulation (GPSR) is a set of rules that requires consumer products to be safe and our obligations to ensure this.

If you have any concerns about our products, you can contact us on ProductSafety@springernature.com

In case Publisher is established outside the EU, the EU authorized representative is:

Springer Nature Customer Service Center GmbH
Europaplatz 3
69115 Heidelberg, Germany

Batch number: 10755284

Printed by Printforce, the Netherlands